Har ~~ng the ~~eelwork of

Wheelwork
of Nature

Harnessing the Wheelwork of Nature

Tesla's Science of Energy

Thomas Valone, Ph.D.

Harnessing the Wheelwork of Nature
Tesla's Science of Energy

ISBN 1-931882-04-5

Printed in the United States of America

Published by
Adventures Unlimited Press
One Adventure Place
Kempton, Illinois 60946 USA

www.adventuresunlimitedpress.com
www.adventuresunlimited.nl
www.wexclub.com

10 9 8 7 6 5 4 3 2 1

Special thanks and acknowledgement to Nikola Tesla, Andrija Puharich, James and Kenneth Corum, Kurt Van Voorhies, Robert Bass, H. W. Jones, Elizabeth Rauscher, William Van Bise, Gary Peterson, Dilettante Press, Mark Seifer and Metascience Foundation, National Public Broadcasting Station, Nikola Tesla Museum, Belgrade, Oliver Nichelson, Toby Grotz, Keith Tutt, Simon & Schuster publishers, David Childress, Adventures Unlimited Press, and my loving wife, Jacqueline Valone, without whom this book would not have been possible.

HARNESSING THE WHEELWORK OF NATURE

Tesla's Science of Energy

by
Thomas Valone, Ph.D.

ADVENTURES UNLIMITED PRESS

OTHER BOOKS BY THOMAS VALONE

Homopolar Handbook: Definitive Guide to Faraday Disk & N-Machine Technologies
Electrogravitics Systems: A New Propulsion Methodology
Energy Crisis: The Failure of the Comprehensive Energy Strategy
Bush-Cheney Energy Study: Analysis of the National Energy Policy
The Future of Energy: An Emerging Science
Future Energy: Proceedings of the Conference on Future Energy, 1999

ABOUT THE AUTHOR

Thomas F. Valone has a Ph.D. in General Engineering from Kennedy-Western University, a Masters in Physics from SUNY at Buffalo, and Professional Engineering license. Dr. Valone is President of Integrity Research Institute The Institute performs scientific testing, patent application review, expert witness and technical consultation for clients including law firms, investment houses and high-tech companies. (www.IntegrityResearchInstitute.org)

Thomas Valone was formerly a US Patent Examiner, specializing in Class 324 Physics, Measuring, and Testing apparatus. He was elected a Board Member of the Patent & Trademark Office Society in 1999. He taught engineering, physics and electronics at a SUNY-accredited college, where he also designed the Instrumentation & Process Control Curriculum and managed engineers and technicians in lab projects. At Scott Aviation, he was a research scientist and Director of R&D. He was responsible for numerous sensor circuit design, instrument design and testing projects.

Memberships include: Institute of Electrical and Electronic Engineers (IEEE), American Association for the Advancement of Science (AAAS), American Institute of Aeronautics and Astronautics (AIAA), and American Physical Society (APS), New York State Society of Professional Engineers, Alliance for Energy and Economic Growth, and formerly, the U.S. Energy Association.

He is listed in *Who's Who in U.S. Writers, Editors & Poets, 1988*

New York, June 20th, 1896.

#46 & 48 E. Houston Str.

Mrs. Elisabeth Porter Gould,

    #100 Huntington Ave.,

    Boston, Mass.

My dear Madam:-

    Please excuse this long delay in replying to your kind letter of June 3rd, as well as this mode of communicating, to which I am compelled to resort for want of time and energy.

    I fully appreciate the honor of being in such excellent company, but it is no inducement for me to forward my picture; quite the contrary, feeling that I would be out of place, I could not grant your wish. But in doing so I apprehend that I would disappoint you, and this consideration compels me to the opposite course.

    The photograph which I am forwarding with this mail has been taken by a new form of electric light, and the objects shown have a meaning which I cannot at present explain. This is the only picture which I would care to have preserved.

    Believe me to be,

        Yours very truly,

        N. Tesla

Note the date at the bottom of this photo of Tesla is June 20, 1896

Table of Contents

The 187-foot
Wardenclyffe
Tower in 1903
which stood
unfinished for
the next 14 years.
The two-story
power plant, by
comparison, is in
the background.

Introduction to Tesla's Science of Energy

Thomas Valone, Ph.D., P.E.

It is a great privilege to present this amazing collection of seminal articles, some of which have never been published before, on Nikola Tesla's science of energy. As I'm finishing my Ph.D. thesis on utilizing zero point energy, I realized that Tesla probably acknowledged the same energy reservoir when he referred to harnessing "the very wheelwork of nature."[1] The visionary scientists who have contributed to this anthology offer a collective argument of what Tesla meant by that phrase. Tesla also recognized that an atmospheric and a terrestrial storage battery exists here on earth, just waiting to be tapped for the good of mankind. Therefore, this is the wheelwork of nature that we want to explore in this book.

The first section of this anthology offers some historical Niagara Falls material and biographical information about the life of Nikola Tesla with the contributions of William Terbo, the grand-nephew of Tesla, Keith Tutt, author of *The Scientist, the Madman, the Thief & Their Light Bulb,* and Dr. Andrija Puharich, whose unpublished biographical manuscript is rich with personal insights. Puharich, a Yugoslavian, also develops with great care, the background and unexpected uses for Tesla's Magnifying Transmitter (TMT). The second section is devoted to Tesla's wireless transmission of electrical power, as distinguished from wireless telegraphy for which he is also famous. It is surprisingly practical, even today, as the brilliant minds in this book prove. Tesla was at least a century ahead of his time, however, so people stole his ideas, left him penniless, and ignored his saintly concern for the human race. I pray that as global community consciousness expands in the 21st century, Tesla's ideas about sharing energy with the whole world will be more understood and appreciated. The third and last section has miscellaneous articles about a few of Tesla's less well-known inventions, including the two-rotor belted homopolar generator and an ozone generator.

Today we are faced with the consequences of the fateful decision in 1905 by J. P. Morgan to abandon Tesla's Wardenclyffe Tower project on Long Island, once he learned that it would be designed mainly for wireless transmission of electrical power, rather than telegraphy. He is reported to have complained that he would not be able to collect money from the customer in any feasible way. This mercenary attitude by the world's richest man forced the nation to pay for thousands of miles of transmission line wires, just so an electrical utility meter could be placed on everyone's house. Today the U.S. Energy Association in Washington, DC trains representatives from the former Russian states how to reliably do the same in their countries.

No one, except for the few great physicists like Drs. Rauscher, Corum, Bass, and Van Voorhies found in this book, has realized that Tesla was very practical when he proposed the resonant generation and wireless transmission of useful electrical power, after returning from his experiments at Colorado Springs in 1900. For example, Professor Rauscher shows that the earth's magnetosphere contains sufficient potential energy (at least 3 billion kilowatts) so that the resonant excitation of the earth-ionosphere cavity can reasonably be expected to increase the amplitude of natural "Schumann" frequencies, facilitating the capture of useful electrical power. Tesla knew that the earth could be treated as one big spherical conductor and the ionosphere as another bigger spherical conductor, so that together they have parallel plates and thus, comprise a "spherical capacitor." Dr. Rauscher calculates the capacitance to be about 15,000 microfarads for the complete earth-ionosphere cavity capacitor. W.O. Schumann

---

[1] "..it is a mere question of time when men will succeed in attaching their machinery to the very wheelwork of nature." –Tesla addressing the Amer. Inst. of Elec. Eng., 1891

is credited for predicting the "self-oscillations" of the conducting sphere of the earth, surrounded by an air layer and an ionosphere in 1952, without knowing that Tesla had found the earth's fundamental frequency fifty years earlier.[2]

In comparison to the 3 billion kW available from the earth system, it is possible to calculate what the U.S. consumed in electricity. In 2000, about 11 Quads (quadrillion Btu) were actually used by consumers for electrical needs, which is equal to 3.2 trillion kWh. Dividing by the 8760 hours in a year, we find that only <u>360 million kW</u> are needed <u>on site</u> to power our entire country. This would still leave 2.6 billion kW for the rest of the world! The really shameful U.S. scandal, unknown to the general public, is that out of the total electrical power generated using wire transmission (about 31 Quads), a full ***2/3 is totally wasted*** in "conversion losses."[3] (See the *Electricity Flow Chart 1999*, which contains US DOE/EIA data, updating the Toby Grotz article in this book.) No other energy production system of any kind in the world has so much wastefulness. Instead of trying to build 2 power plants per week (at 300 MW each) for the next 20 years (only to have a total of additional 6 trillion kWh available by 2020), as some U.S. government officials want to do, we simply need to ***eliminate the 7 trillion kWh of conversion losses*** in our present electricity generation modality. Tesla's wireless transmission of power accomplishes this goal, better than any distributed generation.

Electricity Flow Chart 1999
(Quadrillion BTU)

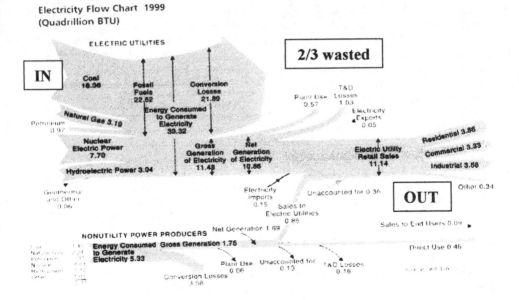

As Tesla himself said,

"In the near future we shall see a great many uses of electricity... we shall be able to disperse fogs by electric force and powerful and penetrative rays... wireless plants will be installed for the purpose of illuminating the oceans... picture transmission by ordinary telegraphic methods will soon be achieved... another valuable novelty will be a typewriter electrically operated by the human voice... we shall have smoke annihilators, dust absorbers,

---

[2] W.O. Schumann, *Z. Naturforsch*, 72, p. 149-154 and 250-252, 1952, (in German)
[3] "National Energy Security Post 9/11" U.S. Energy Association, June, 2002, p. 34

sterilizers of water, air, food and clothing…it will become next to impossible to contract disease germs and country folk will go to town to rest and get well…"

"*If we use fuel to get our power, we are living on our capital and exhausting it rapidly. This method is barbarous and wantonly wasteful and will have to be stopped in the interest of coming generations*. The inevitable conclusion is that water power is by far our most valuable resource. On this humanity must build its hopes for the future. With its full development and a perfect system of wireless transmission of the energy to any distance, man will be able to solve all the problems of material existence. Distance, which is the chief impediment to human progress, will be completely annihilated in thought, word, and action. Humanity will be united, wars will be made impossible, and peace will reign supreme."[4]

The same article which contains this prophetic quotation from Tesla also notes that his

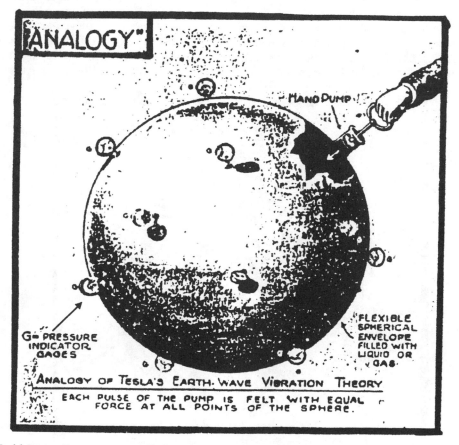

"World System" was conceptually based on three inventions of his:
1. The Tesla Transformer (Tesla coil)
2. The Magnifying Transmitter (transformer adapted to excite the earth)
3. The Wireless System (economic transmission of electrical energy without wires)

---

[4] Nikola Tesla, 1900, as quoted in "Great Scientist, Forgotten Genius, Nikola Tesla" by Chris Bird and Oliver Nichelson, *New Age*, #21, Feb. 1977, p. 42

Tesla states, "The first World System power plant can be put in operation in nine months. With this power plant it will be practicable to attain electrical activities up to 10 million horsepower (7.5 billion watts), and it is designed to serve for as many technical achievements as are possible without due expense."[5] (Note that Tesla's calculated power levels are conservatively estimated, compared to Rauscher's calculations.)

The essay by Toby Grotz on the wireless transmission of power is a great introduction to this wireless power system of Tesla. It contains all of the details for a preliminary test of the system. His Figure 5 also illustrates the transmission of a high voltage pulse of electricity equally around the world where it rebounds at the opposite side and returns to its source, repeating the cycle many times. Grotz also worked with Dr. Corum on "Project TESLA," which was a business venture designed to implement the wireless transmission of electricity.

Dr. Corum notes in his introductory article on the ELF (extremely low frequency) oscillator of Tesla's that the tuned circuit of Tesla's magnifying transmitter was the whole earth-ionosphere cavity. His second article presents probably the most complete article on Tesla's magnifying transmitter that has ever been written. He explains in great detail the meaning of magnification as Tesla intended, with examples and equations. Even if not an engineer, I believe the reader will still appreciate the enthusiastic style with which the Corums describe

Tesla's developments regarding the TMT.

---

[5] Ibid., p.74

There are two diagrams produced at the turn of the century to help explain in simple terms Tesla's wireless transmission of electrical power. The first is a mechanical "Analogy" that is described in Corum's ELF disclosure article. The second is the "Realization" which illustrates the usefulness of the power transmission concept.

Credit: Metascience Foundation

Tesla wrote, "That electrical energy can be economically transmitted without wires to any terrestrial distance, I have unmistakably established in numerous observations, experiments and measurements, qualitative and quantitative. These have demonstrated that it is practicable to distribute power from a central plant in unlimited amounts, with a loss not exceeding a small fraction of one per cent in the transmission, even to the greatest distance, twelve thousand miles – to the opposite end of the globe."[6]

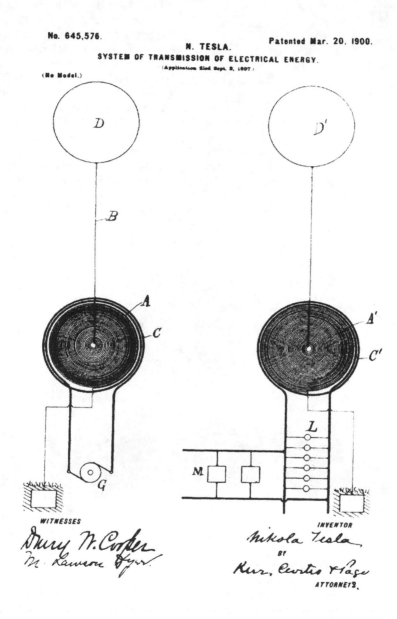

No. 645,576.

N. TESLA.

Patented Mar. 20, 1900.

SYSTEM OF TRANSMISSION OF ELECTRICAL ENERGY.

(Application filed Sept. 2, 1897.)

(No Model.)

WITNESSES

INVENTOR

Nikola Tesla

BY

Kerr, Curtis & Page

ATTORNEYS.

---

[6] Nikola Tesla, "The Transmission of Electrical Energy Without Wires as a Means for Furthering Peace," *Electrical World and Engineer.* Jan. 7, 1905, p. 21

Nikola Tesla

As Tesla experimented with a 1.5 MW system in 1899 at Colorado Springs, he was amazed to find that pulses of electricity he sent out passed across the entire globe and returned with "undiminished strength." He said, "It was a result so unbelievable that the revelation at first almost stunned me."[7] This verified the tremendous efficiency of his peculiar method of pumping current into a spherical ball to charge it up before discharging it as a pulse of electrical energy, *a "longitudinal" acoustic-type of compression wave*, rather than an electromagnetic Hertzian-type of transverse wave.

It is also understood that Tesla planned to include stationary resonant wave creation as part of the wireless transmission of power. Examining the pair of 1900 patents #645,576 and #649,621 each using the same figure on the first page, we find in the first patent that Tesla has

Sections of the Earth and its Atmosphere

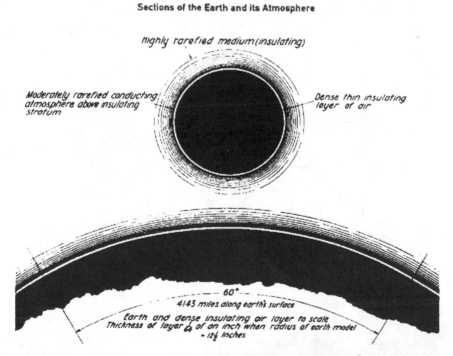

designed a quarter-wave antenna (50 miles of secondary coil wire for a 200 mile long wavelength). More importantly is the sphere on the top which is supposed to be a conductive surface on a balloon raised high enough to be radiating in "rarefied air." As Tesla states,

"That communication without wires to any point of the globe is practical with such apparatus would need no demonstration, but through a discovery which I made I obtained

---

[7] Nikola Tesla, "World System of Wireless Transmission of Energy," *Telegraph and Telephone Age*, Oct. 16, 1927, p. 457.

absolute certainty. Popularly explained it is exactly this: When we raise the voice and hear an echo in reply, we know that the sound of the voice must have reached a distant wall, or boundary, and must have been reflected from the same. Exactly as the sound, so an electrical wave is reflected, and the same evidence which is afforded by an echo is offered by an electrical phenomena known as a 'stationary' wave – that is, a wave with fixed nodal and ventral regions. Instead of sending sound vibrations toward a distant wall, I have sent electrical vibrations toward the remote boundaries of the earth, and instead of the wall, the earth has replied. In place of an echo, I have obtained a stationary electrical wave, a wave reflected from afar."[8]

It is also worth calling attention to Corum's disclosure article on the operation of an ELF oscillator, he proposes that Tesla's x-ray patents were designed for the switching of high voltages in the charging and discharging of the dome of the Wardenclyffe tower (patent #1,119,732). Dr. Bass' article elaborates on the details of <u>longitudinal waves</u> that would be created by such discharges. They have <u>superior properties of transmission</u> which normal radio and television waves today do not possess. Nikola Tesla was very familiar with their benefits.

---

[8] Nikola Tesla, "The Problem of Increasing Human Energy," *Century*, June, 1900

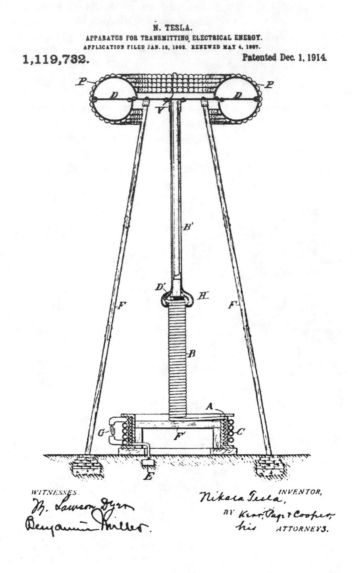

N. TESLA.
APPARATUS FOR TRANSMITTING ELECTRICAL ENERGY.
APPLICATION FILED JAN. 18, 1902. RENEWED MAY 4, 1907.

1,119,732.

Patented Dec. 1, 1914.

WITNESSES

INVENTOR,
Nikola Tesla,
BY Kerr, Page & Cooper,
his ATTORNEYS.

176

Tesla states, "As to the transmission of power through space, that is a project which I considered absolutely certain of success long since. Years ago I was in the position to transmit wireless power to any distance without limit other than that imposed by the physical dimensions of the globe. In my system it makes no difference what the distance is. The efficiency of the transmission can be as high as 96 or 97 per cent, and there are practically no losses except such as are inevitable in the running of the machinery. ***When there is no*** *receiver there is no energy consumption anywhere. When the receiver is put on, it draws power. That is the exact opposite of the Hertz-wave system.* In that case, if you have a plant of 1,000 horsepower (750 kW), it is radiating all the time whether the energy is received or not; but in my system no power is lost. When there are no receivers, the plant consumes only a few horsepower necessary to maintain the vibration; it runs idle, as the Edison plant when the lamps and motors are shut off."[9]

These incredible facts are explained by Dr. Corum and Spainol elsewhere, "...the distinction between Tesla's system and 'Hertzian' waves is to be clearly understood. Tesla, and others of his day, used the term 'Hertzian waves' to describe what we call today, energy transfer by wireless transverse electromagnetic (TEM) radiation...no one wants to stand in front of a high power radar antenna. For these, E and H are in phase, the power flow is a 'real' quantity (as opposed to reactive – Ed. note), and the surface integral of E x H (Poynting vector – Ed. note) is nonzero. The case is not so simple in an unloaded power system, an RF transformer with a tuned secondary, or with a cavity resonator. In these situations, the fields are in phase quadrature, the circulating power is reactive and the average Poynting flux is zero – *unless a load is applied*. They deliver no power without a resistive load. These are clearly the power systems which Tesla created. The polyphase power distribution system was created by him in the 1880s and inaugurated at Niagara Falls in 1895. The RF transformer was invented and patented by him in the 1890s. Terrestrial resonances he experimentally discovered at the turn of the century. And, for the next 40 years he tried to bring through to commercial reality this global power system. Today, millions of us have working scale models of it in our kitchens, while the larger version sits idle."[10]

Receiving coil a great distance from the transmitter lighting a light bulb (white spot) in a test of Tesla's wireless transmission of power in 1899.

---

[9] Nikola Tesla, "Minutes of the Annual Meeting of the AIEE," May 18, 1917.
[10] Corum, Corum, and Spaniol, "Concerning Cavity Q," *Proceedings of the International Tesla Symposium*, 1988, p. 3-15

September, 1917

THE ELECTRICAL EXPERIMENTER

H. GERNSBACK EDITOR
H. W. SECOR ASSOCIATE EDITOR

Vol. V. Whole No. 53   September, 1917   Number 5

U. S. Blows Up Tesla Radio Tower

SUSPECTING that German spies were using the big wireless tower erected at Shoreham, L. I., about twenty years ago by Nikola Tesla, the Federal Government ordered the tower destroyed and it was recently demolished with dynamite. During the past month several strangers had been seen lurking about the place.

Tesla erected the tower, which was about 185 feet high, with a well about 100 feet deep, for use in experimenting with the transmission of electrical energy for power and lighting purposes by wireless. The equipment cost nearly $200,000.

The late J. P. Morgan backed Nikola Tesla with the money to build this remarkable steel tower, that he might experiment in wireless even before people knew of Marconi. A complete description, revised by Dr. Tesla himself, of this unique and ultra-powerful radio plant was given in the March, 1916, issue of THE ELECTRICAL EXPERIMENTER. Everyone interested in the study of high frequency currents should not fail to study that discourse as it contains the theory of how this master electrician proposed to charge this lofty antenna with thousands of kilowatts of high frequency electrical energy, then to radiate it thru the earth and run ships, factories and street cars with "wireless power."

Most of our readers have, no doubt, read about the famous Tesla wireless tower, which structure involved the expenditure of a vast sum of money and engineering talent. From this lofty structure, which was designed some 20 years ago by Dr. Tesla and his associates, there was to be propagated an electric wave of such intensity that it could charge the earth to such a potential that the effect of the wave or charge could be felt in the utmost confines of the globe.

Further, it may be said that Tesla, all in all, does not believe in the modern Hertzian wave theory of wireless transmission at all. Several other engineers of note have also gone on record as stating their belief to be in accordance with Dr. Tesla's. More wonderful still is the fact that this scientist promulgated his basic theory of *earth current* transmission a great many years ago in some of his patents and other publications. Briefly explained, the Tesla theory is that a wireless tower, such as that here illustrated and specially constructed to have a high capacity, acts as a huge electric condenser. This is charged by a suitable high frequency, high voltage apparatus and a current is discharged into the earth periodically and in the form of a high frequency alternating wave. The electric wave is then supposed to travel thru the earth along its surface shell and in turn to manifest its presence at any point where there might be erected a similar high capacity tower to that above described.

A simple analogy to this action is the following: Take a hollow spherical chamber filled with a liquid, such as water; and then, at two diametrically opposite points, let us place, respectively, a small piston pump, such as a bicycle pump, and an indicator such as a pressure gage. Now if we suck

some of the water into the pump and force it back into the ball by pushing on the piston handle, this change in pressure will be indicated on the gage secured to the opposite side of the sphere. In this way the Tesla earth currents are supposed to act.

The patents of Dr. Tesla are basically quite different from those of Marconi and others in the wireless telegraphic field. In the nature of things this would be expected to be the case, as Tesla believes and has designed apparatus intended for the *transmission of large amounts of electrical energy*, while the energy received in the transmission of intelligence wirelessly amounts to but a few millionths of an ampere in most cases by the time the current so transmitted has been picked up a thousand miles away. In the Hertzian wave system, as it has been explained and believed in, the energy is transmitted with a very large loss to the receptor by electro-magnetic waves which pass out laterally from the transmitting wire into space. In Tesla's system the energy radiated is not used, but the current is led to earth and to an elevated terminal, while the energy is transmitted by a process of *conduction*. That is, the earth receives a large number of powerful high frequency electric shocks every second, and these act the same as the pump piston in the analogy.

Quoting from one of Tesla's early patents on this point: "It is to be noted that the phenomenon here involved in the transmission of electrical energy is one of *true conduction* and is not to be confounded with the phenomena of *electrical radiation*, which have heretofore been observed, and which, from the very nature and mode of propagation, would render practically impossible the transmission of any appreciable amount of energy to such distances as are of practical importance."

In the same "Cavity Q" article, the authors also settle the most common criticism of the Tesla wireless power system regarding biological effects. Calculating the circulating reactive power, they find a density of a microVAR per cubic meter at 7.8 Hz to be quite small, while it is well-known that the frequency is very biologically compatible. The authors also look at the present 100 V/m field and again find that raising it by a factor of 4 to 10 will pose no ill effects. (Thunderstorms do it all of the time around the world.)

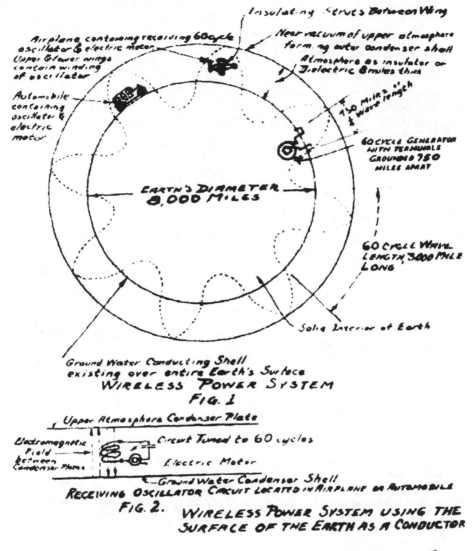

Insulating Struts Between Wing

Near vacuum of upper atmosphere forming outer condenser shell

Atmosphere as insulator or Dielectric 8 miles thick

Airplane containing receiving 60 cycle oscillator & electric motor

Upper Glower wings contain winding of oscillator

750 miles = ¼ wave length

Automobile containing oscillator & electric motor

60 cycle Generator with terminals grounded 750 miles away

EARTH'S DIAMETER 8,000 MILES

60 cycle wave length 3,000 mile long

Solid Interior of Earth

Ground Water Conducting Shell existing over entire Earth's Surface

WIRELESS POWER SYSTEM
FIG. 1

Upper Atmosphere Condenser Plate

Electromagnetic Field between Condenser Plates

Circuit Tuned to 60 cycles

Electric Motor

Ground Water Condenser Shell

RECEIVING OSCILLATOR CIRCUIT LOCATED IN AIRPLANE OR AUTOMOBILE

FIG. 2. WIRELESS POWER SYSTEM USING THE SURFACE OF THE EARTH AS A CONDUCTOR

SKETCH A.

In 1925, an electrical engineer, John B Flowers, developed a proposal to test and implement Tesla's Wireless Power System. He drafted the entire scheme for the Wardenclyffe project and presented it to H. L. Curtis, physicist, and J. H. Dillinger, head of the Radio Laboratory at the Bureau of Standards in Washington, DC. In a carefully worded 10-page document, complete with schematic drawings of the earth imbued with Tesla standing waves, Flowers unveiled a plan for operating cars and planes powered by wireless electricity (Sketch A). The plan was declined even though the mechanical test in Sketch B actually worked. Below is a report on the test results of the mechanical model of Tesla's wireless system:

"Using the concepts in Sketch B, a mechanical oscillator arm was fastened to the tied opening of a rubber balloon 20 inches in diameter. The oscillator arm was operated with an

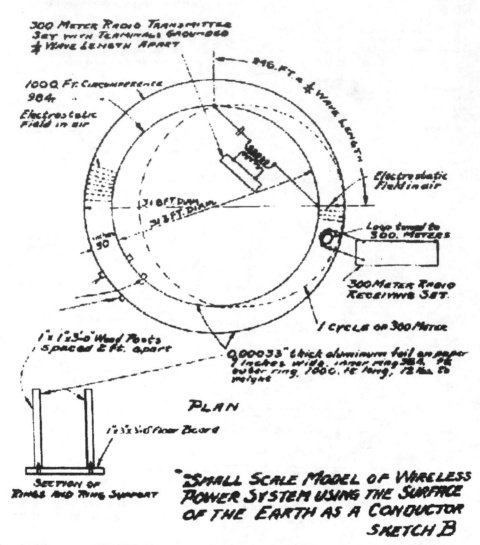

300 METER RADIO TRANSMITTER
SET WITH TERMINALS GROUNDED
¼ WAVE LENGTH APART

1000 FT. CIRCUMFERENCE
984 ..
Electrostatic
Field in air

246.FT = ¼ WAVE LENGTH

Electrostatic
Field in air

Loop tuned to
300. METERS

300 METER RADIO
RECEIVING SET.

313.8 FT DIAM.
313 FT. DIAM.

1 CYCLE OR 300 METER

1" x 1 x 3'-0" Wood Posts
spaced 2 ft. apart

0.00033" thick aluminum foil on paper
9 inches wide, inner ring 984, 76
outer ring 1000, 16 long, 12 in. to
weight

PLAN

1"x 3"x 3'-0" floor Board

SECTION OF
RINGS AND RING SUPPORT

"SMALL SCALE MODEL OF WIRELESS
POWER SYSTEM USING THE SURFACE
OF THE EARTH AS A CONDUCTOR
SKETCH B

electrical motor at 1750 RPM by means of an eccentric on the motor shaft. The balloon hung free in the air. The rubber surface of the balloon represented the earth's conducting surface and the air inside its insulating interior. The waves were propagated in the rubber surface at the rate of 51 feet per second, the frequency of transmission was 29 cycles per second and the wavelength was 21 inches. The mechanical oscillator was used in place of Tesla's electrical oscillator as it presents an almost perfect analogy. Standing or stationary waves of the rubber surface replace the electromagnetic waves of Tesla's system. By the test of this analog, the operation of Tesla's system can be forecast. When the oscillator arm was set in motion by operating the motor, there were three standing waves having six loops on the 'earth's surface' all having the same amplitude of vibration! When the finger was pushed against one or more loops, all the loops were reduced in amplitude in the same proportion showing the ability to obtain all the power out at one or more points! The waves extended completely around the 'world' and returned to the sending station."[11]

Toby Grotz reports in his article that, in the 1980's, about 1/3 of the generated electrical power in this country was lost in transmission. Today, a couple of decades later, we have shamefully doubled our dependence on foreign oil and also doubled our electrical transmission grid inefficiency. From 31 Quads generated, a full *2/3 is totally wasted* in "conversion losses."[12] (This is being repeated for emphasis.) No other energy production system of any kind in the world has so much wastefulness. Instead of trying to build 2 power plants per week (at 300 MW each) for the next 20 years (only to have a total of additional 6 trillion kWh available by 2020), as some U.S. government officials want to do, we simply need to eliminate the *7 trillion kWh* of conversion losses in our present electricity generation modality. ***This book scientifically proves that Tesla's wireless transmission of power will accomplish electrical distribution, better than centralized or even, dispersed generation.***

Tesla discovered the evidence for charge clusters (as patented by Ken Shoulders and Hal Puthoff), the overunity effects of air arcs (as experimentally verified by Dr. Peter Graneau and George Hathaway), and the overunity effects of plasma glow discharge (as experimentally verified and patented by Dr. Paulo Correa).

Many believe it's time for Westinghouse, General Electric, and the J.P. Morgan Foundation to generously support a non-profit vehicle, such as a "Nikola Tesla Institute," to make amends for the billions that they reaped from Tesla's inventions. Several are considering a class action lawsuit, on behalf of Tesla's living descendants, to establish a trust fund. The reasons for legally attacking the profiteers of Tesla's inventive genius to fulfill Tesla's fondest dream of wireless transmission of power are the following. (This is a short list.)

1. General Electric, 1884: "Although Tesla had an antipathy toward the use of direct current motors, he worked to improve Edison's dynamos. He was sure he could increase the output, lower the cost, and decrease the maintenance. Edison replied, 'If you can do this, young man, it will be worth $50,000 to you.' This would mean the realization of a laboratory for Tesla and the means for a life of scientific exploration. This was what he had visioned as the meaning of America's golden promise. He set to work harder than ever, driving himself beyond his endurance, and as a result came up with the design of twenty-four different types of standard machines, short cores, and uniform patterns which were to replace the old ones. Edison was delighted with the results, but there was no $50,000 in Tesla's pay envelope and after some time, Tesla approached him for the money. It is said that Edison replied, "Tesla, you don't understand our American humor." Tesla didn't."[13] Tesla himself states this incident more succinctly, "For nearly a year my regular hours were from 10:30 AM until 5 o'clock the next morning without a day's exception. Edison said to me: 'I have had many hard-working assistants but you take the cake.' During this period, I designed twenty-four different types of standard machines with short cores and of uniform pattern which replaced the old ones. The Manager had promised me fifty thousand dollars on the completion of this task but it turned out to be a practical joke. This gave me a painful shock and I resigned my position."[14] The legal team will also detail all other legal nightmares caused by Edison, who continued to torture Tesla for years. Such examples include the court order to prevent Tesla from using GE light bulbs for

---

[11] J. B. Flowers, July 16, 1925, as quoted in *Exotic Research Report*, July, 1999, p. 48

[12] "National Energy Security Post 9/11" U.S. Energy Association, June, 2002, p. 34

[13] Hunt and Draper, *Lightning in His Hand, The Life Story of Nikola Tesla*, Omni, 1981, p.42

[14] Nikola Tesla, *My Inventions, The Autobiography of Nikola Tesla*, Hart Brothers, 1982, p. 72

the Pan American Exhibition of 1901 in Buffalo, NY and the egregious lies about a 'debt-ridden company' spread by Edison to depress Westinghouse stock, not to mention the electrocution of dogs at state fairs by Edison to show the dangers of AC electricity. It is ultimately possible that Edison can be implicated in the burning of Tesla's NY laboratory in <u>March, 1895</u>, while he was out of the city. The motivation for the crime was overwhelming: Edison (General Electric) lost the Columbian Exposition light bulb contract to Westinghouse in 1892 to the tune of $400,000. General Electric also lost the generator contract for the three initial 5,000 horsepower generators at Niagara Falls in 1893 and was forced to secure a license for the use of Tesla patents. Tesla (Westinghouse) completed the powerhouse in 1895 and residents of Niagara Falls turned on the lights in <u>April, 1895</u>, proving the superiority of AC electricity. A year later and 20 miles away, Buffalo, NY would be the first city in the world to have electric street lamps. Meanwhile, GE lawyers could only repeatedly file petty lawsuits to wear down Westinghouse, so that eventually, it was called, "The War of the Currents." The uneducated Edison led the groundless and unscrupulous battle by scaring the public with words like, "Just as certain as death, Westinghouse will kill a customer within six months after he puts in a system of any size. He has got a new thing and it will require a great deal of experimenting to get it working practically. It will never be free from danger."[15]

2. <u>Westinghouse, 1888:</u> Tesla was awarded patents on the AC system of motors and generators in May, 1888. "Within a few months, Westinghouse acquired the patented American rights and hired Tesla at a salary of $2,000 a month to work in Pittsburgh on the development of the polyphase system. Tesla's system for the transmission and distribution of alternating current, including the induction motor, was covered by 40 historic patents. His motor was the missing link for today's alternating current system of centralized electric generating stations capable of efficiently and economically distributing electricity over long distances. It is widely believed that Tesla received a million dollars for his patents and that Westinghouse was to pay Tesla $1 per horsepower for each AC motor produced. However, according to Westinghouse historical records, the contract specified that Tesla was to receive about $60,000 and earn $2.50 per horsepower for each motor produced. *Four years after* the contract was signed, it was rumored, the accrued royalties totaled approximately $12 million. Westinghouse was advised to get rid of the royalty contract when his form was in financial trouble and the fate of his company was at stake. So Westinghouse told Tesla he did not think he could honor the royalty clause... The 1897 annual report of Westinghouse shows that *Tesla was paid $216,600 for outright purchase* of the polyphase system patents."[16] Over 110 years later, the amount of profit that Westinghouse has realized from Tesla's polyphase system has to be embarrassingly huge amount of money. Today, in comparison, pharmaceutical companies routinely ask Congress to extend the term of their patents beyond 20 years, just so they can "recover" more profit.

3. <u>J. Pierpont Morgan, 1901:</u> "It has been stated that Morgan simply gave Tesla $150,000 with no strings attached. Actually, there were plenty of strings attached. Morgan delayed his check for a few months. Finally it came with the stipulation that fifty-one percent of the patents relating to wireless telephony and telegraphy, not only those to

---

[15] John Shatlan, "Tesla: Scientific Superman who Aided Westinghouse Industry," *Pittsburgh Business Journal*, July 19, 1982 and the *Tesla Journal*, 1986, p. 60

[16] Ibid., p. 60 (Ed. note: the same facts are also found in O'Neill's biography, *Prodigal Genius*.)

be used in the present but the ones to be developed – all were to be in Morgan's name. The $150,000 was well-secured...On March 1, 1901, Tesla sent to Morgan his contract, *signing over the fifty-one percent interest in his patents and inventions and in any future ones relating to electric lighting and wireless telegraphy or telephony*...Morgan's $150,000 was woefully inadequate when Tesla considered all that must be done, but it was a start. He secured a tract of land on Long Island, about sixty miles from New York City, though an arrangement with James S. Warden. Tesla had pictured to Warden a glowing and convincing real estate boom in that site, employing several thousand people who would build their homes on the adjacent land. Warden cooperated to the extent of offering two hundred acres of land for the use of the scientist, twenty acres already cleared and with a well one hundred feet deep. By July 23, 1901, work had started on the project with the roads cleared and the right of way in order. Thus, within a little less than five months after the contract with Morgan was signed, work was started on Tesla's giant project."[17] The rest of the horror story is history, as only the tower frame was erected in the next year. No more money was forthcoming for the project that Morgan initiated, even when the equipment cost alone cost about $200,000. Morgan believed that he would "have nothing to sell except antennas (and refused) to contribute to that charity."[18] Tesla tried and tried for years until in 1917 the U.S. government blew up the abandoned Wardenclyffe tower because suspected German spies were seen "lurking" around it. With Edison as his willing ally, Morgan even publicly discredited Tesla's name, so that all of the five school textbook publishers of the time removed any reference to him. Any wonder why even today, 100 years later, hardly anyone knows who Tesla is?

Upon reading the rest of this book, all of us who contributed to this book know that the engineers and physicists of the 21st century will come to appreciate the benefits of the tremendously efficient (about **95%**) wireless transmission of power. In terms of today's systems theory, Tesla understood that it is vital to "increase human energy" in order to maximize the quality of life worldwide.[19] (See Puharich article for a detailed analysis of this Tesla theme.) In terms of economic theory, many countries will benefit from this service. At first, receiving stations will be needed. Just like television and radio, only an energy receiver is required, which may eventually be built into appliances, so no power cord will be necessary! Just think, monthly electric utility bills will be optional, like "cable TV."

Tesla was an electrical genius who revolutionized our world in a way that DC power could never have accomplished, since the resistance of any transmission lines, (except perhaps, superconductive ones), is <u>prohibitive</u> for direct current. He deserved much better treatment from all three of the tycoons described above, than to spend the last 40 years of his life in abject poverty. However, he was too much of a gentleman to hold a grudge. Instead, regarding the magnifying transmitter, Tesla wrote in his autobiography, "I am unwilling to accord to some small-minded and jealous individuals the satisfaction of having thwarted my efforts. These men are to me nothing more than microbes of a nasty disease. My project was retarded by laws of nature. *The world was not prepared for it. It was too far ahead of time. But the same laws will prevail in the end and make it a triumphal success*."[20]

---

[17] Hunt and Draper, p. 136
[18] H.W. Jones, "Nikola Tesla, Generator of Social Change," *Proc. of Inter. Tesla Sym.*,'86, p.1-89
[19] Nikola Tesla "The Problem of Increasing Human Energy" *The Century Illustrated Monthly Magazine*, June 1900, p. A-109-A-152
[20] Nikola Tesla, *My Inventions*, p. 91

This book is being published in time for the <u>Wardenclyffe Tower Centennial</u>, (1903-2003) which to many, signifies an extraordinary cause to remember and resurrect. Let us fulfill this prophesy of Tesla, making it a triumphal success, by supporting a philanthropic, international wireless power station to benefit the whole world. The scientists who contributed to this anthology are available to make such a global wonder a reality. The benefits, immediately alleviating electric power shortages everywhere, are too numerous to count. (For example, in Tesla's homeland, the Electric Power Company of Serbia will raise their monthly rates by 50% on the day this book goes to the publisher.) Are you willing to help make a world of difference?

(Editorial comments are inserted in many of the following contributed articles. They represent my scientific viewpoints, which may help other researchers. You can recognize these additions by the familiar ending: "– Ed. note.")

All patents cited in this book are available from www.uspto.gov or even better, from www.GetThePatent.com where a free viewer is available http://www.catesianinc.com/products/cpcviewax/install/ or at last resort, send $3 to the USPTO, Box 9, Washington, DC 20231 with the patent number. – Ed. note

Thomas Valone can be reached through Integrity Research Institute, a nonprofit, 501(c)3 corporation, located at 1220 L St. NW, Suite 100-232, Washington, DC 20005, www.IntegrityResearchInstitute.org and iri@erols.com

SECTION I

History

"This coil…shown in my patents Nos. 645,576 and 649,621, in the form of a spiral, was, as you see, in the form of a cone…in an inductive coupling which was not close – we call it now a loose coupling – but free to permit a great resonant rise. That was the first single step, as I say, toward the evolution of an invention which I have called my 'magnifying transmitter.' That means, a circuit connected to ground and to the antenna, of a tremendous electromagnetic momentum and small damping factor, with all the conditions so determined that an immense accumulation of electrical energy can take place." – Nikola Tesla (*Nikola Tesla on His Work with Alternating Currents*, Leland Andersen, Editor, Sun Publishing, 1992, p. 72)

2 Reflections from Tesla's Descendent

William H. Terbo

Reprinted from *Proceedings of the International Tesla Symposium*, 1990

This is the fourth biennial Tesla Symposium. I've had the pleasure of making some opening remarks at each one. My appearance here has a twofold purpose. First, as a descendant of Nikola Tesla, I hope to provide a link between the man we honor and those of us who are here to honor him. Second, as a representative of the Tesla Memorial Society and its Honorary Chairman, I want to reaffirm the cooperation between the International Tesla Society and the Tesla Memorial Society in moving toward our common aims. Simply put, they are honoring and perpetuating the memory and ideals of Nikola Tesla through appropriate cultural and academic activities. The Tesla Symposium is a worthy representation of these aims.

This morning I'd like to touch briefly on three topics. First, to re cap some of the events and accomplishments of the two years since our last symposium which have reflected positively on the name and reputation of Nikola Tesla--plus mention of two or three current projects. Second, I'd like to share some historical detail on the original Tesla/Westinghouse power generating system at Niagara Falls. And lastly, I'd like to provide some personal thoughts about the private character of Nikola Tesla.

Before I begin, let me ask the Executive Secretary of the Tesla Memorial Society, Nicholas Kosanovich, to stand and be recognized. Nick, more than anyone, is responsible for the continued success of the Society. He does yeoman work. I don't know how he does it, I certainly haven't been able to get into harness that way.

Recent Events and Accomplishments

First, let me talk about some of the events and accomplishments that have happened recently. They all tend to promote the name of Tesla. What we are trying to do is gain the recognition for Tesla that he deserves and these all work to build toward that aim.

Pennsylvania and New York have issued proclamations naming July 10 as **Nikola Tesla Day**. At least six U. S. Representatives and Senator Carl Levin of Michigan have made speeches in Congress commemorating Tesla's July 10 Birthday. When we get copies of the Congressional Record, we will certainly make them available to the International Tesla Society. There may be other recognition of this day and we will just have to wait on them. Sometimes we only hear about the proclamations of various States after the fact.

Another item that has been a long time in the works and has come to fruition is a plaque of Tesla at the United Engineering Center Headquarters [345 West 47th Street, New York, NY 10017] near the United Nations in Manhattan (Figure 1). That is also the International Headquarters of the Institute of Electrical and Electronics Engineers. They occupy three or four floors of that building. This is a large plaque, more than one person can lift. It has been put in one of the most desirable locations in the building. It's in a hall of other awards and plaques that connects the lobby and the first floor, the most desirable and prestigious location there. It was originally a gift of the Yugoslav Government and meant to be placed on the New Yorker Hotel, where Tesla died. The New Yorker Hotel, of course, now is a dormitory for Dr. Moon's Unification Church, and there has always been some difficulty in getting cooperation on that account. So I think a better solution was to have it installed at the Headquarters of the IEEE and, with their assistance, this was done. Another plaque was just installed in Belgrade, about 100 years after the fact, to commemorate Tesla's visit to that city in 1892.

The IEEE Power Engineering Society has as their principal annual award, the Tesla Award and Medal (see A-4), which has been issued each year since 1976. In 1989 the recipient was Dietrich R. Lambrecht, an engineer working on turbines for Siemens in Germany. In 1990, just recently, the winner of the reward was Gordon R. Slemon [A-4].

In June 1990, the IEEE and the New York Power Authority dedicated the Adams Hydro-Electric Generating Station in Niagara Falls. They designated it as an "Engineering Milestone," as this is the original plant built to create Alternating Current from Niagara Falls in 1895. The power was first generated in 1896. This is a further indication of the support that we are getting from the IEEE. It certainly is welcome, because having the scientific community support the name of Tesla is always very important.

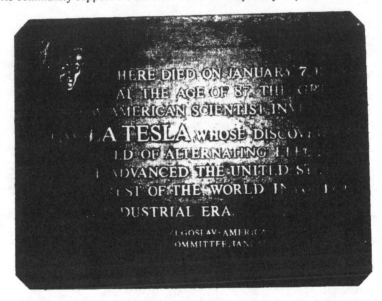

Figure 1. A plaque honoring the achievements of Nikola Tesla. Now located in the United Engineering Headquarters in Manhattan, it was originally to be placed at the Hotel New Yorker, where Tesla had been a long time resident. -- William Terbo

John Wagner has been one of our most active members and is now taking charge of the Youth Division of the Society. He's done a number of things that are worth remarking. First, he's had a bust of Tesla created (Figure 2). It is a very good likeness of Tesla in his prime, in his late thirties. The bust has been purchased by donation and is available to be given to the Smithsonian Institution in Washington, DC. It presently is in the Engineering Library at the University of Michigan in Ann Arbor. It is a fairly costly item and money has been raised by a number of parties including the rock group that goes by the name "Tesla." It's a wonderful spectrum of support that we get. Here is a group that looked upon Tesla as being an outsider who had difficulty in making his way and I think heavy metal rock groups might be considered outsiders having a difficult time making their way. So, they've taken the name, it's not an ethnic connection and so this is really quite generous on their part. They're quite successful, incidentally.

Trying to get the Smithsonian Institution to recognize Tesla has been difficult because their tendency is toward Edison, that is, whenever there is a controversy between Edison and Tesla, the Smithsonian tends to side with Edison. In conjunction with this, John has been instrumental in starting a campaign of signatures of people to make a petition. His aim is to gather 100,000 signatures. Although it sounds like a formidable number, he's got 37,000 already. These signatures are gathered not only by professional institutions, colleges and universities, but by people who are attending the rock concerts by the "Tesla" group. They've actually been responsible for 70 to 75 percent of those signatures. The way they're going, they WILL collect the 100,000.

Figure 2. John Wagner's bust of Tesla in his late thirties. -- John Wagner

Another element John has developed for promoting the name of Tesla and for trying to get the attention of the Smithsonian are top quality sweat shirts and T shirts he has had made that say on them "Bust the Smithsonian." He has done this because the Smithsonian has not yet agreed to accept the Tesla bust. Incidentally, over 250 of these items have already been sold, with all proceeds over actual costs going into a fund for the purchase of additional busts.

Further, John has been carrying on an effort to enlist the support of university level educational institutions to properly recognize the scientific contributions of Nikola Tesla. He has encouraged his students in a letter writing campaign that has produced suggestions to University Physics and Electrical Engineering Department Heads as well as the CEO's of 51 power generating Utilities.

Tesla Biographies

Although its schedule was previously unannounced, I can now mention that Yugoslav TV is now midway through a process of making six, 40-minute biographical episodes on Tesla's life. Those are going to be in Serbo-Croatian, but probably will be subtitled in English. In addition, a 60-minute fully English language film version is planned to be available. I understand that three of the six, 40-minute segments are already in the can and that they are aiming for a 1991 screening date which will commemorate his 135th birth date anniversary. In addition, Henry Golas, who has the film rights to Margaret Cheney's book "Tesla: Man Out Of Time," is still proceeding with his efforts to get that into production. I know of at least two other partially funded videos for public TV that are in the pre-production stage.

In the area of books, Margaret Cheney has indicated that a new hard bound edition of her book "Tesla: A Man Out Of Time" will be available shortly. Another book is being written by Dusko Doder, a very prominent newspaperman. He was the "U S News and World Report" Bureau Chief in Beijing at the time of the unrest and is now based in Belgrade, acting as Bureau Chief for Moscow and Belgrade. He has already published a book which has been quite successful, called "Gorbachev: Heretic In The Kremlin" which has been well received and is now starting on a biography of Tesla. A third book, by a young woman, Carol Costa, is a biography that is oriented toward youngsters, that is now in the process of being published.

"Nikola," a play written by one of our members, a young professional playwright, Karen A. Klami, is a dramatization of crucial events Tesla's younger life. It has had several professional readings and a commercial production is being organized. I don't know whether it will appear on Broadway, but it will appear someplace.

As far as the Tesla Memorial Society is concerned, we've released our Tesla Journal double issue for 1989/90. It is available in the bookstore. It's a good solid piece of scholarship.

The Nikola Tesla Museum

Dr. Marincic will be speaking later in the program and will discuss the Tesla Museum in Belgrade. Certain renovation has been done so it's in really top shape. I think this is important because the archives are there and climate control in that building has to be very carefully monitored. I think that's the essence of the renovation that they've done. It's a beautiful mansion from the turn of the century, like one of the three or four story granite mansions that were once seen on 5th Avenue in New York City (Figure 3). Also in Belgrade, the power generating companies of Yugoslavia have organized their new institute for research. It's in a new building and named "The Tesla Institute," as are most of the power companies in Yugoslavia named after Tesla.

Long Term Projects

The Tesla Memorial Society has three long term projects that are not expected to happen right away and are going to require a lot of long term effort. First, there is solid planning going ahead for a museum at Niagara Falls which will incorporate some of the subterranean tunneling that was done for the original power installation (Figure 4). There are still remnants of the old power plant down there and it can make an interesting tourist attraction, almost like investigating tombs or something of that sort. It has all been sealed up for decades. These tunnels can be the basis for a tourist attraction that will certainly mention Tesla prominently.

Shoreham, Long Island, is the site of the Wardenclyffe Tower that Tesla built in the early part of this century. The foundation of the tower is still there and the old laboratory building next to it is intact and being used for storage. We are trying to get a formal dedication of that site. The building and surrounding complex is privately owned by Agfa-Gevaert, the big Belgian photochemical company, and is no longer used for production. It's a large industrial complex of many acres, and the site occupies only one acre or so on the side, so we may have a possibility of getting their cooperation. It won't happen overnight, that's for sure.

Of course we're looking forward to 1993, the 50th Anniversary of Tesla's death, as an appropriate time to have an exhibition at the Smithsonian. Because of the bent of the Smithsonian, that's going to take a lot of work, but I think something will happen and we're marshalling our forces.

Figure 3. The Nikola Tesla Museum in Belgrade, Yugoslavia. -- Nikola Tesla Museum

The Tesla/Westinghouse Power Generating System

Let me continue now with just a few words about Niagara Falls. I happened to acquire a copy of the very enlightening 1901 promotional brochure from the Niagara Falls Power Company, that was made as a presentation in getting new subscribers to their system. The real age of electric power, in the modern sense, started in Niagara Falls on November 16, 1896, when the first power generated by the Tesla/Westinghouse AC system was delivered to Buffalo, New York. This was the culmination of three years of design and construction. By 1901, according to this booklet, the Niagara Falls Power Company (the predecessor of the Niagara Mohawk Company) had about 45,000 Effective Horsepower capacity and was marketing this new source of power to industries and municipalities in the area. In their brochure, they reviewed the quantity of power used and the applications for each of about 35 customers that they identified by name. By today's standards the quantities seem trivial but, as with the dollar, in 1901 a Horsepower WAS a Horsepower.

Some of the companies listed were:

The Pittsburgh Reduction Company, which is the predecessor of Alcoa. They were using 5,000 horsepower for the electrolytic production of metallic aluminum. They were the largest commercial user (as opposed to municipalities).

Carborundum, who used 2,000 horsepower for electrolytic production of abrasives.
A company called Castner Electrolytic Alkali Company (there is a chemical process called the Castner Process) used 2,400 horsepower for electrolytic production of pure caustic soda.

Union Carbide was making calcium carbide, using the power for electric furnaces.

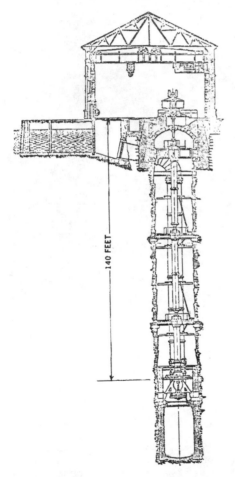

140 FEET

Figure 4. The Canal, Penstock, Turbine, Dynamo and Tunnel of the Niagara Falls Power Companies, Adams Plant Number One Powerhouse. -- Radio Electronics

Natural Food Company, which later became the Shredded Wheat Company used 2,500 horsepower for baking and for motors and for lighting in the plant. Their plant is still standing there very close to the Falls.

About 1,500 horsepower was being carried to Buffalo and to nearby points, principally used either for lighting, electric railroads or electric traction companies. A small portion was also being sold to manufacturing companies.

It made such a change. That availability of cheap and clean electric power alone provided jobs for about 10,000 Niagara Falls people (besides the people outside of Niagara Falls). Those jobs didn't exist before because the processes that were possible with this electrical power were not economical earlier. It turned Niagara Falls from a town of about 10,000 people into a city of about 40,000. Most of the companies that were in distant areas, instead of using electrolytic processes where, what were then, massive quantities of electrical energy were needed, were only using from 20 to 200 horsepower apiece, principally for running industrial machinery in factories. This was, of course, the first practical large scale application of the Tesla Alternating Current concept and was an event that was heard around the world.

Demand was terrific. By 1916, it had already exceeded 200,000 horsepower. The Adams Power Station, the first power station when it was completed in 1896, was by far the largest producer of Alternating Current in the world. It is the remnants of that plant that were dedicated as an Engineering Milestone just last June.

Personal Thoughts

Let me talk for just a moment about the character of Tesla. Much has been written about the public persona of Nikola Tesla, usually dealing with the creation and implementation of his inventions and discoveries. They show a man of intellect, dedication and perseverance, but an over abundance of these quantities creates a picture of a somber and sober person with a single-mindedness that seems to preclude a more human side.

Much has also been speculated on the private side of Nikola Tesla. This speculation was much influenced by his fastidious dress, his formal manners and his precise writing style. Aside from the more outrageous speculations, the conclusion was usually drawn that this was an introverted and driven workaholic, without a fully developed personal side. However, put in the context of the times, dress, manners and precision in writing were really hallmarks of the decades on either side of the turn of the century. We have very few instances of verifiable anecdotes from his closest friends because such gossip rarely found its way into print. (What a change from today's celebrity bashing.) But, in context or out of context, these characterizations of Tesla's private personality have been perpetuated.

Figure 5. The 1901 Promotional Brochure of the Niagara Falls Power Company. -- Niagara Falls Power Company

Now, over a period of time, I have thought of this and it occurred to me that I may very well have a unique insight into Tesla's private personality through a comparison with my father. The parallels in their lives far exceed the common gene pool formed by their blood relationship. (My Grandmother was Tesla's sister, Angelina.)

Except for being 30 years apart, the list of parallels is really staggering:

Lika, the Croatian county of their birth in what is now Yugoslavia, had a special status and responsibility for Serbs as it was part of the Austro-Hungarian military frontier. The Ottoman Empire's boarder was that close to Vienna, a situation that lasted for a period of hundreds of years.

Both Tesla and my father were sons of Serbian Orthodox priests and that's a very severe and demanding faith. (I don't know if anybody here is Serbian Orthodox, I'd like to say that I'm Episcopal for that very good reason.)

Both received a technical education far from home, which was necessary in those days. There was no higher education readily available in the provinces. You had to go, as in my fathers case, to Budapest.

Both became engineers and worked for similar lengths of time for the Budapest Telephone System.

Both emigrated to the United States at 28 years of age.

And, as inventors, made their most important inventions in their middle to late thirties.

Both exhibited qualities of dedication and patience and modesty and a philosophical turn of mind.

Both were strong and vigorous and died at the age of eighty seven.

But, both suffered from a certain naivete, particularly in business.

Both acquired, but let slip from their hands, a considerable fortune.

Now, I knew my father pretty well, even better from the perspective of the years since his passing And I see a human being with human qualities and frailties, and it seems reasonable to attribute to Tesla some measure of these same qualities and frailties.

This brings me to an unexpected point. My father had a sense of humor and therefore, Tesla must have had a sense of humor that has been overlooked in the myth of his private self. (There certainly aren't any books that are titled "Favorite Jokes of Nikola Tesla".)

My father was a story teller, and usually was recalling things that happened when he was a boy. I heard them all dozens of times. I can't repeat them because they'd lose their essence in his way of telling them but my father could never finish these stories without collapsing in laughter. He would get into the memory of some story and would just get helpless with laughter. We all knew what the end of it was and he'd struggle through ultimately. It was an infectious laugh and we didn't mind the same old stories. I still remember them clearly.

And so, in closing, I just want to say that I've taken a long way around, and had a chance to remember my father. But the next time that you think of Nikola Tesla, that discoverer of great concepts, think of him convulsed in gales of laughter, trying to tell some boyhood yarn.

About the Author

Born April 10, 1930, to Nicholas J. and Alice H. Terbo in Detroit, Michigan, William H. Terbo is a Founding Director and Honorary Chairman of the Tesla Memorial Society. His father, Nicholas Terbo (Nikola Trbojevich), a world known research engineer, mathematician and inventor, was nephew and friend of Nikola Tesla. Mr. Trbojevich held nearly 200 U.S. and foreign patents, principally in the field of gear design, including the basic patent for the Hypoid Gear - used on nearly every automobile in the world.

Mr. Terbo's father modeled his professional life after Nikola Tesla, a man 30 years his senior. He was the only family member to join Tesla in the United States, immigrating in 1914, 30 years after his uncle. With such a family history in science and engineering, Mr. Terbo's higher education was a matter of "which engineering school" rather than "what area of concentration." He graduated from Purdue University with a degree in Mechanical Engineering, an area close to his father's specialty.

Mr. Terbo began his professional career as a Stress Analyst in Detroit before moving to Los Angeles where he specialized in computer oriented systems for the Space Program. Since 1973, he has been involved in Strategic Planning and Market Research for the Telecommunications Industry. Working as the Manager of Market Planning and Research for RCA Global Communications, Inc., and more recently as the Senior Staff Member for Corporate Development with MCI International, Inc.

William Terbo can be reached at 21 Maddaket, Southwyck Village, Scotch Plains, NJ 07076

3 Nikola Tesla – Electricity's Hidden Genius

Keith Tutt Copyright © Keith Tutt, 2001
Reprinted from *The Scientist, the Madman, the Thief & Their Light Bulb*, Simon & Schuster Pub.

"I have harnessed the cosmic rays and caused them to operate a motive device."
Nikola Tesla, *The Brooklyn Eagle*, 10 July 1931

"Ere many generations pass, our machinery will be driven by power obtainable at any point in the universe. Is this energy static or kinetic? If static, our hopes are in vain; if kinetic – and this we know it is for certain – then it is a mere question of time when men will succeed in attaching their machinery to the very wheelwork of nature."
Nikola Tesla, "Experiments With Alternate Currents of High Potential and High Frequency," 1904

In 1884 a young Croatian immigrant stepped ashore at the Castle Garden Immigration Office in Manhattan, New York. He was a sharp-featured 27-year-old with a glamorous shock of black hair, named Nikola Tesla. In his coat pockets he carried a few coins, some papers with drawings and calculations on them and, perhaps most importantly of all, a letter of introduction to Thomas Alva Edison, the incumbent king of electricity.

Behind Tesla there was already an extraordinary past filled with invention, hardship and a series of near fatal accidents and afflictions. Ahead of him lay a future in which many of the things he had already imagined would come to pass for the benefit of the world. And yet his greatest wish – of freely available electrical energy for all – would be denied him.

Blessed with an extraordinary mind capable of extravagant and yet detailed visual imagination, Tesla was a complex prodigy who suffered from strange over-sensitivities and symptoms of what we would now call an obsessive compulsive disorder. As well as one of the most highly developed forms of photographic memory, Tesla claimed to possess a superhuman, almost supernatural, power of hearing which enabled him to hear conversations hundreds of yards away and – in a few instances – to hear thunder up to 500 miles away. During a teenage nervous breakdown Tesla could hardly go out of his home, as he had become hyperaware of sounds, atmospheric pressures and sunlight. He seemed to feel the impact of natural phenomena directly within his body. His compulsive side brought long periods of needing to count physical actions he performed – steps along a road, mouthfuls of food, even breaths: he behaved like a self-monitoring machine, a mobile laboratory which his psyche had decided to investigate. Later, when he was able to bring the exercise of his will power to bear over these compulsions, he would make good use of this internal observation.

Invention came naturally to Tesla from an early age. When he was five he modelled a waterwheel which worked without the use of any conventional blades; he was later to recall this when he designed his bladeless turbine [1] He designed a device in which imprisoned beetles powered a wheel with the flapping of their wings. He tried to fly from the top of the family house using an umbrella – a feat which nearly killed him. He tried to take apart and reconstruct his grandfather"s clocks, a skill which had its limits: "In the former I was always successful, but often failed in the latter."[2]

In 1875, at the age of eighteen, he enrolled at the Austrian Polytechnic School in Graz, Austria, where he studied mathematics, physics and mechanics. He was determined to complete the two-year course in one year, and worked most days from three in the morning until eleven at night. One aspect of his compulsion was a need to complete anything he had started.

While it later became a helpful force within his creative production, it often drove him to despair. At college he had started to read the works of Voltaire when he discovered that there were nearly one hundred volumes in small print. Such was the strange conscience of his psyche that he could not rest until all were read.

It was during his time at Graz that his ideas about alternating current first started to surface. Professor Poeschl, a German, was Tesla's inspirational teacher of theoretical and experimental physics. One day Poeschl showed the class a new electrical machine that had just arrived from Paris: called a Gramme Machine, it could function as both a direct current (DC) motor and a dynamo. Tesla reported later that he felt strangely excited by the machine"s arrival. When it was operating the machine"s brushes sparked wildly. Tesla suggested to his teacher that the machine could be improved if the commutator were done away with, and if it were to run instead by alternating current. He didn"t know how this might be done, and yet he had an instinct that somehow the answer might lie within his own mind. The professor was less confident: "Mr Tesla may accomplish great things but he will never do this. It would be equivalent to converting a steadily pulling force, like gravity, into a rotary effort. It is a perpetual motion machine, an impossible idea."[3] However, Tesla's need to complete things would not let this idea rest: "With me it was a sacred vow, a question of life and death. I knew that I would perish if failed."

With this motivation burning away inside him, it was a few more years before finally a burst of creativity hit the young Tesla. He was taking a walk in Graz"s city park with Anital Szigety, a mechanic friend, at the same time reciting a passage from Goethe"s Faust. Then, as Tesla reported it: "The idea came like a flash of lightning, and in an instant the truth was revealed." Tesla started to draw in the dirt with a stick for his friend to see: "See my motor here; watch me reverse it! [4]

He had hit upon a whole new system of electrical operation based on the totally novel concept of producing a rotating magnetic field by running two or more alternating currents out of phase with each other. The rotating magnetic field completely did away with the need for the conventional brush contacts and commutator of the normal DC motor. In his creative flash he had discovered multiphase alternating current (AC) – a leap forward which would make possible the high-voltage widescale generation, transmission and distribution of electricity that is still the worldwide standard today. In that same moment he had also shown Professor Poeschl the error of his skeptical ways. Over the next days, Tesla designed most of the new machines and devices required by the multiphase AC system: particularly the induction motor and all the equipment required for the generation and supply of AC electricity. He wrote of his work: "It was a mental state of happiness about as complete as I have ever known in life. Ideas came in an interrupted stream, and the only difficulty I had was to hold them fast." His work also provided an example of his extreme gift of visualization: "The pieces of apparatus I conceived were to me absolutely real and tangible in every detail, even to the minutest marks and signs of wear. I delighted in imagining the motors constantly running."[5]

As well as an extraordinary intuitive gift for new technological ideas, Tesla was blessed with this extreme form of "mental practicality", by which he was able to save himself many hours of wasted effort in engineering time. Instead of building real, physical devices, he would usually design and construct them in the workshop of his creative imagination. In this virtual testbed, he would set them running, later returning to see what had happened, what had worn or broken down, what had functioned correctly or incorrectly. He would then make imaginative improvements in order to make the devices more efficient or effective, before continuing this refining process. When he was absolutely happy with his mental creation, he would then, and only then, commit his idea to physical reality. It was this gift above all others that enabled him to be so prolific as an inventor.

When in 1884 the confident Tesla set off for America, however, with the AC system and its components firmly embedded in his mind, he had little idea of the difficult path that lay between him and acceptance of his technology – a path that threatened to both make and break the young Tesla.

The War of the Currents

Straight off the ship in New York, Tesla headed for the offices of the Edison Electric Company, where he found the 32-year-old dynamo of the new world Thomas Edison. Already the inventor of hundreds of products and the owner or co-owner of many electrically related companies, Edison was a self-educated genius with the street smart of an alley cat. Tesla presented his letter of recommendation to the short-tempered Edison – a letter from Charles Batchelor, one of Edison's trusted officers in Europe. The note, addressed to Edison, was entirely flattering: "I know two great men and you are one of them; the other is this young man."

Within moments Tesla was attempting to explain his new induction motor and the development of the multiphase alternating current, but was stopped dead in his tracks by an angry Edison. His response was short and sharp: "Spare me that nonsense. It's dangerous. We're set up for direct current in America. People like it, and it's all I'll ever fool with."[6]

Edison was totally opposed to anything but his own DC system, believing, erroneously as it turned out, that his incandescent light bulbs would not work with AC current. Nevertheless he offered the crestfallen Tesla a job on his workshop crew. It was hardly the last he was to hear of Tesla's AC breakthrough. Once Tesla left his employ – following a broken promise over a $50,000 bonus owing to Tesla – he would team up with George Westinghouse, the Pittsburgh business magnate. While Tesla was a scientific genius of the highest level, he faced a continual challenge to fund the great, but expensive, plans that his imagination provided. When he joined George Westinghouse in 1888 to bring AC electricity to the whole of America, he signed a contract, which gave him royalties of $2.50 for each horsepower of generating capacity licensed. The War of the Currents – the battle to electrify America – had begun in earnest.

While Edison had managed to electrify the wealthier parts of New York with a series of local coal- and steam-driven generating stations, his stubbornness could not allow him to think that there might be a more electrically efficient and more cost-efficient solution. With the backing of Pierpont Morgan, one of the wealthiest and most ruthless businessmen of his time, Edison had pinned his colors firmly to the DC mast, and there was no turning back. For him it was a battle to the death – although the fatalities were, in the end, innocent and unlikely victims.

In the War of the Currents Edison became a sinister P.T. Barnum figure: dogs and cats were collected off the streets and publicly electrocuted by Edison to demonstrate that AC electricity was dangerous – even lethal. Edison even convinced the New York State prison service to employ early AC electrical equipment in the world's first electrocution of a convicted murderer. AC was so dangerous, he contended, that all it was good for was killing.

Despite Edison's propaganda, the 1893 Chicago World's Fair saw Westinghouse and Tesla emerge as victors in the War of the Currents, with a combination of showmanship and technical superiority. The same year Westinghouse was awarded the contract to manufacture the generating equipment for the electrification of Niagara Falls, and Tesla was to be in charge of the design. In a compromise, General Electric, which had taken over the Edison Electric Company, was to supply the transmission and distribution lines for the twenty-six miles from Niagara to Buffalo – the nearest major city. Yet even General Electric's proposal was now based on alternating current technology. For Tesla this was a double triumph: not only had alternating current been accepted for its technical superiority, but he had also been given a strange confirmation of the power of his mind.

At the time he had modeled his first waterwheels, while in school in Gospic, Croatia, he had seen some pictures of Niagara Falls in a school book. He had experienced a powerful reaction, and – as often – further associated creative pictures had appeared in his mind. He saw a huge wheel with water cascading over it. He told his uncle that one day he would travel to America and make this waterwheel. Some thirty years later his prophecy had come true.

By 1897 his royalties from AC were already worth some $12 million, and had they continued they could have reached billions. Tesla would have been the Bill Gates of his day. It was not to be. Westinghouse came under pressure from his commercial enemies. The General Electric Company managed a dirty tricks campaign that lowered the Westinghouse Company's stock and made it close to impossible for it to continue independently. George Westinghouse had to go back to Tesla and ask him to forego all his royalties -past, present and future – in order that the company could survive independently. Tesla, who believed that Westinghouse could still fulfill his dream of AC for all, gave up his right to the millions he was due, and accepted a single payment of just $216,600 for the outright purchase of all his AC patents. A large sum, perhaps, but not enough to independently fund Tesla's researches into the even more radical energy technologies that were already spinning around his mind.

Westinghouse survived to fight another day with General Electric over the country's seemingly infinite energy needs, even though court fights over patents would sap the company financial reserves for many years to come. From that time on it would be others who would benefit from Tesla's genius.

Forgotten Genius?

To demonstrate the genius of Tesla, we only need to list some of his patented inventions apart from those related to AC electricity: the arc light; the speedometer; the first radio-controlled boat; superconductivity; and the first tube light. He also laid the ground for radar, cryogenics, wireless radio and telephony, the use of X-rays and our understanding of the sun's cosmic rays. Cosmic rays were at the heart of some of Tesla's later ideas about energy production. In his own time, though, there were few who could accept his concept that the sun threw out showers of tiny, highly energetic, fast-moving particles. Although no record remains of his methods he claimed that he had measured their energy at hundreds of millions of volts. [7] Thirty years after he first aired his controversial theories, two Nobel laureate physicists, Dr Robert A. Millikan and Arthur H. Compton, admitted their debt to Tesla's work, even though they disagreed violently about the nature of the rays – whether they were in fact photon (light) rays or, as Tesla had believed, charged particles. Millikan, though, managed to measure their potential at 64 million volts, close to Tesla's figure. We now know that cosmic rays, which are many and varied, result from the formations, decays and collisions of many different kinds of particles – some from the sun and some from other, more distant stars, novae and supernovae. Nevertheless, Tesla's principal concept was closer to the truth than any of his contemporaries knew.

Many of Tesla's discoveries and inventions are often mistakenly attributed to better-known names. While most lay people still believe that Marconi perfected the transmission and reception of radio waves, there is no longer reason to believe this: in June 1943 the US Supreme Court ruled that Tesla's patents predated Marconi's claims on the prize of radio. Popular history is, though, still slow to catch up. Errors committed in print can take many years to correct. The just do not always get to write the history books, and even during his lifetime Tesla became an object of ridicule and derision for his "outlandish ideas."

There were times when he may have contributed to this -for instance when he agreed with Lord Kelvin in 1902 that Mars was trying to make contact with America. (It is now believed he may have been the first person to have measured – without realizing its origin – the pulsing of distant stars.) However, Kelvin and Tesla also agreed on a further, more prophetic point: that the world's non-renewable resources – such as coal and oil – should be conserved and that wind and solar power should be developed [8] Tesla's creative scientific skills seemed to know few boundaries; yet many who saw him work were scared by his radical approach to natural forces. In public demonstrations he would often wreathe himself in sparks and crackling bolts of high-voltage electricity without ever seeming to do himself harm:

"I still remember with pleasure how, nine years ago, I passed the discharge of a powerful induction-coil. through my body to demonstrate before a scientific society the comparative harmlessness of very rapidly vibrating electric currents, and I can still recall the astonishment of my audience. I would now undertake, with much less apprehension than I had in that experiment, to transmit through my body with such currents the entire electrical energy of the dynamos now working at Niagara -forty or fifty thousand horsepower. I have produced electrical oscillations which were of such intensity that when circulating through my arms and chest they have melted wires which joined my hands, and still I felt no inconvenience." [9]

A famous photograph of Tesla captures him sitting on a chair in the laboratory he built at Colorado Springs in 1899. From the huge electrical coil in the centre of the room, white arcing sparks – some over twenty feet long and as thick as a man's arm – squirm and leap around him. With millions of volts of electrical charge appearing to surround his posing figure, he seems perfectly, archly, "at home" – and to prove it he is calmly reading a book. It is a seminal image of the man who was more comfortable with the awesome power of natural electricity than perhaps anyone else – either before or since. The image is, in fact, a double exposure, a flashy kind of hoax; nevertheless, it demonstrates a key part of Tesla's personality – his love of showmanship.

Transmission Without Wire

While many of Tesla's dreams were achieved, his most ambitious visions remained unfulfilled during his lifetime. It is a matter of some considerable speculation, given his great achievements, as to why some of his plans did not reach fruition. While Tesla had gained great respect as an engineer and inventor, there were always those – like his professor in earlier times – who did not believe that his imaginings could really come to anything. There were others who were in commercial and technological competition with Tesla – Edison, for example – who were willing to ridicule him and to diminish his standing as a way of promoting their own interests. And then there were the backers, the moneymen, who both fed and starved him according to their preference. Tesla's individual wealth was never enough to finance his own projects, and when his projects cost more than expected, as they inevitably did, he would throw himself on the mercy of a series of investors and benefactors. Throughout his life Tesla's finances swung from copious amounts of cash -which were soon invested in new machinery and inventions – to mountainous debts.

In early 1899 Tesla secured new investment from a number of wealthy individuals including Col. John Jacob Astor, owner of New York's Waldorf Astoria Hotel. With this money he set up an elaborate laboratory in Colorado Springs, where he unleashed artificial lightning discharges of several million volts (blowing up the local generating station in the process). Tesla was convinced that he could transmit radio signals hundreds, even thousands of miles around the globe. In the 1890s he had secured patents on many aspects of radio transmission. In late 1900 Tesla needed a large investment if he were to get his Worldwide Wireless Telephone Transmitter to deliver its promise. After false starts with a number of investors he approached J. Pierpont Morgan, who had been Edison's backer during the early days of Edison's DC developments. Morgan's habit was to own 51 per cent of everything he became involved in, and when Tesla approached him with plans for his worldwide radio broadcasting system, the magnate Morgan was happy to forward him $150,000 secured on 51 per cent of Tesla's interests in his own radio patents.

Tesla did not tell Morgan his hidden agenda, which he had earlier confided to the now unsupportive Westinghouse:

"You will know of course that I contemplate the establishment of such a communication merely as the first step to further and more important work, namely that of transmitting power. But as the latter will be an undertaking on a much larger and more expensive scale, I am compelled to first demonstrate such feature to get the confidence of capital." [10]

Through his experiments he had become convinced that there were ways to transmit unlimited amounts of electrical energy to any point on the globe without using any conventional transfer medium such as copper cable. Writing later in 1900, he described how he had developed his ideas:

"For a long time I was convinced that such a transmission on an industrial scale could never be realized, but a discovery which I made changed my view. I observed that under certain conditions the atmosphere, which is normally a high insulator, assumes conducting properties, and so becomes capable of conveying any amount of electrical energy." [11]

But in order to carry out all the experiments, he needed to first put in place the worldwide radio broadcasting station. He had already proved to his own satisfaction that he could broadcast and receive signals over seven hundred miles, and now he offered Morgan the possibility of both transatlantic and transpacific radio communication. Tesla quickly purchased 200 acres of Long Island, which he christened "Wardenclyffe". The money was soon being spent on the transmitting tower that would be Tesla's landmark, the symbol of his life's vision. Wardenclyffe tower was 187 feet high and topped with a massive fifty-five-ton mushroom-like dome. This contained Tesla's most important component – the magnifying transmitter capable of generating oscillating signals of some hundreds of millions of volts.

In the two years or so that it took Tesla to build the transmitter he had developed two major problems. With escalating costs and long delays he was now in desperate financial straits. His second problem was Marconi, who had, on 12 December 1901, sent the first wireless signal from Cornwall, England, to Newfoundland. What Morgan, and many others, did not know was that Marconi was using Tesla's radio patents, which were to become the focus of much dispute before Tesla's primacy was established in 1943.

Nor did Morgan appreciate how Marconi was able to achieve this with much less equipment and cost than Tesla was employing. He also didn't know, but was about to find out, Tesla's hidden power agenda. Tesla had already filed a patent relating to the wireless transmission of power (US Patent No. 787,412 "Art of Transmitting Electrical Energy through the Natural Medium") and would later apply for a more important US Patent, No. 1,119,732 "Apparatus for Transmitting Electrical Energy", based on his work at Wardenclyffe. In his comprehensive vision every person on the planet would have a receiver, which, just like a radio, they could tune to receive unlimited, unmetered power.

When, on 3 July 1903, Tesla made his final plea for more finance, he threw himself on Morgan's mercy, a quality that the magnate had never shown in any abundance: "If I could have told you such as this before, you would have fired me out of this office ...Will you help me or let my great work – almost complete – go to pots?" [12]

Morgan"s reply came on 14 July: "I have received your letter ...and in reply would say I should not feel disposed at present to make any further advances."[13]

In a Promethean display of anger, the next night saw the skies around Wardenclyffe tower lit up with massive streaks and bolts of Tesla's artificial lightning, powered by the magnifying transmitter. But it was to be the last show of its kind. Neither Morgan nor Westinghouse, and none of the other big money people, were willing to start a new electrical revolution when they were still reaping the profits of the first revolution that Tesla had played his part in.

In the end, Wardenclyffe tower was demolished for scrap and Tesla moved on to more "acceptable" projects. Yet his desire to make energy freely available would never go away.

Tesla's Free Energy Devices

The wireless transmission of power was, essentially, a distribution technology. It still relied on a conventional power generation method such as coal and steam turbine to produce the enormous amounts of power it would have required. Since many years earlier, however, Tesla had been fascinated by the idea of new, untapped energy sources. In one of his famous lectures of 1892 he told an astounded audience:

"Ere many generations pass, our machinery will be driven by a power obtainable at any point of the universe... Throughout space there is energy. Is this energy static or kinetic? If static, our hopes are in vain; if kinetic – and this we know it is, for certain – then it is a mere question of time when men will succeed in attaching their machinery to the very wheelwork of nature." [14]

In June 1900 in *The Century Illustrated Magazine* Tesla wrote what he considered to be the most important of all his articles, "The Problem of Increasing Human Energy." The article was radical, even sensational, in its ideas and caused a significant controversy amongst both scientists and the general public at the time of its publication.

"Whatever our resources of primary energy may be in the future, we must, to be rational, obtain it without consumption of any material. Long ago I came to this conclusion, and to arrive at this result only two ways appeared possible – either to turn to use the energy of the sun stored in the ambient medium, or to transmit, through the medium, the sun's energy to distant places from some locality where it was obtainable without consumption of material." [15]

Among many ideas for energy generation in the future, Tesla put forward a radical thought experiment:

It is possible, and even probable, that there will be, in time, other resources of energy opened up, of which we have no knowledge now. We may even find ways of applying forces such as magnetism and gravity for driving machinery without using any other means. Such realizations, though highly improbable, are not impossible. An example will best convey an idea of what we can hope to attain, and what we can never attain. Imagine a disk of some homogeneous material turned perfectly true and arranged to turn in frictionless bearings on a horizontal shaft above the ground. This disk, being under the above conditions perfectly balanced, would rest in any position. Now it is possible that we may learn how to make such a disk rotate continuously and perform work by the force of gravity without any further effort on our part: but it is perfectly impossible for the disk to turn and do work without any force from the outside. If it could do so, it would be what is designated scientifically as a "perpetuum mobile," a machine creating its own motive power. To make the disk rotate by the force of gravity we have to invent a screen against this force. By such a screen we could prevent this force from acting on one half of the disk. and rotation of the latter would follow. At least, we cannot deny such a possibility until we know exactly the nature of the force of gravity. Suppose that this force were due to a movement comparable to that of a stream of air passing from above toward the centre of the earth. The effect of such a stream upon both halves of the disk would be equal, and the latter would not rotate ordinarily; but if one half should be guarded by a plate arresting the movement, then it would turn. [16]

A screen against gravity? Even now such an idea delights and tantalizes – as does his other assertion that all we needed for free energy was a magnet with one pole, or else a way of shielding magnetism. This assertion has led to much experimentation into "permanent magnet motors" – motors that have no motive force apart from that of their own magnetism. In the 1920s Werner Heisenberg, one of the fathers of quantum mechanics, and the progenitor of the Uncertainty Principle, put forward the idea that we would indeed use magnets as a power source, despite the conventional theory that says magnets are incapable of doing physical work.

One of Tesla's many patents (No. 685,957 filed on 21 March 1901 and granted on 5 November 1901) was for an "Apparatus for the Utilization of Radiant Energy" – a machine to capture the sun's cosmic rays and turn them into electricity. The concept for the device was relatively simple, and involved putting an insulated metal plate as high as possible into the air. A second metal plate is inserted into the ground. Wires are run from both into a capacitor.

The sun, as well as other sources of radiant energy, throws off minute particles of matter positively electrified, which, impinging upon [the upper] plate, communicate continuously an electrical charge to the same. The opposite terminal of the condenser being connected to ground, which may be considered as a vast reservoir of negative electricity, a feeble current flows continuously into the condenser and inasmuch as the particles are charged to a very high potential, this charging of the condenser may continue, as I have actually observed, almost indefinitely, even to the point of rupturing the dielectric.[17]

This simple design for capturing a large electrical charge, and potentially an electrical current, may well have been the starting point for T. Henry Moray (see Chapter 3 of Keith's book – Ed. note) and those who have followed his work to turn "radiant energy" into electrical current. (In Chapter 9 I look at how the radiant energy or "ether" concept has now been updated in the light of modern physics.)

Another fuelless energy device Tesla mentioned in his *Century Illustrated* article "The Problem of Increasing Human Energy" was a mechanical oscillator, which first appeared in public at the Chicago World's Fair in 1893. "On that occasion I exposed the principles of the mechanical oscillator, but the original purpose of this machine is explained here for the first time."[18] Tesla describes how large amounts of heat can be extracted from i the ambient medium using a high-speed oscillator, a steam-driven engine used for producing high-frequency currents.

"My conclusions showed that if an engine of a peculiar kind could be brought to a high degree of perfection, the plan I had conceived was realizable, and I resolved to proceed with the development of such an engine, the primary object of which was to secure the greatest economy of transformation of heat." [19]

Tesla envisioned the mechanical oscillator as part of a technology to capture differentials in energy – a form of energy pump – but he was, it appears, finally defeated not just by the complexities of the other components that would be required, but also by the economics of the project:

"I worked for a long time fully convinced that the practical realization of the method of obtaining energy from the sun would be of incalculable industrial value, but the continued study of the subject revealed the fact that while it will be commercially profitable if my expectations are well founded, it will not be so to an extraordinary degree." [20]

One of the initial spurs for his work on "energy pumps" had been Lord Kelvin, who had stated that it was not possible to build a machine which could extract heat from its surrounding medium and utilize the energy gained to run itself. In one of his many thought experiments Tesla pictured a very tall bundle of metal rods, extending from the earth to outer space. Since the earth is warmer than outer space, heat would be conducted up the metal rods together with an electric current. All that would be required to capture the current would be a very long power cable to connect the two ends of the metal bar each to an electric load such as a battery or motor. A motor should keep running continuously, Tesla believed, until the earth had cooled to the temperature of outer space – something which, depending on the size of such a device, might never happen: "This would be an inanimate engine which, to all evidence, would be cooling a portion of the medium below the temperature of the surrounding, and operating by the heat abstracted." [21] By such means, Tesla contended, such a machine could produce energy without "the consumption of any material" – his key ideal.

Tesla and Faraday's Unipolar Dynamo

Michael Faraday, discoverer of the laws of electromagnetic induction, was the inventor of the first electric motors in the 1830s. One of his stranger, and often neglected, devices was the unipolar dynamo (discussed in Chapter 4 of Keith's book – Ed. note), consisting of a metal disk rotating between magnets in order to produce electrical current. Tesla's involvement with the unipolar, or homopolar generator, led him to believe that it might be capable of acting as a "self-activating" generator. Indeed, in 1889 he filed and received a patent for the "Dynamo Electric Machine" based on Faraday's original design, but with an improved design intended to increase its efficiency by reducing its drag or back torque. Tesla was postulating that if the back torque could be engineered to work in the direction of movement, rather than against it, then the machine could be made self-sustaining. While Tesla was not able to achieve such a feat in his lifetime, his, and Faraday's, ideas were to be picked up by a number of researchers including Bruce DePalma – inventor of the N-machine – in the 1970s and '80s.

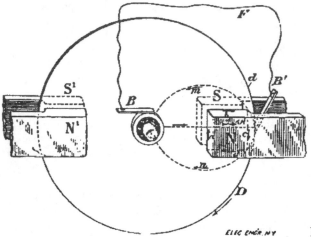

 Early Unipolar Dynamo

These are not the only attempts Tesla made to develop a fuelless energy generator, but just how far he got in his quest is far from clear. Tesla himself clearly stated that he had achieved energy generation from a new energy source on a number of occasions, although he was not always forthcoming about the technology behind his claimed achievement. On 10 July 1931, for instance, *The Brooklyn Eagle* carried an article in which Tesla was quoted: "I have harnessed the cosmic rays and caused them to operate a motive device. "More than twenty-five years ago I began my efforts to harness the cosmic rays and I can now state that I have succeeded."

On 1 November 1933 Tesla made a similar claim in the New York American, under the headline "Device to Harness Cosmic Energy Claimed by Tesla": "This new power for the driving of the world's machinery will be derived from the energy which operates the universe, the cosmic energy, whose central source for the earth is the sun and which is everywhere present in unlimited quantities."

These two articles, written during Tesla's later creative phase, demonstrate his concern to solve "the energy problem" which he saw before him. While he had been critically responsible for the expansion of electricity use, he also felt a passionate need to conserve the coal reserves for future generations.

In November 1933 he was asked by a journalist from the *Philadelphia Public Ledger* whether his fuelless technologies would upset the present economic system. "Dr Tesla replied, "It is badly upset already." He added that now as never before was the time ripe for the development of new resources."

Summary

So why haven't we seen any of these free energy technologies working? There is little doubt that Tesla was one of the great scientific geniuses not just of his own time, but perhaps of the entire twentieth century as well, but the reasons why his technologies were not developed may be complex.

Some researchers have claimed that, like Leonardo da Vinci, he was not just fifty or a hundred years ahead of his time, but perhaps many hundreds of years in advance of contemporary thinking. Scientific and technological ideas need support, both intellectual and financial, if they are to thrive.

Is it possible, then, that new generations of scientists have not been able to develop his visionary ideas into physical technologies? This question bears on the notion of genius in science, as opposed to genius in the arts and other fields of endeavor. While we accept that no one else could have written Beethoven's symphonies or Shakespeare's plays, it seems harder to accept that science is subject to the same vagaries of human beings. Even though Galileo Galilei, Michael Faraday and Albert Einstein possessed unique minds, we often assume that if they hadn't "come up with" their discoveries someone else would have done the same pretty soon after. Perhaps that assumption is erroneous, or at least, highly limited. If it hadn't been for Tesla it is quite possible that we would have developed a much more primitive and limited electrical system based on small generating stations every few miles.

Once Tesla had brought about one electrical revolution, the world was not ready for another, even more radical development of electrical power. The commercial powers that controlled the electrical landscape – based as it was on a distributed network of copper cable – had no interest in throwing away their investment in favor of the wireless, and potentially costless, transmission and reception of electricity. They seem to have had even less interest in Tesla's ideas of free-energy technologies. T. Henry Moray, who adopted some of Tesla's ideas in his radiant energy device (see Chapter 3 of Keith's book – Ed. note) faced many of the same oppositions that Tesla faced. While we can thank Tesla's genius for bringing distributed AC electricity to most of the world, we have yet to receive the gift he really wanted to give. In his more enlightened times Tesla himself maintained a balanced view:

"I anticipate that many, unprepared for these results, which, through long familiarity, appear to me simple and obvious, will consider them still far from practical application. Such reserve, and even opposition, of some is as useful a quality and as necessary an element in human progress as the quick receptivity and enthusiasm of others... the scientific man does not aim at an immediate result. He does not expect that his advanced idea will be readily taken up. His work is like that of the planter – for the future. His duty is to lay the foundation for those who are to come, and point the way." [22]

Eventually on 7 January 1943 Tesla ended his days, alone and poor in a shabby New York hotel where only a few pet pigeons shared his thoughts.

***

(With a Foreword by Arthur C. Clarke, Keith's book is a fascinating collection of scientific stories on pioneers as Tesla, Moray, Faraday, and many others, along with chapters on Swiss ML Converter, cold fusion, Blacklight Power, zero-point energy, an energy primer, and Tesla patents. – Ed. note)

Keith Tutt can be reached through Street Farm, Topcroft, Bungay, Suffolk, NR35 2BL, UK
keith@thesearchforfreeenergy.com

4 Tesla's History In Western New York

Thomas Valone, M.A., P.E.

Reprinted from *Proceedings of the International Tesla Symposium, 1986*

Introduction

Home of the "Tesla Memorial Society of the US and Canada", Western New York has a rich history of being touched and brought to life by Nikola Tesla:

1) Niagara Falls, New York was the first city in the world to have commercial alternating current generation of electricity, owing to Tesla;

2) Buffalo, New York was the first city in the world to receive electric power generated from a long distance away (22 miles) because of Tesla;

3) Buffalo, New York was the first city in the nation to have electric street lights thanks to Tesla.

Spending most of my life in Buffalo, and lots of enjoyable hours in Niagara Falls, it is my great yet humble pleasure to bring to you tonight an historical travelogue; a trip back in time, to the 1890's when the excitement and thrill of Tesla's fame was felt by everyone in Western New York.

Sit back, relax, and get ready to experience what it was like to live in that period, just as the transformation of these two cities was taking place. For the first time, articles from the <u>Niagara Gazette</u> (1893-1897) have been uncovered, including an interview with Tesla on the occasion of his first visit to the newly erected Adams Plant. Let's see how electrical power was extracted from Niagara Falls, from the Adams, to the Schoellkopf, and finally the Robert Moses Power Plant. We'll examine the mistakes that were made as well as how Niagara Falls, New York, has tried to commemorate the "greatest electrician that ever lived," Nikola Tesla.

To begin with, I'll read a passage from <u>Tesla's Magnifying Transmitter</u> by Dr. Andrija Puharich, a book that is in the hands of Dell Publishers at this time.

> In 1890, the Morgan financial group had started a company to try to develop the electric power potential of Niagara Falls. It was called the Cataract Construction Company, and its president was Edward Dean Adams. An International commission was chaired by Lord Kelvin. The commission found that none of the plans were feasible. So the Cataract Company asked for plans and bids from companies in order to get the work under way. It was an incredible personal triumph for Tesla when his concept and plan were adopted by the commission. In October, 1893, two companies were awarded the contracts to electrify Niagara Falls: Westinghouse won the contract to build the AC power plant at the Falls; and General Electric, using licensed Tesla patents, was awarded the contract to build the transmission lines and distribution systems to Buffalo, New York, 22 miles away. Tesla had set huge ideas and forces in motion years ago, and now the tide of industrial civilization was lifting them higher and higher to the thundering crest of Niagara Falls' worldwide reputation ...

On the night of March 13, 1895, a fire broke out in the basement of 33 South Fifth Avenue and swept through the entire structure -- including Tesla's laboratory. All of his hundreds of invention models, plans, notes, plans, notes, laboratory data, tools, photographs -- all, all were destroyed ... Even as the disaster of his laboratory loss was still ashes in his heart, the power from Niagara Falls began to flow in August, 1895 - Tesla's greatest triumph to date. The builders and backers of this biggest of all electrical power plants on the planet were highly pleased with the success of the Tesla polyphase system. [1]

We have three dates to keep in mind:
1893 - start of the Adams Plant, contract awarded
1895 - power turned on at the completed Adams Plant
1896 - power sent 22 miles to Buffalo, for the first time.

Figure 1. Nikola Tesla. -- Introducing Nikola Tesla.

Now starting off with a picture of Tesla, from a small rare booklet found in the Niagara Falls library entitled Introducing Nikola Tesla by Thomas L. Richardson of the Tesla Research Headquarters of Canada, let us look at some of the articles of Tesla's time from the same library.

Biggest Motor in the World

An article entitled, "Biggest Motor in the World" appeared in the Niagara Gazette, August 24, 1897. In 1897 they installed a 400 horsepower motor in Niagara Falls.

The caption under the title says "the biggest alternating current machine will operate on the regular 2,000 volt current, from the Power House generators, without the use of a convertor -- built for the Electric Light Plant." The article goes on to report:

> Anything new in electrical machines is interesting in Niagara Falls where so much electricity is flying around, and consequently the news of a new motor being installed in the power house of the Buffalo and Niagara Falls Electric Light and Power Company, that is the largest of its kind in the world, is not without its interesting features.

> When it is said that the new motor is 'the largest of its kind' it is an essential fact in this story. There is a motor larger, but not of the most modern type.

> The new machine is what is familiarly called a Tesla Induction type motor. It is built on the design of Tesla, who discovered a method of using an alternating two-phase electric current for operating a motor after other electricians had tried and failed to discern this method. It was considered one of Tesla's greatest achievements to discover this method for many reasons, but chiefly for its economical features....

Tesla's Renown

The earliest article to be examined is one entitled, "Tesla's Renown" from the Niagara Gazette, May 29, 1893:

The subtitles read, "A young man who is becoming known to the world as the greatest living electrician. Niagara power over the sea. To perform this feat is one of his dreams -- power enough to drive every railroad, propel every ship and produce every article manufactured." The rest of this short article is reproduced in its entirety:

> Nikola Tesla has been called by scientific men, who do not award praise freely or indiscriminately, 'the greatest living electrician.' At the recent convention of electricians held in St. Louis a well-known electric journal issued daily bulletins or 'extras' giving a list of the delegates and distinguished attendants as they arrived at the convention, a program of the day's proceedings and a special article of immediate interest having reference to the current discussions of the convention.

> One evening during the session of the convention, Mr. Tesla lectured on some of his recent experiments. The bulletin for that day contained a brief account of his life. So great is the interest taken in this young man that over 4,000 copies of the journal containing this biographical sketch were sold on the streets of St. Louis -- something unprecedented in the history of electrical journalism -- and in the evening his lecture, in the Grand Music Entertainment Hall, was listened to by a larger audience than had ever been gathered together before in the United States on an occasion of this kind. Many were glad to purchase complimentary tickets at $4 or $5 apiece.

> It is singular that this remarkable man comes to us from one of the smallest and least known nations of Europe, Montenegro; that he is a young man (only 30 years of age), and that the best scientific minds believe that he has only begun to give the world the result of valuable researches.

> The following is taken from an interview with Mr. Tesla by a New York Herald reporter: "In this great country (of which I am proud to be a citizen) we have water powers which, in aggregate amount, are sufficient to supply all the needs -- in fact, far more than the needs -- of the whole human race. Take, for instance, Niagara. This famous waterfall is estimated variously all the way from five to six millions horse-power. Now 1,000,000 horse-power economi-

cally directed would light every lamp, drive every railroad, propel every ship, heat every store, and produce every article manufactured by machinery in the United States.

"It will not be long before we can transmit that power under quite practical conditions by means of wires with the alternating system over distances as great as 1000 miles. Engineers now object to the use of very high pressure which would be necessary in such transmissions of power. But I believe the time will come when we shall transmit that energy without any wire.

"Since I have experimentally proved that we can get back electric impulses over one single wire without any return we may avail ourselves of the earth as a medium of transmission as one difficult obstacle to overcome. In fact, the only serious objection to this scheme is to find a means to concentrate the energy of vibrations spread over a great area on one spot.

"If this power is to be transmitted across the ocean, it will of course involve the expenditure of an enormous quantity of energy. It has been suggested that I can produce a set of lenses made of asphaltum or gutta percha or any other good so-called non-conductor of electricity, and can concentrate these rays, or waves, to a focus where their effect would be powerful.

"This plan if at all practicable, could be applied as well across the Atlantic as it could at shorter distances on land.

"Electricity is becoming more and more an important factor in our daily life and more and more closely connected with our comfort. I think, after a considerable lapse of time, it will become practically necessary for our existence. For instance, there is the question of light. The advantages of the electric light are so great that even with the present wasteful methods we have been able to succeed in making practical use of it.

"But what will be our success when we shall be able to produce a hundred times as much light as we do at the present day? To do this is merely a question of time. Electric power is obtained by the use of dangerous, cumbersome and complicated appliances. But we have electric machines now, which require no attention whatever, and which will, in a few years, supplant all other motors, simply because of their higher efficiency and ideal simplicity.

"Even now the cost is very great. Eventually we will very likely be able to heat our stoves, warm the water and do our cooking by electricity, and in fact, to perform any service of this kind required for our domestic needs.

"It has been said that it will be unpracticable to heat our houses by means of electricity on account of the great cost, but as I have said we are now looking for other methods of getting electrical energy cheap. Even with the present methods any rich man certainly prefers, instead of a stove in his room, to have it warmed by electricity. The method is expensive but ideal.

"Electric energy can be applied to bicycles, carriages and all sorts of vehicles. It will certainly be applied to rowboats and will probably be so cheap that any man in ordinary circumstances can own a boat and propel it by this means. It would be a gloomy prospect indeed for the world if we did not think that this great power will be used to the advantage of the vast majority of the human race and its benefits will not be confined merely to the wealthy.

"Some years ago I demonstrated that a lampate filament could be made to glow from a current from the human hand. The light coming from the hand is produced by the agitation of the particles of molecules of the air. I charge my body with electricity, and from an apparatus which I have devised, I can make the electricity vibrate at the rate of a million times a second. The molecules of the air are then violently agitated, so violently that they become luminous; and streams of light then come out from the hand.

"In the same manner I am able to take in the hand a bulb of glass filled with certain substances and make them spring into light; I make light come to an ordinary lamp in a similar way, simply by holding it in the hand.

"When I was in London I had the pleasure of performing one of these experiments privately before Lord Rayleigh. I shall never forget the eagerness and excitement with which that famous scientist saw the lamp light up. I can only say that the appreciation of such men simply repays me for that pains I take in working out such phenomena."

Tesla's Dream

Another article entitled "Tesla's Dream" is taken from the <u>Niagara Gazette</u>, November 22, 1893. This is still when plans were getting under way at Niagara Falls and they had just signed the contract. The excitement is building. Tesla has a plan to send Niagara Falls power through New York "by electricity's aid." "All eyes on Niagara Falls. Superintendent of Public Works, Hannan feels that the Niagara power ... can wire the Canal and propel the boats -- Governor Flower given some advice," read the subtitles.

This is an interesting sideline to the Adams Plant project that never actually took place but let's read about it:

An Albany special has the following in regards to the recent test of the possibilities of electric propulsion of canal boats; Edward Hannan, the Superintendent of Public Works, is highly gratified over what he considers the successful demonstration on the Erie Canal at Brighton last Saturday that electricity can be used in the propulsion of canal boats. Speaking about the experiment today he said:

"I think it was clearly made evident at Brighton that canal boats, by using the trolley system of supplying themselves with electricity for electric motors, can get sufficient power to be driven with their ordinary cargoes from one end of the Erie Canal to the other. As to the cost of putting trolley wires along the Erie Canal and of supplying electricity, as well as the cost of putting electric motors in the canal boats -- this will have to be learned by many computations."

"Do you favor the suggestion that the State should put up the trolley wires?"

"I do not. The governor asked my opinion of the State's undertaking such a work and I told him that in my judgement it was not feasible for the State to put up electrical wires or go to any expense of like nature; that it ought to be left to private enterprise. In my opinion this Niagara Falls Electric Power Company will eventually be able, if it chooses to do so, to put up trolley wires all along the line of the Erie Canal and supply electricity for the canal boats."

Commenting on the declaration of the superintendent, George Westinghouse, Jr., when asked by the writer on Saturday what distance it would be commercially profitable to send electricity, replied: "I think that now it can be sent with profit a distance of 200 miles."

Nikola Tesla, the eminent electrician, said to the writer: "I have plans for sending electricity from Niagara Falls to New York City -- plans which I believe will ultimately be accepted by capitalists and carried out."

Note how Mr. Westinghouse was already conceiving of a distance 10 times greater than the distance from Niagara Falls to Buffalo, which still had not been accomplished at that time.

Tesla's Great Ideas

One day later, another article appeared in the Niagara Gazette entitled, "Tesla's Great Ideas," November 23, 1893. The subtitle reads, "W. R. Rankine talks of them -- Nothing would surprise him -- thinks that possibly some of his plans will be brought to a practical reality -- satisfied with progress of the work." Then the article begins (reproduced in its entirety):

> William R. Rankine, secretary and treasurer of the Cataract Construcion Company, arrived in the city this morning from New York looking extremely well and happy.
>
> Mr. Rankine dined at the Prospect House today and was interviewed by a GAZETTE representative on matters in general.
>
> "How is work progressing on the works, Mr. Rankine?"
>
> "Very satifactorily indeed! The contracted work is being pushed along and is progressing as rapidly as one would wish."
>
> "What is your opinion of this matter of electricity on the canals?"
>
> "I think it gives the newspapers a fruitful topic for discussion and the public something to think about."
>
> "What about Tesla's project of transmitting electricity from Niagara Falls to New York?"
>
> "Tesla is always ahead of the procession and I have come to that point where there is nothing astonishing to me in anything this remarkable man may propose. It would not be surprising to me to see some of his wonderful ideas brought to a practical reality in the near future."

Nikola Tesla, An Accurate Sketch

Now we are getting to 1894 just before the power is turned on at Niagara Falls. The article is entitled, "Nikola Tesla, An Accurate Sketch of the Wonderful Serbian Wizard Who Deals in Electricity." Subtitled, "His remarkable genius. Sees 'the low lights flickering on tangible new continents of science' -- inherits his inventive turn from his mother -- early history of a romantic life."

Right next to this article is found another of a decidedly less scientific nature: "Malaria and Epidemics Often Avoided by Partaking of Hot Coffee in the Morning." We can see the state of discernment of the scientific method in those days.

I'll just quote the first paragraph here since it is available elsewhere:

> The readers of the GAZETTE will appreciate the following sketch of Nikola Tesla, the famous electrician who has frequently visited here. It is taken from the February Century and is by Thomas C. Martin. Nikola Tesla was born in Serbia, a land so famous for its poetry that Goethe is said to have learned the musical tongue in which it is written, rather than lose any of its native beauty. There is no record of any one having ever studied Serbian for the sake of Serbian science; and indeed a great Slav orator has recently reproached his one hundred and twenty million fellows in Eastern Europe with their utter inability to invent even a mousetrap. But even racial conditions leave genius its

freedom, and once in a while nature herself rights things by producing a men whose transcendent merit compensates his nation for the very defects to which it has long been sensitive ...

Electric Lighting

In the next article, where Tesla is interviewed, we note that he will refuse to discuss his new invention, the electric arc lamp, that was keeping him busy while the Adams Plant was being completed. Here is an interesting article that reveals efficiencies for incandescent lamps that still have not been surpassed today. Let's take a look at a short article that appeared in the Niagara Gazette, May 22, 1896, entitled, "Electric Lighting." "Nicola Tesla Has a New Scheme Which Will Revolutionize The Present System" reads the subtitle.

New York, May 22. Nicola Tesla has solved the problem which he set before himself many years ago and which may revolutionize the system of electric-lighting. It is, electrical experts say, the nearest perfect adaption of the great force to the use of man.

In Mr. Tesla's laboratory in Houston Street is a bulb not much more than three inches in length, which when the current turns into it, becomes a ball of light. The heat is almost imperceptible. With it a very large room is so lighted that it is possible to read in any corner. Yet this is done without the attachments necessary in existing lights.

The rays are so strong that the sharpest photographs may be taken by them.

No new dynamo is required to produce the current. The bulb is attached to a wire connected with the street current. There is no danger of harmful shock in its use.

Mr. Tesla has been working for many years on his theory of the necessity and practicability of the conservation of electrical energy. The present incandescent light gives only three per cent of illuminating power. The other 97 per cent is wasted in heat.

The bulb which he has perfected gives 10 per cent of light and loses 90 per cent of energy. He declares that he will, with the aid of a few more experiments, be able to produce 40 per cent of light, so that the waste will be reduced to only 60 per cent, or 37 per cent less than at present.

This article is no less than amazing because today our incandescent bulbs still check in at about 3 per cent efficiency. Where did Tesla's invention go?

Nikola Tesla, An Interesting Talk

Now we get to what I believe is the most exciting article of all. Here is an actual interview with Tesla just after the power is being turned on at the Falls and Buffalo is just about to get some of the power (not reprinted or available anywhere else in the literature). Here Tesla is visiting the Niagara Falls Adams Plant to inspect the work that has been finally finished according to his design. The article, from the Niagara Gazette, July 20, 1896, is entitled, "Nikola Tesla, An Interesting Talk with America's Great Electrical Idealist." The subtitles read, "Remarkable personality. The dreamer in science was in the city yesterday, inspecting the wonders which had been achieved in harnessing Niagara. He had but little to say. Mr. Tesla was here with George Westinghouse, President Adams of the Cataract Construction Company, Commodore Melville of the United States Navy, Mr. William R. Rankine, and other distinguished men." The article, a real gem, is quoted in its entirety:

Nikola Tesla, the brilliant Serbian electrician who believes that ultimately electricity, generated by flying atoms, will be pumped out of the ground for use anywhere, was a visitor at Niagara Falls yesterday.

He was accompanied by Edward D. Adams, president of the Cataract Construction Company; George Westinghouse, president of the Westinghouse Electric Company; his son, Herman H. Westinghouse of New York; Thomas D. Ely, superintendent of motive power of the Pennsylvania Railroad; Commodore George W. Melville, chief engineer of the United States navy; Paul D. Cravath, counsel for the Westinghouse Company, and William R. Rankine, secretary of the Cataract Construction Company.

It is a difficult thing to interview Nikola Tesla, but to sit down and talk with him, man to man, is a privilege to be enjoyed and remembered. One seldom meets a man more free from affections and self-conciousness. He does not like to talk about himself and when the subject comes up he is sure to steer away from it as soon as possible.

With due apologies to Mr. Tesla for so much personality, it may be said that he has the same cast of countenance as Paderewski -- long and thin, with fine, clean cut features, long forehead, and a certain gleam of the eye that denotes what might be called spirituality. Anyone who has read of the personality of Edgar Allan Poe and who has also had the pleasure of a talk with Tesla, would feel instinctively that the unhappy inspired child of Parnassus and the Serbian electrician would have found much in common if they had ever met.

Tesla is an idealist, and anyone who had created an ideal of him from the fame that he has won, will not be disappointed in seeing him for the first time. He is fully six feet tall, very dark of complexion, nervous, and wiry. Impressionable maidens would fall in love with him at first sight but he has no time to think of impressionable maidens. In fact, he has given as his opinion that inventors should never marry. Day and night he is working away at some deep problems that fascinate him, and anyone that talks with him for only a few minutes will get the impression that science is his only mistress and that he cares more for her than for money and fame.

He had one of his rare moments yesterday when he could be induced to talk of science and when asked of the advances made in the problem of transmission, with earnest face and eyes fairly ablaze, he said, "There is no obstacle in the way of the successful transmission of power from the big power house you have here. The problem has been solved. Power can be transmitted to Buffalo as soon as the Power Company is ready to do it."

As the famous electrician grew enthusiastic he gestured with his hands which are apparently trustworthy indicators of his nervous condition. They trembled a little as he held them up and the conclusion to be drawn from them was that their possessor was a man of tremendous nervous energy.

Mr. Tesla is a man between 38 and 39 years of age and looks even younger. He was born in a town called Smiljan in Serbia on the borderland of Austria-Hungary. His father was an eloquent preacher of the Greek Church, and his mother was a woman of remarkable ingenuity. He had an inherited taste for mechanics, and it is his mother's blood that makes him what he is.

Tesla Interviewed

The article continues:

A squad of Buffalo and local newspaper men greeted the visitors as they emerged from the dining room of the Cataract House yesterday afternoon and Secretary Rankine courteously introduced the reporters to his distinguished guests.

Mr. Tesla's first visit to this city made him the object of much interest, and while decidedly backward in interviews he was a most agreeable talker. He said, "I am just off of a sick bed and not very strong yet," when first greetings were over. "Yes this is my first visit to Niagara Falls and to the power house here. Oh, it is wonderful beyond comparison; these dynamos are the largest in the world. It always affects me to see such things. The shock is severe upon me."

"What do you think of the project of transmitting power to Buffalo?" he was asked.

"It is one of the simplest propositions," he said. "It is simply according to all pronounced and accepted rules, and is as firmly established as the air itself."

"Do you think that the cost will be less for power transmitted than for using steam power?"

"Certainly. Even if steam was as cheap as electricity, it would be a full steam plant and never be reduced in quantity to be less than 25 per cent of the full power no matter how small the quantity is that you use, while electricity the moment you shut it off, costs nothing."

"What is your opinion of Buffalo's prospects with such great power so near it and so easily obtained?"

"It is an ideal city with a great future, a wonderful future before it." Further on he said: "Niagara Falls has the greatest future of all. For here it will be the cheapest to obtain power and its limit is hard to imagine." In regard to transmission, Tesla asserted that it is cheaper to transmit power in large quantities than in small quantities; the larger the force the less the loss in transmission, and in this connection Secretary Rankine stated that power would be transmitted to Buffalo not later than November of this year. The contract for completing the pole line would be let this week, and by November the company would send all the power they could spare to that city. This would not exceed 1,000 horsepower. Next year, when the new dynamos are ready, this amount would be increased as rapidly as the demands for it came in.

Mr. Tesla said that he was not prepared to talk on his latest invention, the new vacuum light. He was devoting his energy and study to the subject of transmission and insulation in order to bring it down to as near a perfect point as possible. He said he was going back to his laboratory from here and begin to work zealously on the important matters referred to.

Mr. George Westinghouse, who was among the group and who stands preeminent in the electrical world, regarded the conversation with much interest and good nature. He spoke to the Buffalo men present in the most flattering manner of the outlook for that city, but of course he said Niagara Falls was bound to receive the first and greatest benefits of the development of power here. "It will be Greater Niagara first," he said, "but Buffalo's possibilities are to be made marvelous as well." From his practical mind the project of power development for this city and Buffalo seemed unlimited.

In regard to the comparative cheapness of power in Buffalo he said that were electricity as high in price as steam it would be cheaper for use, as there was nothing required in the way of skilled labor to use it. Anyone could shut it off and turn on an electric current, but only a few could run a steam engine. Then the convenience of electrical power over steam power in manufacturing was so great that its value was manifold in this direction. The cost, however, in Buffalo for electric power transmitted from this city, he did not know as he was not connected with the power company. Secretary Rankine came to his aid here and said, "You can say it will cost one half what steam power cost there."

"Mr. Tesla, what is your opinion of the effect of this development of power on Buffalo and Niagara Falls," was asked of the great inventor as he was turning away.

"The effect will be that both cities will stretch out their arms until they meet," he said in an enthusiastic manner, which indicated the true characteristics of the man so clearly.

Secretary Rankine stated that the object of the party here was purely one of a personal nature. The company has adopted Tesla's system of a two phase current for transmitting power and they also use two of Tesla's motors for starting the big dynamos and Mr. Westinghouse has made all of the machinery for the company and consequently both men were interested in the plant here. The visit of Commodore Melville of the navy was one of inspection. That officer is deeply interested at present in improving in every way possible the electrical machinery on the new warships now being built. He was the guest of the Power Company's officials and took great interest in all he saw here.

The visitors departed yesterday afternoon on the West Shore for New York at 5 o'clock.

Celebration

The last article that I discovered was printed during the time of celebration of the great accomplishment of AC power generation and transmission to the distant city of Buffalo. Dated January 11, 1897, this article from the Niagara Gazette is entitled, "Are Coming to This City" with subtitles, "Many prominent men who are interested in the big power development; Important meeting to be held; The directors of the Cataract Construction Company will probably take some important steps regarding new contracts. The visitors will attend a great banquet in Buffalo tomorrow night."

Knowing the historical value of this last article to mention Tesla, let me take the liberty of quoting it in its entirety:

Tomorrow morning a special car will bring to this city from New York nearly all the directors of the Cataract Construction Company, also officers of the Power Company and some of the most noted electricians in the world. A meeting of the Cataract Construction Company is to be held here, an inspection of the work in progress made and some important steps are to be taken regarding new contracts, etc.

Tomorrow night Buffalo will formally celebrate the coming to that city of electric power for commercial purpose. The celebration is to be in the form of a banquet given at the Ellicott Club, and to which many distinguished guests are invited and will be present. This banquet is the only method Buffalo has of celebrating and to those who are to be present it is a glorious way. The menu is to be fine, in fact it is to be the very best that any 350 men ever sat down to, and the main feature of the occasion will be the toasts and addresses made by some of the greatest men of the day in advancing electrical science and turning it into practical and commercial benefits. Among those who are to attend are such men as Thomas A. Edison, Nicola Tesla, Frank Spragde, the inventor of the trolley system, Elihu Thompson, inventor of the arc electric light; also E.J. Houston, an electric light system inventor; Charles F. Brush the original electric light man; George Westinghouse and a host of others.

The officers of the Niagara Falls Power Company are the only representatives from this city, with the exception of Albert H. Porter, who was formerly resident engineer of the Cataract Construction Company.

The list of toasts had not been completed on Saturday night, but all will be ready today. One of the speakers is to be Tesla, that is sure, and others will probably do some talking too."

The Power of Niagara

To get a feel for the untamed power and energy of Niagara Falls, which Tesla revered even as a child, I will take you on a helicopter ride over the falls.

Figure 2. Bird's eye view of Niagara Falls, includes American Falls, Goat Island, and the Canadian Falls. -- Thomas Valone.

We begin at the lower side of the Rainbow Bridge and gradually approach the American and Canadian Falls respectively. The Niagara River has an average of 202,000 cubic feet per second water flow. The thundering power of this rushing water is so loud when one stands next to the falls that we can easily understand why Tesla was so intent on trying to tap some of it for the large scale generation of electricity.

Swinging around the Canadian Falls, also referred to as the "Horseshoe Falls" because of their shape, we see the land mass between the Falls called "Goat Island" where the Tesla Statue stands today. Notice also the tour boats, called, "Maid of the Mist" boats, which go right up to the base of the Horseshoe Falls and spray all of the passengers with water, while they experience the most magnificent rainbows in the world.

Generating Stations

A plaque has been placed at the site of the earliest power generating station at Niagara Falls. Located downstream from the American Falls, (very near the spot where the helicopter started from), the Schoellkopf Hydro-Electric Power Station was inaugurated on December 14, 1881 by the Niagara Falls Hydraulic Power and Manufacturing Company, predecessor of the Niagara Falls Power Company. From a data book supplied by the "Power Authority of the State of New York," we note that the Schoellkopf plant has been documented as "the first public demonstration of electricity at Niagara Falls." It involved DC generator arc light machines using the 86-foot drop of a paper company mill shaft. Supplying "the light of 2000 candles" to a few companies in the local vicinity, it awakened everyone to the potential of cheap electricity from Niagara Falls.

Figure 3. Aerial view of the Canadian (Horseshoe Falls) at Niagara Falls –Pana-vue

Figure 4. Aerial view of the American Falls at Niagara Falls. -- Pana-vue.

The Edward Dean Adams Hydro-Electric Power Station Number One was inaugurated on August 26, 1895 by the Niagara Falls Power Company. Currently the Station's nameplate and the entire archway of the entrance to the building stands on Goat Island directly in back of the Tesla Statue. The Adams Plant Number One contained ten 5,000 horsepower generators yielding 37,000 kilowatts. A second Adams Plant (Number Two), doubled that output. The original plant was designed for 25 Hz only, though "subsequent expansion included conversion to 60 Hz." [2]

The Schoellkopf Power Generating Station #3A in 1914 had a total output of 130,000 horsepower. It was razed in 1958. Schoellkopf Power Stations #3B and #3C, completed in 1920 and 1924 respectively, produced a total of 322,500 horsepower. Unfortunately, these two plants were destroyed in an unanticipated rock slide which occurred in 1956. A beautifully worded plaque is mounted about 20 feet from the Schoellkopf plaques. Erected by the Niagara Falls Power Company in 1922, it says, "To the engineers financiers scientists whose genius courage and industry made possible here the birth of hydro-electric power and created the first five thousand horse power water turbines directly connected to alternating current generators and inaugurated in America long distance transmission of power by electricity." Tesla's handiwork made it all possible!

rockslide

Figure 5. Aerial view of Schoelkopf plant rockslide. – Tom Valone

Figure 6. Two million kilowatt Robert Moses Power Plant -- Tom Valone

What really happened in 1956 that devastated most of the Schoellkopf plants? Well, a book entitled, Colossal Cataract shows the before and after pictures. A tremendous collapse of the cliff above the #3B and #3C plants occurred. The wall was never finished with the fine masonary work that still covers the #3A cliff on the left to this day. The entire Schoellkopf facility was rated at 365,000 kilowatts before disaster struck and part of it was restored to 95,000 kilowatts for a couple of years afterwards.

Within three years after the Schoellkopf Plant #3A was razed, the Robert Moses Niagara Power Plant was opened, with a capacity of 1,950,000 kilowatts, enough to supply a city the size of Chicago today with electricity. Its thirteen generators are the largest of their kind ever constructed by an American manufacturer. For comparison, the Grand Coulee generators are rated at 108,000 kilowatts. Tesla was right when he foresaw the enormous power potential of Niagara Falls. The Robert Moses Plant required 3,650,000 cubic yards of concrete and 284,000,000 pounds of reinforcing steel. Power is produced at 13,800 volts and stepped up to the current high voltage limit of 365,000 volts for efficient long-distance transmission. The Power Plant structure is 1840 feet long and 390 feet high. No rock slide could ever disturb this installation!

Adams Plant Number One

Now that we have seen the entire progression of electrical power development at Niagara Falls, let's go back now to the first Adams Plant and examine some of its details. In Figure 7, we see the actual plant. The next figure shows the interior of the plant with Board of Directors of the Cataract Construction Company, all wearing the same style hat, including Edward Adams himself, (the shortest man with the biggest moustache). Notice the relative size of just one of Tesla's generators, towering above the men.

Figure 7. Adams Plant Number One. -- Niagara Falls Power Company.

Figure 8. Board of Directors of the Cataract Construction Company. – Niagara Falls Power Company

Figure 9. Interior of Adams Plant Number One Powerhouse. – Niagara Falls Power Company

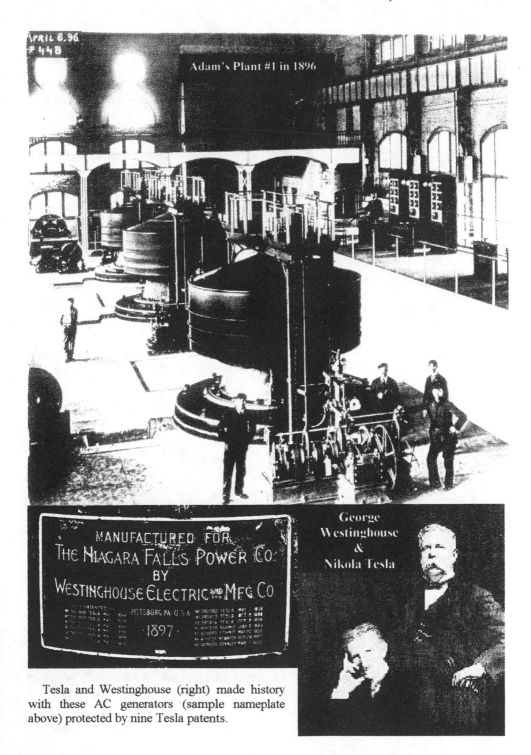

Tesla and Westinghouse (right) made history with these AC generators (sample nameplate above) protected by nine Tesla patents.

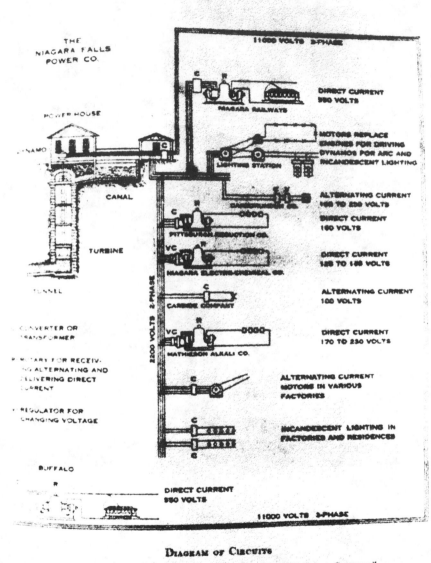

DIAGRAM OF CIRCUITS

from a paper on "The Installation of The Niagara Falls Power Company,"
Engineers' Club of Philadelphia, April 17, 1897

Figure 10. Diagram of Adams Plant One Power Distribution. – Niagara Falls Power Company. Notice that 11,000 volts of 3-phase AC power (top) is being sent to Buffalo, mostly converted into 550 volts DC for railways and the first street lights in the nation.

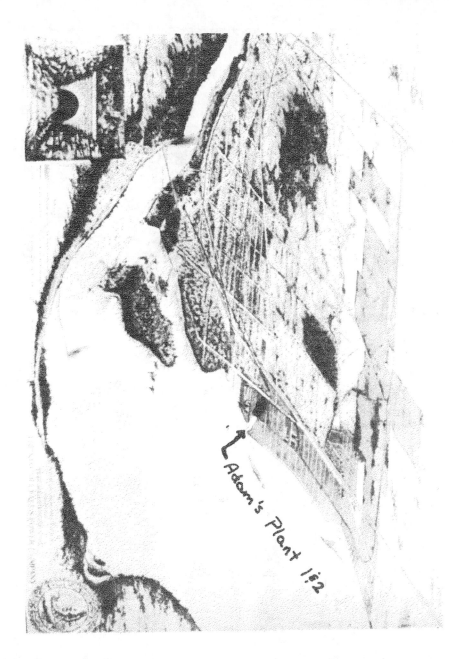

Figure 11. Aerial viewmap of Niagara Falls, circa 1900. – Niagara Falls Power Company

Figure 12. Street map of Niagara Falls showing location of the Adams Plant.

Figure 13. Huge arc formed when opening early DC circuit breaker. – NF Power Co.

These illustrations were taken from various historical books that provide a detailed record of the Niagara Falls Power Company from its inception. In Figure 9, we can see a beautiful shot of the interior of the completed Adams Plant #1, with all of the 5000 hp generators visible. The text noted that two plants were planned each with a 50,000 hp capacity. Figure 11 shows an aerial view from an old Niagara Falls Power Company map. (Niagara Falls was officially incorporated as a city only a few years earlier in 1892.) The Falls are close by. Both Adams Plants are shown in the street map, though I've highlighted the Number One plant. Figure 10 is especially interesting since we can see the 140 foot drop below the Adams Plant where the long generator shafts had to reach, as well as who received the electrical power. Besides a few companies in Niagara Falls, we see that Buffalo, at the bottom, received 11,000 volt, 3 phase power, as well as some DC power too.

Ever wonder what it is like trying to stop 11,000 volts at a few thousand amperes? Well, Figure 13 shows the results. A huge arc is generated by the circuit breaker, used on the early Buffalo circuit. Figure 14 shows the transmission lines to Buffalo, following the tradition of telegraph lines, the only example available at the time.

Figure 14. Repair wagon and first Buffalo transmission line. -- Niagara Falls Power Company.

Adam's Plants 1, 2, & 3

The more I looked at the old Niagara Falls map, showing the Adams plant site, the more I realized that I could probably find it if I went looking. The librarian at the Niagara Falls library also mentioned that one of the plants was still there as well. So I set out to photograph the site. In Figure 15 we see the opening to the canal and the Robert Moses Expressway that now passes over it.

Walking toward the highway and inland we can see how wide the canal is, as we look toward the spot across the canal where Adams Plant Number Two once stood. Crossing the highway, I am now standing on the site of the original Adams Plant Number One (Figure 16). How many people realize that it actually was there? There are no signs commemorating the site, which was quite surprising. I started to pick up a few rocks on the ground, knowing that they probably once were a part of the building that housed Tesla's generators.

Figure 15. Robert Moses Expressway over existing Adams Plant canal – Tom Valone

Figure 16. Site of the original Adams Plant Number One – Tom Valone

Looking across the property of the Sewage Treatment Facility adjacent to the canal, I spotted a building that turned out to be Adams Plant Number Three. In Figure 17, we see the only remaining building of the almost 100-year old trio comprising the world's first AC power stations. It is simply fenced off, again with no sign advertising the extraordinary significance of the building. The Niagara Falls Power Company is now called Niagara Mohawk. An amazing article was discovered from the February, 1962 issue of the <u>Ontario Hydro News</u>, page 13. In 1961, when the Robert Moses Power Plant was opened, the original Tesla generators, which kept working right up until then, were shut down. It was noted in the article that the Niagara Falls Historical Society, which doesn't exist today, was trying to keep the Adams Plant as an "electrical museum." The director of the society said, "It will be a crime if the place is destroyed. The original generators are still there, and it is a natural setting for an electrical museum." Since no money was obtained to buy the buildings, both Adams Plants were razed. I am told that at least one of the generators will be placed in the Smithsonian Museum in Washington, DC.

Figure 17. Modern day site of the original Adams Plant Number Three with building intact. -- Thomas Valone.

Tesla Statue

The large oversized statue of Nikola Tesla stands on Goat Island in Niagara Falls, with the only remaining part of the Adams Plant, the entrance archway, in the background. It is the only full figure statue of Tesla in the world. Created by a Yugoslavian sculptor, it was unveiled on July 23, 1976, commemorating the 120th anniversary of Tesla's birth. He looks sad as we see him from the side, studying his notes, his fingers worn from all of the kids that climb up on his lap. Most of the kids have no idea who Tesla was, but take advantage of the statue.

Figure 18. Tesla at 79 and his statue on Goat Island at Niagara Falls, NY.

In conclusion, as the world consumes about 70 million barrels/day of oil (47 million gal/sec), it is amazing to find that this is about 1/3 of the American Falls water flow (150,000 million gal/sec). The Niagara Falls Historical Society worked to preserve the first Adams Plant (see next page) and failed. Today, we still have a chance to make the third Adams Plant a beautiful commemorative site. What better tribute than to preserve the site of the first generation of AC power in the world? We have here a giant who walked among men. Let us commemorate his memory in the minds of everyone by at least establishing a Tesla Museum in the city that benefited the most from Tesla's invention of the AC generator. We are the future now, half a century since Tesla left the earth. As he himself said, "Let the future tell the truth and evaluate each one according to his work and accomplishments. The present is theirs; the future, for which I really worked, is mine."

References

[1] Puharich, Andrija. *Tesla's Magnifying Transmitter*, 1985, p.69. Private manuscript. The first five chapters are reprinted elsewhere in this anthology.

[2] *Radio Electronics*. August, 1983, p. 52

[3] Ibid., p. 52

ADAMS POWER PLANT TRANSFORMER HOUSE, 1895,
Buffalo Ave. near Portage Rd.,
Niagara Falls, N.Y.

This fine building owes some of its distinction to its having been designed by Stanford White of McKim, Mead and White, America's most prestigious architectural firm at the turn of the century. Then at the height of its power, the firm was also engaged in the design of the two opulent Williams mansions on Delaware Ave. and North St. in Buffalo.

Of rock-faced, Niagara limestone construction, one-and-a-half stories high, the building was erected for the production of electricity from the Niagara River. It pioneered successful experiments in cheap long distance alternating current transmission and thus helped to catalyze the Niagara Frontier's industrial development.

The building, which was saved from demolition about 1965, has certain Richardson Romanesque elements, such as the segmental-arched, double-door entrance with flanking flat-arched openings and blind roundels. Pilasters articulate the end bays with their large arched openings and also articulate the projecting side bays with their recessed spandrels. The building boasts a belt course, impost banding, and a molded cornice. Its appearance suggests its function - power.

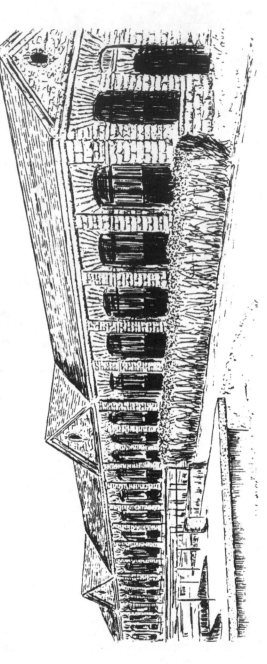

Source: *Designated Landmarks of WNY*

The Alternating Current Induction Motor

Tesla's brilliance is shown by the fact that he designed not only the generators that gave the world alternating current electricity, *but also the machinery that would use AC*. He presented the world with a complete ready-to-use "package" of inventions.

The alternating current induction motor is a good example of these inventions. Electric motors have to rotate in one direction, but alternating current changes direction dozens of times every second. How could changing currents, and the changing magnetic fields they produce, be converted into "one-way" motion?

Tesla's motor uses electromagnets. The magnets do not move, but the magnetic fields they produce attract the rotor and spin it around an axis. By connecting the electromagnets to the generator in a special way, Tesla converted the constantly-changing electric current into a a series of magnetic fields that rotate *in one direction*. His motor made it practical to use AC, thus helping to bring this efficient form of power transmission within reach of everyone in the world.

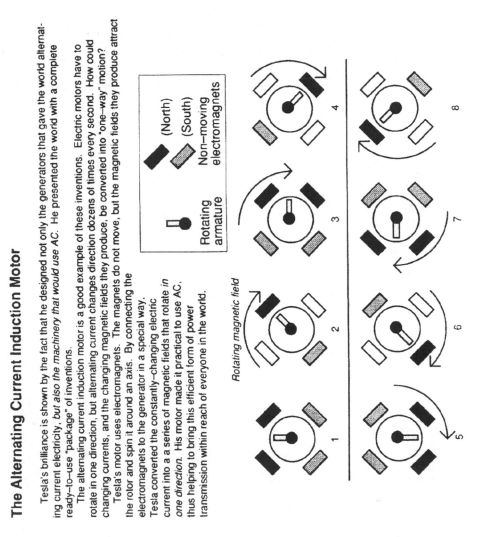

Explanation of the AC motor by Daniel Dumych. (Published by the Niagara Falls Convention and Visitors Bureau)

5 Niagara Falls Electricity Centennial

Thomas Valone

After preparing and presenting the preceding paper, "Tesla's History in Western New York" to the 1986 International Tesla Symposium as a special Saturday night slide show, I felt an obligation to celebrate the 1996 centennial of Tesla's greatest achievement, in gratitude for the electricity that my hometown of Buffalo, NY received from his work at Niagara Falls. Therefore, carrying copies of my paper in booklet form, I made the effort in 1992 to meet personally with Councilman John Accardo in Niagara Falls, who also was the Chairman of the Board of the City Council of Niagara Falls. We discussed the possibility of a city-wide event in 1996 involving the NY Power Authority, Niagara Falls Chamber of Commerce, and the City Council. We envisioned a new plaque, at least three banners across major streets of Niagara Falls, and a few signs around town. The Tesla Memorial Society in Lackawana, NY, Dorothy Rolling and Dan Dumych, the local historians were all very interested in participating, as was William Terbo, the grand-nephew of Tesla. Of course, my burning question of what will happen to the remaining Adams Plant Three could not be resolved by Councilman Accardo, nor even by Niagara Mohawk. It is apparently embarrassing to all of them that only the archway of the Adams Plant Number One was preserved.

Soon afterwards, perhaps in 1994, I began discussions with Steve Brady, Public Affairs representative for Niagara Mohawk and Chairman of the local Foundation Committee. He told me that Niagara Mohawk, the company which took over the original Niagara Falls Power Company, "wants to be a part of it in any way or another." Chris Mierzwa at Niagara Mohawk was also very interested, according to Professor Dollinger from SUNY at Buffalo, though I never talked to him directly. Dollinger told me he wanted to include a tour of the new Robert Moses Power plant that is north of the Falls, with a photo shoot at Goat Island, where Tesla's statue is, as well as a lunch. However, he said that the local IEEE had no funds for such an event. There was some discussion about also including Ontario Hydro, the Canadian electric company that also has a generation station at the Canadian side of Niagara Falls

The most encouraging meeting I had was with Cathleen Barber, the Senior Community Relations Representative of the New York Power Authority which runs the Niagara Power Project at Niagara Falls. It is their decision, for example, to send most of the electricity generated at Niagara Falls to New York City nowadays. Cathy was willing to support any date I would select and promote it, with professors from the local universities invited. We guessed that a one-day event with a plant tour, presentations, lunch and dinner would cost $3000 to $5000, but it was not clear who would pay for it.

I then had a meeting with Frederick Caso, Jr. the Executive Vice President of the Niagara Falls Chamber of Commerce, who was willing to offer mailing labels of members. He suggested Congressman LaFalce, who agreed to a *Congressional Record* insert, much like the one Congressman Henry Nowak created on April 28, 1981 (V.127, No.62) commemorating the 125th anniversary of Tesla's birth. Surprisingly, Buffalo's interest was minimal from discussions I had with the Buffalo mayor's office and the Niagara Parnership. As 1996 approached, with no one offering assistance for the Centennial, I sent a letter to Don Glynn, a reporter at the *Niagara Gazette*, to alert him to the historically significant event. Then, without advance notice except in Niagara Falls, the **Niagara Mohawk Power Corporation** chose to celebrate the Centennial themselves without historians, professors nor authors. Only industrial CEO's and politicians were there to present sterilized information and an exaggerated absurdity about being responsible for the "energy of life." Witness the following publications.

Large-scale electrification began 100 years ago—right in our backyard! In 1895 a Niagara Mohawk predecessor company opened the Adams station for the generation of electricity at Niagara Falls. When the Cataract Construction Company—under the leadership of Edward Dean Adams—was formed for the purpose of harnessing the power of the Falls in 1886, there was no consensus on how that power would be transmitted. It took five years of study before electricity was selected over pneumatic and mechanical means.

The construction of the immense tunnel that would carry water for more than a mile under the town of Niagara Falls was the largest engineering project of its day, and a risk of enormous proportions. The reward was the revolutionizing of modern life.

Compared to direct current, alternating current was easier and cheaper to transmit over long distances, an important consideration for remote generating plants.

Alternating current won the day, and George Westinghouse won the contract to build the generators, basing his design on several theories and patents of Nikola Tesla. On November 15, 1896, the switches were thrown to transmit electric power 26 miles from Niagara Falls to Buffalo—history's first long-distance transmission of electricity.

Thus, companies that would become Niagara Mohawk gave birth to the business of long-distance electrical transmission and distribution. For 100 years, every minute of every day, our energy has brought life to your world.

Niagara Mohawk

For 100 Years
WE'VE BROUGHT YOU
the Energy of Life!

On November 15, 1896, the switches were thrown to transmit electric power 26 miles from Niagara Falls to Buffalo—history's first long-distance transmission of electricity.

Life changed from that moment on. On the Niagara Frontier. And around the world.

Niagara Mohawk played a leading role, along with local businessmen, Wall Street financiers and engineers from around the world including Nikola Tesla, George Westinghouse and Thomas Edison to harness the incredible power of the Falls.

From the Adams Hydroelectric Station, built at the side of the Falls, Niagara Mohawk brought to life the dreams of many on that night 100 years ago.

You are invited to a reenactment of this historic event on Friday, November 15 beginning at 5:30 p.m. outside Niagara Mohawk's Electric Building on Washington Street, downtown Buffalo.

Events will include the lighting of Buffalo Place's Holiday Tree, fireworks, entertainment and ice skating at Rotary Rink.

For 100 years, every minute of every day, our energy has brought life to your world.

Adams Hydroelectric Station No. 1

Niagara Mohawk

CENTENNIAL CELEBRATION
November 15, 1896 - 1996

Dear Conference Participants:

Today we are gathering to mark the centennial of one of the more momentous events in the history of industrial America, the day switches were thrown that put the mighty power of Niagara Falls to work for the City of Buffalo in the form of electricity. Flipping those switches began an era of remarkable progress, as electric energy became universally available to homes and businesses throughout the country.

Over the past century, the wonder that people felt on that November day in 1896 has become a casual acceptance of electricity as part of everyday life. The sense of awe has moved on to new phenomena, from television to space flight, that have themselves been superseded in the public's imagination. That is human nature. The urgency of the present and the promise of the future occupy our thoughts.

On an occasion such as this we should stop to measure and appreciate progress and the extraordinary individuals who were its engines. This centennial is especially significant, because today we find ourselves at the brink of another era of dramatic change, as impending competition fundamentally changes the relationship between customers and their electricity providers.

On behalf of Niagara Mohawk Power Corporation, Westinghouse Electric Corporation, General Electric Company, and the New York Power Authority, I am pleased to welcome you to this Centennial Celebration. It is our hope that you will gain today a greater appreciation of this past and the lessons it holds for the future.

William E. Davis

William E. Davis

Chairman and CEO
Niagara Mohawk Power Corp.

| Niagara Mohawk Power Corporation's |
| Centennial Celebration Program of Speakers |

CELEBRATING A CENTURY OF ELECTRICAL ENERGY, NOVEMBER 15, 1996

8 a.m. to 8:30 a.m.   *Opening reception*

8:30 a.m. to 9:45 a.m.   *Bridging the Past to the Future*
William E. Davis
Chairman and CEO
Niagara Mohawk Power Corporation

Thomas A. Christopher
Power Generation Business Unit
Westinghouse Electric Corporation

Ronald R. Pressman
GE Power Systems
General Electric Company

10 a.m. to 11:15 a.m.   *What is the Future of Electrical Energy in New York State?*
Panel Participants:
Albert J. Budney Jr.
President
Niagara Mohawk Power Corporation

C.D. "Rapp" Rappleyea
Chairman and CEO
New York Power Authority

Louis R. Tomson
Deputy Secretary to Governor George Pataki

The Honorable Paul Tonko
Chairman
New York State Assembly Energy Committee

11:30 a.m. to 1 p.m.   *Luncheon*
Welcoming Remarks:
Charles P. Steiner
President and CEO
Niagara Falls Area Chamber of Commerce
and
The Honorable James C. Galie
Mayor
The City of Niagara Falls

Keynote Speaker:
William E. Davis
"Electricity: Appreciating the Past, Anticipating the Future"

1 p.m. to 1:30 p.m.   *School Awards Presentation and Ceremonial Reenactment*

Attendees are invited to take a few minutes during the conference to look at exhibits put together by high schools from around Western New York. The exhibits highlight art, history and technology related to the 100-year anniversary event.

Buffalo News
Nov. 16, 1996

100TH ANNIVERSARY

Ceremonies celebrate Niagara electricity

By MIKE VOGEL
News Staff Reporter

Talk of a brighter future came immediately true for power industry leaders Friday in downtown Buffalo as they threw a switch that lit the city's holiday tree and touched off a fireworks display celebrating the centennial of Niagara Falls electricity.

Tensions flickered just below the surface, though, as utility executives and power policy setters used the 100th anniversary celebrations to reflect on a future of deregulation and restructuring for the service that Niagara first provided to the world and to downtown Buffalo 100 years ago.

By the turn of the century, Niagara Mohawk chairman and chief executive William E. Davis noted, utility customers probably will be picking electricity suppliers much as they now choose telephone companies.

"We hope to be among the more successful of those callers," he added.

But while the atmosphere was definitely electric in both Buffalo and Niagara Falls for the celebrations, the future of electrical service still glows only dimly.

Legislation that will unleash competition still is being considered, and seven utility companies in the state still are reviewing each other's transition plans — most of them unveiled only last month.

"The move toward competition is going to be a complex and time-consuming process," said Clarence D. Rappleyea, chairman of the New York Power Authority.

On a day that mingled discussion of future prospects with remembrances of the past, many electrical experts found strong comparisons in the uncertainty facing the industry today and the mixture of hope and apprehension facing the region's hydropower pioneers a century ago.

The ceremonies and a power conference marked the throwing of switches that completed the world's first successful long-distance transmission of electricity — 26 miles from Niagara Falls to Buffalo — a few seconds after midnight on Nov. 15, 1896.

"It was in the middle of the night, and the ceremony was unadvertised, just in case something went wrong," Rappleyea quipped.

The harnessing of hydropower at Niagara changed the world, providing a source of energy that would power a revolution of technology in industries and households alike. That change started when Niagara Falls Power Co. executive William Rankine

See Power Page C4

MIKE GROLL/Buffalo News
Niagara Falls Mayor James C. Galie re-enacts the throwing of the switch as Niagara Mohawk Chairman William E. Davis watches.

Power: Niagara Falls electricity was to run Buffalo's streetcars

Continued from Page C1

closed a switch in the Cataract City and an engineer watched by Buffalo Mayor Edgar Jewett did the same at the new Buffalo power house.

The results were less than spectacular. The intent had been to use Niagara to power Buffalo's streetcars, but a missing voltmeter kept cautious engineers from unleashing the new power genie on the city system as planned.

At the power house that night, success was measured simply in the humming of a small rotating armature. Soon, though, the voltmeter was installed, and electricity was running not only the streetcars but Buffalo's new street lights.

The victory also marked the end of the "Battle of the Currents," which pitted Nikola Tesla's alternating current equipment against an earlier empire that Thomas Edison had built on his hopes for direct current.

Tesla, backed by George Westinghouse, carried the day — but a hundred years later, at least one Centennial Conference questioner noted that technical advances finally may make multivoltage DC a better technology.

Tesla, born in Croatia to Serbian parents, still garnered the lion's share of honors Friday as a genius whose work still powers achievements today.

"Nikola Tesla was a superstar a hundred years ago, much as athletes and entertainers are today," said New Jersey aerospace and telecommunications data engineer William H Terbo, a great-nephew of the inventor and head of the Tesla Memorial Society.

But Tesla's celebrity faded with time. His lasting legacy was his inventiveness.

"He was so innovative in so many areas, that some of his patents are just now coming into use," Terbo said. "He often thought of himself as a failure, because so many of the things he thought of he was never able to produce."

The dawning of the age of electricity that took place here a century ago "literally changed the world," about 350 conference participants were told at the Niagara Falls Convention Center.

"In a way, it's almost impossible to appreciate the true magnitude of what took place at Niagara Falls a century ago, as we take electricity for granted," Westing-house Power Corp. general manager Thomas A. Christopher said.

What was once amazing is now commonplace, as electric light and power have become an everyday miracle.

Before participating with Niagara Falls Mayor James C. Galie and Buffalo Mayor Masiello in separate commemorative switch-throwing ceremonies Friday, utility executives and public officials spent even more time pondering the future.

Long a system of protected and highly regulated monopolies, the power industry is moving toward an era of deregulation designed to drive down costs by promoting competition. Spurred by federal action in 1992, New York is now among 46 states planning a competitive environment.

The state's power supply system will be divided into a regulated core of transmission lines and a deregulated power generation and marketing industry, paralleling a current system of telephone companies and long-distance lines.

State Sen. Paul Tonko, head of the Senate Energy Committee, said the issue should be a priority in next year's legislative sessions. New York industries and residents need the relief from electricity costs that are "among the highest in the nation," he said, but the legislation also will have to protect the environment, work force, system reliability and safety.

"Customer choice will lower costs more effectively that any state or federal regulation," said Rappleyea, who advocated Power Authority ownership of the state's high-voltage lines as a "public electric Thruway."

Power companies will have to react to a business environment in which "customer service will be more important than ever," said Niagara Mohawk President Albert J. Budney Jr.

General Electric Co. Vice President Ronald R. Pressman said technical advances will continue to be a key factor as utilities seek "to retain their customers in what we expect to be a brutally competitive deregulated market in the years ahead."

U.S. energy systems have to stop lagging behind other nations in efficiency if they want to expand into global markets or just protect their own turf, he added.

*Niagara
Gazette*
Nov. 16,
1996

Power industry looks to future

MILESTONE: *While
celebrating an anniversary,
electric utilities are
preparing for competition.*

By Don Glynn
Niagara Gazette

A century after Niagara celebrated the first long-distance delivery of hydropower, the electric power industry faces massive structural changes that should end the current monopoly and create more competition to benefit consumers.

"If you give customers the power to choose, they will lower costs more effectively than any rule or regulation from Albany or Washington," said C.D. "Rapp" Rappleyea, chairman and chief executive officer of the New York Power Authority.

Addressing 200 electric, industry representatives and business leaders attending a conference marking the centennial of power

power Nikola Tesla's invention of the alternating current. Rappleyea said a recent study by a Washington-based research group estimated that the electric bills for a typical New York industry would be cut by 26 percent as a result of competition.

In May, the state Public Service Commission ordered the state's utilities to deregulate their operations, including the generation, transmission, distribution and service of hydropower. Some utilities filed lawsuits in the state Supreme Court challenging the PSC order.

"Another study by an energy analyst shows that a penny off the average electric rate would produce more than $1 trillion in economic activity in the United States by 2010, putting nearly $1,000 more in the pockets of each resident," Rappleyea said.

Under the PSC directive, utilities are required to separate their

markets to wholesale competition by 1997 and to retail by 1998, steps already viewed as complex and time-consuming.

William E. Davis, chairman and chief executive officer of Niagara Mohawk Power Corp., predicted that by the turn of the century many electricity users will be getting calls from would-be electricity suppliers offering services and incentives, much like telecommunications firms today.

"Customers will be able to choose services tailored to their specific needs, in contrast to the one-size-fits-all approach offered under utility regulation," he said.

Citing the Tesla milestone of Nov. 15, 1896, Assemblyman Paul D. Tonko, D-Amsterdam, chairman of the Assembly Standing Committee on Energy, said he thinks New York can regain its pre-eminence in the century-old power industry.

"But our time is short and the consequences of failure to act are

severe," he said.

The assemblyman said he is confident the Legislature plan — known as Competition Plus/Energy 2000 — will take

the industry and move it toward the next millennium.

"It offers a prescription for change and offers the tools to do it," he said.

James Neiss/*Niagara Gazette*

POWER CONFERENCE: Paul D. Tonko, left, chairman of the state Assembly Standing Committee on Energy, and William E. Davis, chairman and chief executive officer of Niagara Mohawk Power Corp., stand in front of information on Nikola Tesla, who invented alternating current

Innovation didn't spark confidence

C.D. "Rapp" Rappleyea, the chairman and chief executive officer of the New York Power Authority, was talking about the future of electrical energy in New York state during a conference Friday at the Convention and Civic Center.

"We might like to believe that 100 years ago, people were very confident about the future and the wonders of new electric technology," Rappleyea said, noting the centennial of inventor Nikola Tesla's development of the alternating current system that made possible the long-distance transmission of hydropower.

"After all, that was an age that enjoyed the scientific genius of Tesla and Thomas Edison...and the bold entrepreneurship of industrial giants such as George Westinghouse and J.P. Morgan," Rappleyea said.

"But apparently, some people in the 1890s were just as skeptical and wary of changes as we are today."

Rappleyea said he had been told that in 1896, the civic leaders in Buffalo were very cautious about Tesla's plan to transmit power 26 miles from Niagara Falls to Buffalo.

"So cautious, in fact, that Buffalo Mayor Edgar Jewett waited until after midnight to pull the switch. And the ceremony was unadvertised, just in case something went wrong," Rappleyea said.

The good news: It all worked out as planned. And the next morning, as people headed out to work, they found that Buffalo's streetcars were running on Niagara's power, transmitted over wires strung by the General Electric Co.

Tesla's discovery — a rotating magnetic field produced by two or more currents alternating out of step — opened a new era of electric light and hydropower.

Among those attending the "Century of Electrical Energy" conference here was William H. Terbo, the honorary chairman of the 3,500-member Tesla Memorial Society Inc., based in Lackawanna.

A LINE OR TWO

Don Glynn

Terbo, who lives in Scotch Plains, N.J., is one of four direct descendants of Tesla. Mostly forgotten for decades, the famous inventor who died in 1943 gained new prominence during the nation's Bicentennial in 1976, Terbo said, delighted that his granduncle was being honored after 100 years.

■ ROAD REPORT: That road

through Joseph Davis State Park, linking Pletcher Road and Lower River Road, will be closed for the winter, beginning Dec. 1.

At Artpark in Lewiston, the Portage Road entrance to the upper park area and the fishermen's trail parking lot also will be closed.

The district state parks commission, which administers both parklands, closes the roadways to reduce costs for plowing and maintenance.

■ QUOTE OF THE WEEK: "That's the Billionaires Club, with a 'b,'" said Dan Shumny, a marketing consultant for Casino Niagara, describing an exclusive lounge on the upper floor of the new attraction ready to open Dec. 9. A reporter wondered if he meant "Millionaires Club."

Shumny shot back: "No, I said 'Billionaires.' Everyone's a millionaire these days."

■ ON THE LINE: People interested in obtaining information from the Western New York Coalition Against Casino Gambling may call 882-4793. The coalition is located at 1272 Delaware Ave., Buffalo, N.Y. 14209-2496.

Don Glynn is a veteran Gazette reporter, editor and columnist. His column appears Wednesday, Friday and Sunday.

Niagara Gazette, Nov. 17, 1996

Edward D. Adams

Nikola Tesla

Samuel de Champlain

Awesome Power In A Setting Of Stunning Natural Beauty

Niagara! Perhaps no other river on Earth is as recognized for its awesome power and breathtaking beauty. While exploring in 1604, Samuel de Champlain recorded the first European reference to Niagara Falls.

As American civilization unfolded, Niagara Falls proved to occupy a strategic position on major land and water trade routes.

Its attraction as a center of commerce and a site for harnessing water power grew by leaps and bounds. In 1841 the earliest calculation of the power of Niagara Falls was made. The tremendous flow was figured to be 374,000 cubic feet per second.

At a height of 160 feet Niagara Falls was capable of imparting a total of 6,800,000 horsepower, of which two thirds could effectively be captured through water wheels.

The Falls' location, majestic splendor and staggering power were a magnet for tourists and developers the world over. Niagara Falls was and still is recognized as one of the few great "Wonders of the World."

An Unknown Scientist with a Revolutionary Theory

At age 28, Nikola Tesla came to the United States. Four years later he announced his invention of the "polyphase alternating current system."

Tesla's discovery that a rotating magnetic field is produced by two or more currents alternating out of step made transmission possible over hundreds of miles. This opened a new age of limitless electric light and power.

Tesla's 1888 announcement before the American Institute of Electrical Engineers captured the interest of George Westinghouse, the Pittsburgh railroad entrepreneur who was systematically buying up electrical patents and developing a full scale frontal attack on Edison's direct current system.

In 1893 Tesla and Westinghouse contracted to install the power and lighting equipment for the Chicago World's Fair. This successful demonstration of alternating current set the stage for its later use at Niagara Falls.

Great Risk- Great Reward

In 1883 Thomas Evershed an engineer for the Erie Canal, suggested building a gigantic tunnel to tame the power of Niagara Falls.

In 1889, a group of New York bankers agreed to put up the money for the tunnel on the condition that Edward Dean Adams, a lawyer engineer and banker personally back their interests.

By the late summer of 1890, Adams' International Niagara Commission decided that power would be generated in a central station in the form of electricity using water diverted from the Niagara River above the Falls.

The tunnel was started in October 1890. Twenty five hundred men removed 600,000 tons of rock and built

Orders went out for four giant turbines larger than the world had ever seen.

By April 1895 the first hydroelectric unit was tested successfully. In 1896 Edison's Industrial Electric Company completed the construction of a 10,000 volt transmission line stretching an incredible 26 miles from Niagara Falls to Buffalo.

Just after midnight on November 15, 1896 electricity produced from the waters of Niagara Falls was transmitted for the very first time to Buffalo the Queen

Niagara Mohawk Centennial brochure

SECTION II

Principles of Wireless Power Transmission

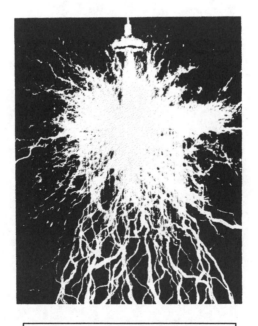

In 1900, world's largest Tesla coil, 25 feet
in diameter, with a 12-million volt discharge

6   Effects of Tesla's Life & Inventions

Andrija Puharich, MD, LLD
circa 1985

SECTION ONE

<u>From Birth to Arrival in the U.S.</u>
1856 to 1884

Nikola Tesla was born under the Austro-Hungarian empire in the village of Smiljan in the region of Lika, in the mountains of present-day northwestern Yugoslavia. The simple little village looks very much today as it did some one hundred and twenty years ago. His father, Milutin, a priest of the Orthodox Serbian Church, and his mother, Djuka, received the newborn Nikola into this world at midnight between 9 July and 10 July 1856. Although young Nikola's life was idyllic up to the age of seven, he later wrote that during this early formative period, he was weak and vacillating, "a slender reed moved around by every emotional breeze." He lived in the great spaces of the mountains and benefited from his background in the "literary" world so that he was able to read and write at a very early age. I put the word literary in quotes because his mother had not been taught to read and write, which was the common lot of women in that day and age.

This phase of his life was abruptly ended by the death, due to injuries suffered by a fall from a horse, of his elder brother, Dane, age 14. So great was the shock to all in the family that Nikola's father could not bear the familiar surroundings of Smiljan and decided to leave the scene of the tragedy. The family moved to a nearby town, Gospic, which was noted as a market center of some 3000 people.

The shock of his only brother's death, and the departure from the cozy familiarity of nature had a profound effect on the seven-year-old Nikola. He suddenly became a recluse and began to live in his father's library, and in the local library, devouring every book that he could read and understand. It was during this unusual and early encounter with books that Tesla first became aware that he possessed unusual mental powers. Much later, he gives us a glimpse of these powers:

"In my boyhood, I suffered from a peculiar affliction due to the appearance of images, often accompanied by strong flashes of light, which marred the sight of real objects, and interfered with my thought end action. They were pictures of things and scenes which I had really seen, never of those I imagined. When a word was spoken tome, the image of the object it designated would present itself vividly to my vision, and sometimes I was quite unable to distinguish whether what I saw was tangible or not. This caused me great discomfort and anxiety. None of the students of psychology or physiology whom I have consulted could ever explain satisfactorily these phenomena. They seem to have been unique, although I was probably predisposed, as I know that my brother experienced a similar trouble.

"The theory I have formulated is that the images were the result of reflex action from the brain on the retina under great excitation. They certainly were not hallucinations such as are produced in diseased and anguished minds, for in other respects I was normal and composed. To give an idea of my distress, suppose that I had witnessed a funeral or some such nerve-racking spectacle. Then, inevitably, in the stillness of night, a vivid picture of the scene would thrust itself before my eyes and persist despite all my effort, to banish it. Sometimes it would even remain fixed in space though I pushed my hand through it.

TIME
The Weekly Newsmagazine

NIKOLA TESLA*
ill the world's his power house.
(See SCIENCE)
From a portrait by Princess Lwoff-Parlaghy.

Volume XVIII

Number 3

July 20, 1931

Fig. 1

"To free myself of these tormenting appearances, I tried to concentrate my mind on something else I had seen, and in this way I would often obtain temporary relief; but in order to get it I had to conjure continuously new images. It was not long before I found that I had exhausted all of those at my command; my "reel" had run out, as it were, because I had seen little of the world-only objects in my home and the immediate surroundings.

92

"As I performed these mental operations for the second or third time, in order to chase the appearances from my vision, the remedy gradually lost all its force. Then I instinctively commenced to make excursions beyond the limits of the small world of which I had knowledge, and I saw new scenes. These were at first blurred and indistinct, and would flit away when I tried to concentrate my attention upon them, but by and by I succeeded in fixing them; they gained in strength and distinctness and finally assumed the concreteness of real things. I soon discovered that my best comfort was attained if I simply went on in my vision farther and farther, getting new inspirations all the time, and so I began to travel -- of course, in my mind. Every night (and sometimes during the day), when alone, I would start on my journeys -- see new places, cities and countries.

"I was about twelve years old when I first succeeded in banishing an image from my vision by willful effort, but I never had any control over the flashes of light to which I have referred. They were, perhaps, my strangest experience and inexplicable. They usually occurred when I found myself in a dangerous or distressing situation or when I was greatly exhilarated. In some instances I have seen all the air around me filled with tongues of living flame. Their intensity, instead of diminishing, increased with time and seemingly attained a maximum when I was about twenty-five years old.

"These luminous phenomena still manifest themselves from time to time, as when a new idea opening up possibilities strikes me, but they are no longer exciting, being of relatively small intensity. When I close my eyes I invariably observe first, a background of very dark and uniform blue, not unlike the sky on a clear but starless night. In a few seconds this field becomes animated with innumerable scintillating flares of green, arranged in several layers and advancing towards me. Then there appears, to the right, a beautiful pattern of two systems of parallel and closely spaced lines, at right angles to one another, in all sorts of colors with yellow, green and gold predominating. Immediately thereafter the lines grow brighter and the whole is thickly sprinkled with dots of twinkling light. The picture moves slowly across the field of vision and in about ten seconds vanishes to the left, leaving behind a ground of rather unpleasant and inert gray which quickly gives way to a billowy sea of clouds, seemingly trying to mold themselves into living shapes. It is curious that I cannot project a form into this gray until the second phase is reached. Every time, before falling asleep, images of persons or objects flit before my view. When I see them I know that I am about to lose consciousness. If they are absent and refuse to come, it means a sleepless night.

"To what an extent imagination played a part in my early life, I may illustrate by another odd experience. Like most children, I was fond of jumping and developed an intense desire to support myself in the air. Occasionally a strong wind blew from the mountains rendering my body as light as cork and then I would leap and float in space for a long time. It was a delightful sensation and my disappointment was keen when later I undeceived myself.

"During that period, I contracted many strange likes, dislikes and habits, some of which I can trace to external impressions while others are unaccountable. I had a violent aversion against the earrings of women, but other ornaments, such as bracelets, pleased me more or less according to design. The sight of a pearl would almost give me a fit, but I was fascinated with the glitter of crystals or objects with sharp edges and plane surfaces. I would not touch the hair of other people except, perhaps, at the point of revolver. I would get a fever by looking at a peach, and if piece of camphor was anywhere in the house, it caused me the keenest discomfort. Even now I am not insensible to some of these upsetting impulses. When I drop little squares of paper in a dish filled with liquid, I always sense a peculiar and awful taste in my mouth. I counted the steps in my walks and calculated the cubical contents of soup plates, coffee cups and pieces of food -- otherwise my meal was enjoyable. All repeated acts or operations I performed had to be divisible by three and if I missed, I felt impelled to do it all over again, even if it took hours."[3]

Tesla discovered at school, The lower Real Gymnasium, when he was ten, that he could call out all arithmetical and mathematical calculations in his head just as clearly as if he were working it all out on a blackboard. This capacity served him like a modern high-speed computer all of his life. In 1870, at the age of fourteen, he graduated from The Real Gymnasium and shortly thereafter had the second major shock of his life. He was swimming in a stream with his friends wherein was anchored a long and large float. In order to surprise his friends, he decided to dive under the float, i.e., to "disappear" and emerge at the far end. He did not realize that he lacked the capacity to swim this length under water. So he swam as long as he could and came up to surface for air -- only to find a wooden bean against his head, and no air. By this time he was getting frantic for lack of air, and the large build-up of carbon dioxide in his blood further aggravated his sense of suffocation. At this point his brain was reeling and he began to sink. Just then a flash of light illumined his mind, and he thought he saw the planks above the beam trapping some air. He floated up to the planks, pressed his mouth against them, and found enough air to inhale. With his lungs and brain ventilated he was able to escape his entrapment. This close escape from death by drowning in his fourteenth year, however, was only the prelude for a sea of troubles that plagued him for the next seven years.

Following graduation and this near-drowning episode in 1870, he was sent to the Higher Real Gymnasium in Karlovac, Croatia, where he lived with his uncle. Here he lost his robust health when he contracted malaria, which racked his body with aches and fevers for the next three years. The only control then known for malaria was quinine, and he did not know which was more deleterious – the malaria or the quinine. We do know today that one of the most common deleterious effects of quinine is damage to the hair cells of the hearing nerves. That Tesla did not suffer from this kind of damage will be made evident from some of his subsequent experiences in hearing.

However, in spite of weakness from malaria and supersensitivity to all stimuli, his three years at Karlovac were the true beginning of his scientific work, the only life he would really know. We begin to see in these painful years the birth of his major ideas. His teacher in physics was Prof. Martin Sekulic who was well-informed as to what was happening on the contemporary scientific scene, as can be gathered from his Communications to the Yugoslav Academy of Arts and Sciences. He particularly emphasized electricity in his physics teaching. The young Tesla was utterly fascinated as he watched Prof. Sekulic vigorously turning the handle of a static electricity machine which developed a charge very much the way rubbing a plastic comb with a piece of wool develops sparks. The spark output of the static machine was then directed toward a small globe made of paper covered with metal foil balanced on the tip of a needle so that it could freely rotate like a magnetic compass. It fascinated Tesla to understand how the static charge is converted into a rotational motion. As he said himself, each such demonstration set off in his mind a thousand echoes of further probing and explorations. But his mind did not stop at the electrical forces acting on the small spinning paper globe. He expanded this idea to include the entire terrestrial globe. He worked out a plan to build a huge ring around the equator which would spin at the speed of some 1000 miles per hour, or as we would say today, in synchronous orbit. He planned to use this as a means for high-speed transport around the earth and toyed with various methods of getting his passengers on and off of his high-speed platform.

He now had two main elements working in his mind, which were to lead him on to his first great discovery: *the rotating magnetic field*. He knew from simple static machine demonstration that he had an experiment, which he could run in his head and put it on his mental display screen and re-run it over and over again. Today we would say that his mind was organized like a super-computer with all powers of analysis, integration, enormous memory bank, and such powerful visual display of all operations that they competed with physical events for reality quality. It was in this same period that another primordial image entered his mind, which he was able to give birth to thirty years later. He visualized the mighty torrent of Niagara Falls in the far-off United States/Canadian border after seeing a postcard picture.

In his mind, he invented a mighty water wheel with which to get mechanical energy from the falling waters. Thus, his mind's eye was endlessly working over three primordial programs that he had to solve:

1) the bulb spun by static electricity;
2) the ring platform suspended around earth's equator;
3) the power of water turning a wheel.

After three years of intermittent illness, and the ecstasy of learning to run his powerful mental computer, Tesla graduated from Karlovac in 1873 at the age of seventeen. Upon his return home to Gospic, this budding genius was to enter a crossroads crisis of his life and endure the third great shock to his entire being. He started his journey home with some sense of foreboding because he had to face his father on the question of his future career. His father desired strongly to have Nikola enter the life of the clergy. Nikola with equal desire and strength of purpose wanted to become an electrical engineer.

The very day that Nikola arrived home for the showdown encounter with his father, he contracted the dreaded cholera. He had been debilitated enough by his three-year bout with malaria, and now cholera. He lay between life and death in bed for the next nine months with scarcely the strength to move. How the pending problem with his father was resolved and his health restored is tersely described by Tesla:

> My energy was completely exhausted and for the second time I found myself at death's door. In one of the sinking spells that was thought to be my last, my father rushed into the room. I can still see his pallid face as he tried to cheer me in tones belying his assurance. "Perhaps," I said, "I may get well if you let me study engineering." "You will go to the best technical institution in the world," he solemnly replied; and I knew that he meant it. A heavy weight was lifted from my mind, but the relief would have come too late had it not been for a marvelous cure brought about through a bitter decoction of a bitter bean. I came to life like another Lazarus to the utter amazement of everybody. [4]

Unfortunately, we do not know what kind of bean he had been treated with. Having survived his third great shock, Tesla had to face another major crisis as he approached the age of eighteen. He was about to be called up as a conscript in the Austro-Hungarian Army. While Tesla does not explicitly mention this episode in his life story, it is known from other sources that he had no intention of becoming a military conscript. [5] This was especially more painful in that both sides of his family had a long list of military careers to their credit, as well as priests, of course. Having survived the family priestly pressures, he now had to survive the family military pressures.

Since we have no reliable data to go on about how Tesla managed to escape being a conscript, we have to reconstruct this period of his life from the historical context. We do know that Tesla states that it was his father's idea that he should disappear into the mountains for a time, to which proposition Tesla states that he reluctantly agreed. It so happens that my own father was born under the Austro-Hungarian Empire, not far from the region where Tesla was brought up. When my father reached the age of eighteen, he faced the same crisis as Tesla with respect to military conscription. He did not want to become a conscript, and his father supported his wishes. So my father disappeared into the Dinaric Alps for a year while his father tried to buy off the military people. In this effort, he was unsuccessful. My father, Franjo Puharich, told me that he had only two choices left since he could not hide forever. The first was to maim himself so that he would be unfit for military service. He tells me that this was a common practice in his day (ca. 1910). There are rumors afloat to this day amongst people still living in Yugoslavia that Tesla attempted such self-mutilation, but this rumor cannot be confirmed. My father told me that his second alternative was to escape abroad and enter some country illegally. He chose the latter course, stowed away aboard a ship and entered the U.S. as an illegal alien. In this way he escaped military conscription and eventually became a U.S. citizen. It was some fifty years before he returned to his native land.

All we know is that Tesla spent a year in the mountains, and when his father had made the arrangements, he enrolled as a student at the Polytechnic School in Gratz, Styria (now Austria) in 1875. Tesla only states that the year in the mountains helped to restore his health and gave him the freedom to pursue his grand "Gedanken," i.e., thought experiments, in his mental laboratory. He must have put much of his life into order because he was determined to get answered as many of his questions as he could. At Gratz, Tesla programmed himself to study every day from 3:00 AM to 11:00 PM twenty hours of work, seven days a week. He not only mastered the foundations of physics, mechanics and mathematics, passing his first year at Gratz with the highest honors, but found time to further his knowledge of French, German, Serbian, and Hungarian. He wanted to learn philosophy and decided to read Voltaire, the great French philosopher, as written in the original French. Having committed himself to complete this task, he found after h had started that Voltaire had written some 100 large volumes in very small print. Having made his bargain with himself, he was bound to keep it, and thus found the "time" to read the 100 volumes of Voltaire in his "spare time." This feat clearly shows us the magnitude of his prodigality at the age of twenty. He saturated himself with the key literary products of European art and science. So intense was his work and learning that he found out later that his professors, who loved him, had secretly written to his parents asking that somehow Nikola should be encouraged to slow down -- lest he kill himself with overwork.

Completing his first year at Gratz was a total triumph of his will over all obstacles. But his second year slowed him down, not because of lack of will or mind power, but because no scholarship aid was available in spite of his brilliance, it seemed that if ever there were a worthy student to receive scholarship aid, it was Nikola Tesla, but the fates conspired to slow him down. He stayed on in Gratz, auditing all the courses he could, but of course did not have to take exams. He read in the library, and attended demonstrations in the laboratories. His professors loved him so much that they allowed him to attend all classes even though they knew he had not paid tuition. It was at one of these demonstrations that Professor Poeschl showed the newly invented Gramme Dynamo, which he had received, from Paris. This was a crude direct current generator, which had a horseshoe shaped magnet for the field, and as a rotor turned inside it, electricity was produced. What disturbed young Tesla was the scientific lack of aestheticism of the Gramme Dynamo It produced electricity with much noise and sparking at the commutator. The commutator was a set of rings on the dynamo shaft that collected the electricity, which the turning rotor collected from passing across the magnetic field of the horseshoe magnet. The principle here is that if a wire is moved through a magnetic field (from the horseshoe magnet) an electric current is produced, and is passed to a ring (the commutator) on the shaft of the rotor, and a sliding contact moving over the ring called the brush picks up the electricity and passes it by wires to the load. So offended was Tesla's deep sense of scientific elegance with the clumsiness of this arrangement that he protested to Prof. Poeschl with the opinion that there must be a better way to accomplish the goal. To this Prof. Poeschl replied with heavy-handed German authoritarianism, looking Tesla in the eye: *"Mr. Tesla may accomplish great things, but he certainly will never do this. It will be equivalent to converting a steadily pulling force, like that of gravity, into a rotary effort. It is a perpetual motion scheme, an impossible idea."* [6]

This was the challenge Tesla needed: to solve an "impossible" problem. From 1876 to 1882 he ran his prodigious mental computer laboratory over and over this problem. He admits that by 1880 (age 24) he was beginning to realize that perhaps Prof. Poeschl might be right -- the problem might be insoluble. More of this later. In spite of his brilliance, Tesla could not solve the simple problem of making a living -- as many a genius has found out. So Tesla turned his powerful mind to an easy way of making a living: gambling. European student life in his day was dominated by drinking, duelling, gambling and sexual adventures. In this environment it was always easy to enter into a gambling encounter to make money, and from this expertise Tesla stayed on and lived at Gratz until 1879 -- the year in which he would have graduated had he been able to pay tuition. We do not know much about these years except that Tesla, in order to support himself by gambling, billiards end cards, had to be "one of the boys", and therefore had his share of personal indulgences. However, he learned from this experience that his system could not tolerate coffee, and eventually abstained from it. Alcohol he was able to tolerate in small amounts, and continued to use it as an after dinner drink for the rest of his life. His passion for gambling during these three years became a fever, but one which he finally conquered.

He left Gratz in 1879 to visit his family in Gospic, and no sooner had he come home then his father died. This sad event imposed further burdens on his already threadbare poverty. As hard and difficult as his life had been, he was determined to continue with his postgraduate education. He felt, even by auditing courses, that he had learned everything possible at Gratz. He decided to go to Prague (Bohemia) and enroll there in the ancient and distinguished University of Prague. To his great disappointment he found that he did not qualify for enrollment because he had not studied Greek in high school. In spite of this setback and continued poverty, he persisted in learning electrical engineering by auditing courses, and, of course, reading everything of interest in the library.

Tesla's family, of course, knew of both his great desire to learn and his equally dire financial straits, and tried to help him. His mother's brother, Pajo Handic, was a military officer stationed in Budapest. Pajo had a friend, Perenc Pukas, who was an executive of the Central Telegraph Office of the Hungarian government. Through this friend, Pajo arranged a job for his nephew. Tesla arrived in Budapest in January 1881 at the age of 24, eager to begin his long awaited career as an electrical engineer. However, he was bitterly disappointed to find out that the only job available was as a draftsman -- work he really disliked.

Fortunately for Tesla, the new telephone invented by Alexander Graham Bell in 1876 had just reached Europe, and the Hungarian government was eager to install an exchange in Budapest. The inspector-in-chief of the Telegraph Company recognized Tesla's mathematical and engineering talents and awarded him the job of designing the new installation. Tesla gives every indication that he was happy with his new work for the first time and his freedom from poverty in Budapest was a joy.

It is puzzling to know that in January of 1882, Tesla suffered a fourth great shock: he had a complete nervous breakdown. What is meant by this phrase needs some explanation. We have no indication that Tesla was frustrated by his work. On the contrary, he states that in the few months before he moved to Budapest, while still in Prague, his "mental computer" was so free-running that he "invented" in his Gedankan experiments all of the motors and dynamos for which he later became famous. But he admits that while he built the mechanical models in his head, the underlying principle escaped him. Perhaps the unending quest for this Holy Grail of electrical first principles haunted him more than he realized. His nervous breakdown was in fact an exact opposite of breakdown in that it was a super sensitivity of senses and of mind organization.

Tesla retreated from the world in that month of January 1882; he found insulation between himself and the noisy world. He describes lying in bed and distinctly hearing the ticking of a pocket watch -- three rooms away! When a fly landed on the table beside his bed, he experienced a dull thud in his ears. The vibrations of a carriage passing over cobblestones several miles away wracked his body. The ground under his bed and under his feet rumbled continuously from any sound; he felt as though he was in a continuous earthquake. If the sun's rays accidentally fell upon him, his brain felt as though it were being clubbed; and if the sun hit him while moving along a road where trees produced a stroboscopic effect, he felt as though he were being engulfed in hammer blows of lightning. His whole body from time to time was convulsed by twitching and Tremors. One could almost say that sensory stimuli were exciting epileptic-type electrical storms throughout his brain and body.

Even today there is no way to describe how his nerves could amplify the weak electrical signals of his sense organs. It was as though his nervous system had gone from normal thermal level electrical conduction to super-cooled typed of electrical superconductivity [7] The only other instance we know historically of such super sensitivity is from the lives of certain saints who, in undergoing a kind of final refinement and purification, would enter an ecstatic state similar to Tesla's condition.

In Tesla's case this condition of general hyper-sensitivity does not seem to have lasted for more than a month, because he recounts that with the aid of his devoted athletic friend, Antal Szigety, he began to recover. Szigety insisted that Tesla get out of bed; he walked him, and exercised him. Tesla later admits that in the recesses of his awesome computer mind was the solution to his quest -- the perfect alternating current motor, but he could not reach it. Perhaps he let his body enter a higher dimension of sensitivity in order to find the solution. But it is Tesla who must describe this ultimate experience culminating the quest of his life to this moment. The climax and recovery of health rapidly came in February 1882; we do not know the exact date:

"A powerful desire to live and to continue the work, and the assistance of a devoted friend and athlete (Antal Szigety), accomplished the wonder. My health returned and with it the vigor of my mind. In attacking the problem again, I almost regretted that the struggle was soon to end. I had so much energy to spare. When I undertook this task, it was not with a resolve such as men often make. With me, it was a sacred vow, a question of life and death. I knew that I would perish if I failed. Now I felt that the battle was won. Back in the deep recesses of the brain was the solution, but I could not yet give it outward expression".

"One afternoon, which is ever present in my recollection, I was enjoying a walk with my friend in the City Park and reciting poetry. At that age, I knew entire books by heart, word for word. One of these was Goethe's *Faust*. The sun was just setting and reminded me of the glorious passage:

"Sie ruckt und weicht, der tag is uberlebt,
Dort eilt sie hin und fordert neues Leben.
Oh, dass kein flugel mich vom Boden hebt
Ihr nach und immer nach zu streben!
Bin schoner Traum indessen sie entweicht,
Ach, *ru* des Geistes Flugeln wird so leicht

Keinen korperlicher Flugel sich gesellen!"

Translation:
"The glow retreats, done is the day of toil:
It yonder hastes, new fields of life exploring;
Ah, that no wing can lift me from the soil,
Upon its track to follow, follow soaring
A glorious dream! though now the glories fade.
Alas, the wings that lift the mind, no aid
Of wings to lift the body can bequeath me!"

"As I uttered these inspiring words; the idea came like a flash of lightning and in an instant the truth was revealed. I drew with a stick on the sand the diagram shown six years later in my address before The American Institute of Electrical Engineers, and my companion understood them perfectly. The images I saw were wonderfully sharp and clear and had the solidity of metal and stone, so much so that I told him: 'See my motor here; watch me reverse it.' I cannot begin to describe my emotions, Pygmalion seeing his statue come to life could not have been more deeply moved. A thousand secrets of nature, which I might have stumbled upon accidentally, I would have given for that one which I had wrestled from her against all odds, and at the peril of my existence."

What did Tesla visualize in his computer that had solved his agonizing problem of how to make an alternating current motor of aesthetic design? While Tesla was the first human being to have the vision of a rotating magnetic field, subsequently many engineers and scientists have been able to have this vision due to subsequent detailed scientific expositions being made of the phenomenon. Let us try to recapture Tesla's vision of February 1882 when he was all of 25 years old. Please refer to Figure 2.

AC-1 shows the representation of a normal sine wave. This can represent the rise and fall (M) of a wafer wave, and its travel to the right; or it can represent the rise and fall of an electric current from positive (+) charge state (up) to negative (-) charge state (down) and direction (➔) of travel; it can represent the swing of a magnetic wave from north pole (+) to south pole (-), and direction of travel. The rise end fall is shown by degrees on a 360° scale Just as in a circle, and one complete cycle is 360°. The rise phase in AC-1 is positive (+) from 00 to 900; then the falling positive (+) phase is from 90° to 180° where it reaches zero value, 0, neither (+) or (-). This completes half a cycle of the sine wave.

From 180° the wave goes from 0 value down to full negative (-) value at 270°; from 270° the value goes from full negative (-) up to zero at 360° This completes a full cycle of action that has the form of a sine wave. Many phenomena in nature follow this cyclical pattern. Of immediate interest is that this is the way an alternating current (AC) is displayed on an oscilloscope in a two-dimensional plane. However, if one saw this AC wave as it exists in nature, and as Tesla undoubtedly saw it in his mental visual display computer, it looks more like a corkscrew in three dimensions. If the direction of travel is from left to right (➔), then components of the AC point in different directions. The magnetic component of the electro-magnetic AC wave points upward in the plane of the paper where the arrow is marked M. The electric component of the electro-magnetic AC wave paints directly down through the paper (perpendicular to the plane of the paper).

What Tesla knew, and other scientists knew, was that if one placed a second alternating current in a circuit, AC-2, leading AC-1 by 90° this is called a phase difference; certain effects would occur which could be used to turn a magnet that was suspended like a compass needle.

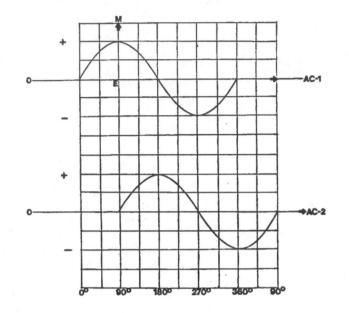

Fig. 2

Let us begin to build up the vision that Tesla had of the rotating magnetic field using the simple elements we have given.

Referring to Fig. 3, place a compass an the center of the circle over the part marked magnetic bar rotor. Line up the compass needle, and the line of the magnetic bar rotor so that they both point north. Now take bar of iron a pocket knife blade will do - and bring it close enough to the north pole of the compass needle so that the needle can be moved. Now move the knife point along the rim of the compass so that the needle moves first to 0° (north) and then to 90° (east). Practice guiding the compass needle so that you can move it smoothly first from 0° to 90° then from 0° to 180°; then from 0° to 270° then the complete circle from 0° to 360°. This in effect is how an alternating current motor works. Your hand 16 the alternating current that goes through a full cycle (or circle) of 360°, and it guides the magnetic component of the alternating current (the iron bar, or knifeblade), i.e., a magnetic field. In such a way that its force produces a torque, or rotation on a rotor (the compass needle). Now this part is easy. What Tesla had to solve was how to produce the magnetic field whirlwind around the circle of the rotor without any mechanical motion to create the magnetic field. The vision he had can now be visualized by us.

Referring to Figure 3, remove the compass and note the two circles, each of which is eccentric to the circle around the magnetic rotor bar: and further note at 90°and 180° that the circles are 90° out of phase with each other. Each of these circle, represents one cycle of an alternating current, AC-1, and AC-2 (as in Fig 2), but now shown as a complete cycle in the form of a circle rather than a sine wave. Now to visualize what Tesla saw: imagine circle AC-1 to be a hula hoop of blue color and watch it go around a person, or better still, watch a child swing a hula hoop an his hips. You can now see and feel the swing of a magnetic loop around a central rotor circle. Referring to Fig. 3, note that the outer circle represents the (+) swing of a sine wave and its perimeter the maximum (900 as in Fig. 1 AC-1) north pole magnetic field strength. The inner perimeter circle shows the (-) swing of a sine wave and represents the south pole maximum magnetic field strength (270° as in Fig. 1 AC-1). Between these two maxima there is a circle, which represents zero magnetic field strength (0° and 180° as in Fig. 1 AC-1).

Fig. 3

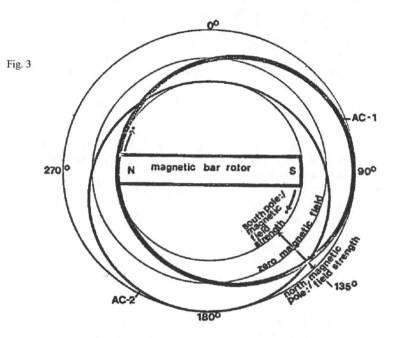

Now we add a second hula hoop (AC-2), which leads the first hula hoop (AC-1) by 90°, and get them both spinning around the hips. See Fig. 3. As long as the two hula-hoops maintain their 90° phase difference, we have the identical condition of magnetic field whirlwind that Tesla saw in his vision. However, using his prodigious calculating capacity, Tesla could plot the magnetic field strength relations for every instant of time, and for every degree of the circle – and compute the field effect on the magnetic bar rotor in producing mechanical rotation and power. For example, if we take an instant of time at 90° to see what the magnetic field strengths are of AC-1 and AC-2 in Fig. 2, we see the following: Since AC-2 peak magnetic field strength is at 180° with respect to AC-1 (see Fig. 3), it will pull the south pole of the rotor clockwise toward it. Since there is inertial, or resistance in the rotor it will lag the maximum magnetic field strength of AC-2 by some degrees. Since AC-1 is going in a falling phase of magnetic field strength (going from 90° to 180° Fig. 2) its pull on the rotor is weakening, allowing the latter to follow AC-2. Since the magnetic field strengths of AC-1 and AC-2 are additive between 90° and 180° (Figs. 1 and 3) or at 135° – maximum north magnetic field strength – the rotor will be found at this part of the circle at this instant. As the two hula-hoops of magnetic field strength sweep around they will pull the rotor with them, just as if the hand were moving a magnet around a compass needle. Not only did Tesla see this immaterial set of magnetic forces spinning around, hut he encased them with the proper mechanicals – the shaft with its mechanical racer; the field coil structure surrounding the rotor. Then he encased these mechanicals with the right materials -- the iron for the magnetic cares, which he wound with the proper copper wire coils. Then he devised the proper geometry and circuits that connected all the coils, which guaranteed the proper phase differences between AC-1 and AC-2. Then he set the motor running in his head, and quietly saw that by throwing one switch he could reverse the phase between AC-I and AC-2 so that AC-1 now leads AC-2 by 90° – and presto – the motor smoothly reversed direction! His grand design was done – he had grasped the principle of the rotating magnetic field, and given it physical embodiment in a working model – all on the visual display computer screen in his mind.

In the fall of 1882 Tesla moved co Paris, hoping to work for the French branch of Thomas Alva Edison's Continental Electric Company. He was able to bring with him his good friend and now master mechanic, Antal Szigety, as an assistant. He readily obtained the job he sought. However, work demands prevented him from building his new invention in the flesh of iron, steel and copper. It was not until the summer of 1883 when he and Szigety were sent to Strasbourg to do some motor repairs chat they had the time and a machine shop in which to build the first physical alternating current motor. The test was an instant success, and followed faithfully in performance the model that Tesla had been running on his mental computer screen. He was now 27 years old with a proven theoretical and practical solution of the mission declared impossible by his erstwhile Professor Poeschl.

Tesla was kept busy in Strasbourg until the spring of 1884, when he returned to Paris. There he enlisted the interest of the Paris manager of the Edison works, Mr. Batchellor. Mr. Batchellor recognized his genius and the value of his invention; and urged him to go to America to work with the great Mr. Edison. Tesla accepted this invitation, and prepared to leave Paris. At the Paris train station, he discovered to his horror that his baggage, his transatlantic passage tickets, and his wallet had been stolen! As the train began to pull out, he had to make a choice: should he chance getting on the train and the boar -- or cancel his journey? He chose to leap aboard the train. When no one showed up to take his place on the steamship, he was allowed to board. And so he was on his way across the ocean to an unknown land, but one to which all immigrants in 1880's believed was the land of promise and of gold. He landed at Ellis Island in New York Harbor on June 6, 1884 with four cents in his pocket. But he had in his pocket a letter from Mr. Batchellor introducing him to Mr. Edison, which said in part:

"I know two great men and you are one of them; the other is this young man."

SECTION TWO

From Darkness to Light

The period from age 28 (1884) where Tesla is penniless and homeless, digging ditches in New York City to survive - to the age of 35 (1892) when he becomes the most famous inventor in the world, is a millionaire, socially lionized, and honored by all of the great scientists in London. It is difficult to find in the life of any contemporary person one who has gone through so many transformations in seven years as did Tesla from 1884 to 1892, age 28 to 35. With four cents in his pocket, and his letter of introduction, he found his way to Edison. This meeting could have been the break that Tesla needed, because Edison, as a patron, could have given him all the opportunity necessary to realize his dreams. It is an irony of history that the two men net at all, because, temperamentally, they were at opposite poles. Tesla with his great scientific mind was systematically seeking first principles in nature end using his inventions as tools for deeper, more refined probing. Edison had the great mind that could find associative relations between material systems, which resulted in new inventions as an end in itself. Tesla's mind was seeking to understand the architecture of nature with geometrical comprehension. Edison followed each clue of nature from point to point with linear reduction to practice.

Much has been written about the nine-month period when Tesla worked for Edison. Suffice it to say that Edison was committed to a future electrical technology based on direct current -- from batteries and direct current generators. He had perfected the incandescent lamp and was trying to create a marker for it -- and to do this he had to supply electricity. In order to do this, he had to create DC power stations aboard ships, in cities, and for factories. In this Edison was successful, and power stations were being built rapidly as the United States became electrified. Furthermore, the financial backing of J.P. Morgan was securely behind the Edison ventures which assured success.

Against this array of industrial growth and financial power stood the tall, thin Tesla with his visions of using alternating current to industrialize and power the earth. He tried to catch Edison's attention by performing meticulously and creatively as an engineer. In this he succeeded and won Edison's highest compliments. But when he tried to open up a conversation about his ideas on alternating current, Edison would not listen. Edison had made up his mind that the future electrification of the world would be with direct current. Tesla became discouraged realizing that he would have to develop his ideas on his own, and left the employ of Edison. Little did Edison realize that his lowly employee would soon become his greatest and most worthy technological and business opponent. The war of the giants was in the making.

Tesla had made such a powerful impression on other Edison employees that they offered to finance his inventions. Accordingly, the Tesla Electric light and Manufacturing Company was founded in March of 1885, in New Jersey, across the Hudson River from New York City. Tesla now had a modest amount of capital, but he could not begin work on his beloved AC motor because his backers wanted him first to develop commercially saleable arclights for street lighting. This he did, and soon he was filing a stream of patents. But his success was short-lived. As the great financial depression of 1886 clamped down on the U.S., his new company was forced into bankruptcy.

Now in the 30th year of his life, just as the possibility of the successful realization of his motor was tantalizingly held before him, Tesla was thrown into a pit of despair. There was no work to be had, although occasionally Tesla was able to get a job digging ditches in New York City, as public works for sewers and water mains continued during the depression. He had no place to live and slept wherever he could get shelter for the night. He later spoke of living "through a year of terrible heartaches and bitter tears." It must have seemed pure torture to toil in a ditch, when he did have work, and at the same time to run his mental computer in the dazzling images of his colorful and poetic rotating magnetic fields. Fortunately, in the ditch with him was his foreman -- also working out of his field just to keep alive. As the two men sat together eating their meager lunches, Tesla poured out his dreams and inventions. Tesla did not realize it then, but his fellow worker was the magical connection to A.K. Brown of the Western Union Telegraph Company. Almost miraculously, under Brown's initiative, the foreman's faith, and Tesla's total conviction in his vision, The Tesla Electric Company was formed in April 1887 within competing distances of the shops of the Edison Company. Tesla's working capital, in the form of a loan, was $500,000. In today's monetary values, this had the purchasing power of ten times that amount.

Tesla had been like pent-up lion for the past five years – ever since he understood the principle of the rotating magnetic field. But whether he was building telephone systems in Budapest, repairing DC motors and generators in Paris or Strasbourg, designing motors for Edison in New York City, or digging ditches, his powerful visual display computer was developing new designs and refining them. Thus, it was that when he obtained his laboratory, he produced an explosion of devices and patents such as had never been seen before in such a short time.

What Tesla did was to design a <u>system</u> to produce alternating current, distribute it over large distances, end utilize it in motors and lamps. He worked out every crucial detail of this system to such perfection that, to this day, his technology and concepts dominate AC systems virtually unchanged. His system is called the: "Tesla Polyphase System" because, following the conception portrayed in Figures 2 and 3 of using the 90° phase difference between two alternating current. to generate the "hula hoop" rotating magnetic field, he added more pairs of hula hoops, up to 384, to the field. This had the same effect as adding more pistons to a gas engine: it gave greater and smoother rotatory power (torque) to the motors. Thus, he built fields around his rotors, which had hundreds of "hula hoops" rotating, each pair with a precise number of degrees of phase difference from its neighbor. See appendix for Tesla patents illustrations.

With the rapid issuance of one patent after another from the U.S. patent Office, the world of electrical scientists suddenly became aware that a new luminary -- a new Faraday of electricity -- was in their midst. Correspondence and interviews suddenly made heavy demands upon Tesla's twenty hour days. A year later on May 16, 1888, Tesla gave the first of his famous lectures at Columbia University, New York City, under the auspices of the American Institute of Electrical Engineers [1] Here he revealed and demonstrated the full sweep of his polyphase alternating current system. The electrical world was amazed at his revelations. The insoluble problem had been solved: alternating current could be transmitted far hundreds of miles for industrial usage -- compared to a mile or two by Edison's system. As Tesla was being applauded for his victory, the Edison forces were preparing interests massive counterattack against this threat to their vested interests.

In this lecture Tesla received not only scientific and academic recognition, but also industrial recognition. About a month after the lecture the founder of Westinghouse electric, George Westinghouse, came to see Tesla at his laboratory. Without much negotiation, Westinghouse bluntly offered to buy up all of Tesla's polyphase system patents for one million dollars in cash. Tesla said he would accept the offer if Westinghouse also paid a royalty of one dollar per horsepower of motors produced. Westinghouse agreed, and the deal was made. Thus, in two tumultuous years, Tesla went from the despair of ditch digging to world renown for his genius and independence.

What did this shift of status and recognition do to Tesla? First, he paid back the half million-dollar loan that had made his success possible. The rest of the money he used to support his laboratory and future research. But, as part of his contract, he was to go to Pittsburgh, Pennsylvania and work with the Westinghouse engineers to develop the production prototypes of the various parts of the polyphase system. However, here Tesla learned something about himself: it was not possible for him to gear down his mind and work to the pace of other mortals. He felt stifled, irritated, and less-than-creative. During this period of self-assessment, Tesla learned still more about himself. He could not accommodate his mind, emotions, or actions to <u>any</u> human being. The drummer that he marched to beat a rhythm uniquely solo to Nikola Tesla. He knew now that total dedication to his calling, and total celibacy, were to be his way of life.

In order to consolidate his new role in life as a wanted, famous person, and his own desire for privacy, he decided to return to Europe after a five-year absence and re-formulate his entire working philosophy. The past five years of poverty and creative effort had exhausted his reserves, and so, when Tesla returned in 1889 to his native bika, he went into retreat at the Gomirje monastery. This was curious since Tesla had fought for years to keep away from organized religion. One wonders if he did not contemplate withdrawal from the world into a monastic life. But his decision must have been to go back to live in the world and yet remain apart – because he returned to New York City in January 1890.

In 1890 Tesla established a pattern of life, which he was to keep up, whether rich or poor, to the end of his days. He moved into a private suite at the best hotel in New York City – the famed Waldorf-Astoria that was then at 34th street – the site of the Empire State Building. His elegant six foot, two inches, 140-pound body was encased in the most fashionable tailored clothes. These he always wore, whether in the laboratory, or at a fancy dinner party. He worked at his laboratory from 9 AM until about 6 PM. At the Waldorf when he appeared at 8 PM he insisted that only the headwaiter should wait on him, and all food was especially prepared under his meticulous directions. He dined alone, unless he was giving a banquet for his friends. Every aspect of the meal had to pass his scrutiny for an almost "Kosher" rigidity of preparation. The headwaiter could not touch any dish or food without the interposition of a fresh napkin between hand and dish. Tesla himself handled everything with a dozen or two dozen napkins during a meal – each napkin being used once and discarded.

Tesla, the former ditch digger, gambler, and billiard professional, could no longer tolerate physical contact with another human. He never shook hands with a person. The mere contact with human skin gave him the sensation akin to an allergic reaction; the energy of others was like poison entering his blood.

Yet in spite of the noxious quality he felt emanating from all persons, and his fear of germ contamination, he was considered absolutely charming and thoroughly sociable in his reserved and aloof way. This paradox gave his personality an incredible attractive power in society. Men trusted him. Women found him attractive and charming with no physical undertone. He was Lionized by the social "400" families of New York at that time. His resolve to be celibate, ascetic, and dedicated to his work was forever under attack. He did occasionally attend parties from 8 PM to 10 PM and then left to go to his laboratory. If he had a new electrical effect to demonstrate, his select friends would troop in evening clothes and gowns to his laboratory just south of Washington Square.

Here he would astound them with his Promethean displays of electrical fire and his light displays -- where he would pick up a long glass tube without wires, and it would light up in his hands. He had a great sense of drama and many thought that his laboratory was the best theater in New York. But the dinners at the Waldorf, the fancy clothes, the Tesla Theater were only small dues, which he paid to his adulating society. The real Tesla inside returned to his laboratory after the 10 PM dinner and went to work until 3 AM. Here in the stillness of his laboratory he pursued his vision secretly. It is only years later that we are able to piece together what was really going on in that most magnificent of laboratories: the mind of Tesla.

What the world saw from Tesla were a series of inventions, each of which would have been the climax of a lifetime of work for an inventor. But what no one knew was that Tesla was only developing new tools in order to explore deeper and deeper into the mysteries of electricity.

What he had in mind was to build the tools that could produce any kind of electricity he desired. First, he wanted to explore the various frequencies of electricity. He knew that electricity must have different qualities when the number of cycles per second changed, or increased. Therefore, he had to invent machines that would produce electricity smoothly over the range from one cycle per second (now called Hertz, or Hz) through tens per second, hundreds per second, and so on up to light frequencies. But he was the only one on the planet earth who had the total scientific approach for producing these frequencies at this time.

First, he invented and built a series of alternating current generators, which allowed him to reach frequencies of tens of thousands of cycles per second. Here he mastered all the problems of building copper coils to produce the magnetic fields that would turn the rotors, which produced alternating current. Here he solved the complex geometries of magnets, coils, and windings that produced the desired frequencies. But he learned the limitations of the production of AC waves by the rotating generator. He found that magnetic coils produced such high self-induction that they damped electromagnetic oscillations so much that the higher frequencies neither could nor be produced. In order to solve this problem, he came upon his next greet invention, which is called the Tesla Coil. The invention is utter simplicity itself, but no one had seen the solution before him.

Since the days of Faraday, experimenters had obtained different frequencies and different voltages of electricity by means of an inductance coil. This device uses a battery as a direct current source. A long wire coming from one terminal of the battery is coiled like a spring upon a round stick of iron, and then the other end of the wire is connected back to the other terminal of the battery. The electric current coursing through this coil (called a primary coil) produces a magnetic field in the iron care. This magnetic field can be used to energize a second call wrapped around the primary coil, and a current will be induced in the secondary coil. The value of such induced currents is that the voltage can be markedly increased in the transference of energy from the primary to the secondary in a definite ratio, which is proportional to the number of wire turns between the secondary and the primary. Far example,

$$\frac{1000 \text{ turns secondary}}{100 \text{ turns primary}} = \frac{10}{1} \text{ ratio} = 10{:}1 \text{ voltage step-up}$$

So if you put 100 volts DC into the primary, you can get 1000 volts DC out of the secondary. Furthermore, to get a pulsed DC current out of the induction coil, one interrupts the DC current with a kind of telegraph-key switch, and one would now have an "interrupted current."

Now Tesla would produce nice sine waves (he called them harmonic waves) out of his AC generator, and then in order to increase their voltage he would use an induction coil to exceed the peak voltage output of his generator. Yes, he would get higher voltages, but he would mess up his nice sine waves of alternating current end get chopped up pulses from the coil output. (See Fig.4.) He solved this problem by using a simple device called the Leyden Jar, which had been around longer than the induction coil. The Leyden Jar can be simply described as an open glass cup whose outside is lined with metal foil. This geometric form of an insulator (glass) lined with foil (as conductor) has the property of being able to store a large amount of electric charge, and furthermore, to discharge it quietly. This Leyden Jar is more formally known as a condenser. Lord Kelvin had studied the properties of the condenser and had described them in an elegant mathematical formulation. Tesla knew this Kelvin formulation. He suspected it had the key to his problem. The details of Tesla's solution are well known to every physics student today, and we will try to explain it in layman's language. Just as Tesla had placed two AC waves 90 degrees out of step to create the hula hoop rotating magnetic field, so he knew that in any single AC wave there were two components within that wave that were 90 degrees out of step: in any AC wave the current lags the voltage by 90 degrees. Now the voltage can be likened to the pressure, which a tall column of water exerts; and the current is the amount of water in that column. Obviously, if we keep the amount of water constant (volume) we can increase or decrease the pressure it exerts by changing the height of the column. Now Tesla further knew that the current in a magnetic coil is 90 degrees out of phase with the current in a condenser, now called a capacitor. It is as though one held two glasses, on in each hand, one of which is filled with water. Now the glass in the left hand is held vertically and is empty. The glass in the right hand is filled with water and is

Fig. 4

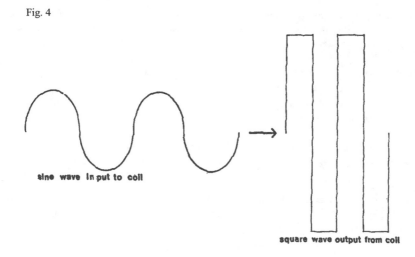

sine wave input to coil

square wave output from coil

tipped so that it is at right angles (or 90 degrees) to the glass in the left hand. As the right hand glass is tipped it will empty water into the left hand glass. If we rock the two glasses (always held at right angles to each other) back and forth, the water will slosh back and forth between the two in a smooth harmonic transfer. The water, of course, is to be likened to the amount of electric charge held first in the magnetic field <u>coil</u> (right hand), and then transferred to the capacitor, or <u>condenser</u> (left hand). The <u>rate</u> at which the electrons can be sloshed back and forth between the coil and the condenser becomes the frequency of the oscillation of the alternating current. It so happens, by the laws of nature, that the timing of the filling up and emptying of the condenser with electrons can be precisely determined by two values. The first is by the electron charge-carrying capacity of the insulator (the dielectric constant); and the second is by the surface area of the metal foil conductor covering the insulator. These two factors give a precise <u>time constant</u> for the charge/discharge cycle of a given capacitor; and when mated to an equivalent time constant for the coil (inductance), Tesla could precisely control the peak frequency of any given coil, and thus produce his smooth harmonic sine wave alternating current Once he had perfected his "Tesla coil", he could produce any frequency of electricity of any potential, or power level, he desired. He now had the cools to find out about the nature of electricity.

He knew the great equations of Clerk Maxwell which said that electric waves were identical to light waves. [3] In 1887 Heinrich Hertz had shown that by using an interruptor spark mil he could produce electric waves that behaved like light waves. But what Tesla wanted to do was to produce elegant sine waves octave upon octave from the very low frequencies (8 Hz) up to light waves (9 million million Hz) and find out what was the deep inner secret of light and electricity. He had two driving insights to check out. The first was his interest in lord Kelvin's theory that all of life's processes are electrical, and that there is an unknown <u>force in the organism</u> that guides and integrates such electrical effects. The second was his endless fascination with his power to visualize all experiences and to create his great inventions upon his mental computer display screen. Between bioelectricity and light, he

107

hoped not only to master the forces of nature, but also to divine the nature of life itself. He hints at these goals in his famous Franklin Institute Lecture (1893): [4]

"In all this vast world, of all objects our senses reveal to us, the most marvelous, the most appealing to our imagination, appears no doubt a highly developed organism, a thinking being.

"Again in all the perfect harmony of its parts, of the parts which constitute the material or tangible of our being, of all its organs and senses, the eye is the most wonderful. It is the one, which is in the most intimate relation with that which we call Intellect. So intimate is this relation, that it is often said the very soul shows itself in the eye.

"But there is something else about the eye which impresses us still more than these wonderful features – an optical instrument of being directly affected by the vibrations of the medium. (In Tesla's day "the medium" meant the aether.) This is its significance in the processes of life.

"A single ray of light from a distant star falling upon the eye of a tyrant in by-gone times may have altered the course of his life, may have changed the destiny of nations, may have transformed the surface of the globe, so intricately, so inconceivably complex are the processes in nature. In no way can we get such an overwhelming idea of the grandeur of nature as when we consider that, in accordance with the law of the conservation of energy, throughout the infinite, the forces are in a perfect balance.

Hence, even a single thought may· determine the motion of a universe. "

Thus, out of the toils and coils of the laboratory, we perceive that Tesla is a really only developing better tool with which to ask more profound questions of nature. Having developed sources of alternating current and the Tesla coils to produce any frequency or voltage of alternating current, Tesla now proceeded to use these tools to produce light. He produced every known form of light in those few years, from low frequency flickers (20-30 pulses per second) of phosphorescent materials, through incandescence of solids, liquids, and gases, to X-rays and, eventually, cosmic rays. He explored the entire spectrum of radiant energy known to man today – all before 1892. He studied the illumination of every kind of gas at his disposal under high pressures, atmospheric pressures, low pressures, and high vacuum. Since he made little attempt to exploit these discoveries commercially, it has long been a puzzle as to why he pursued these arcane studies.

We find the clue in his third great lecture in 1892 in London: [5]

"Such discharges of very high frequencies, which render luminous the air at ordinary pressures, we have probably often occasioned to witness in nature. I have no doubt that if, as many believe, the Aurora Borealis is produced by sudden cosmic disturbances, (such as eruptions at the sun's surface, which set the <u>electrostatic charge of the earth in an extremely rapid vibration)</u> the red glow observed is not confined to the upper rarefied strata of the air, but the discharge traverses, by reason of its very high frequency, also the dense atmosphere in the form of a <u>plow,</u> such as we ordinarily produce in a slightly exhausted tube. <u>If the frequency</u> were very low the dense air would break down as in a lightning discharge."

Now we begin to see what Tesla is doing. He has calculated in his superb mental laboratory the various gases that exist around the earth at various altitudes, and their respective pressures. He has discovered that he can duplicate on the laboratory bench, effects at any altitude around the earth -- lightning, Aurora Borealis, clouds, night glows, etc. He has discovered that high frequency alternating current under vacuum pressures becomes super-conducting, and that therefore he can use the ionosphere as a conduit. He has discovered, on the contrary, that very low frequencies neither do nor conduct well, especially in the stratosphere and clouds, and that here they build up their energy into accumulated static electrical charge. He is beginning to think of an attempt to control the gigantic forces of nature -- weather modification! Tesla is beginning to wonder if the earth has an electrical charge. Does the sun have an electrical charge? Suppose, he thinks, if these too bodies are charged, I can perhaps modulate the electrostatic force of the ball of the earth with low frequencies, and this may in turn draw energy from the sun by sympathetic vibration. He is already thinking of drawing more energy from the solar system, from the cosmos. That he is so thinking we know from his later (1900) article "Talking with the Planets", but we are not yet ready to understand the far reaches of his mind.

While Tesla is exploring the ionosphere laboratory equivalent with high frequency currents, he is exploring the earth as a conductor for very low frequencies. He hints at many experiments using the earth se a conductor. He is beginning to think that the earth has a resonance to electrical waves somewhere around 10 Hz, but he has not yet explored it. But now he wonders what the effect of such extra-low frequencies (ELF) will be upon living things -- if he is to use the earth as a conductor. He begins to explore the effects on himself. He repeatedly demonstrates that he can pass over a millions volts of high frequency current through his body with safely, even though his body will be engulfed in a complete sheet of electrical fire. [7] Since he knows the limits of safety here, he explores the low frequencies

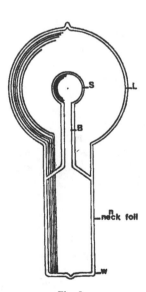

Fig. 5

that may be used in earth conduction of electricity. He does many experiments in the ELF range (8-20 Hz), the VLF range (20-40 Hz), and in the low frequency range (40-100 Hz). He discovers that ELF (8-20) can be painful, and could be dangerous. He discovers that a visual flicker-fusion of frequencies occurs at 16 cycles; and that very low frequencies (20-40) are less harmful than ELF. He finds that the low frequencies (40-100) are safe and insists that the standard for alternating current power transmission be set at 60 Hz -- which has prevailed in the U.S. to this day. But he waits for the day when he can build a generator big enough, and then to use it as a test probe for the whole earth. He makes one more notable invention as he seeks tools to explore the nature of electricity and light, the earth and its ionosphere, and the electrical relations between planers. He develops a model of the earth when it is electrically charged. He first revealed it in his London Lecture of February 1892, at age 35. He called it a rotating brush bulb. This is a method of illumination, which, at first appearance, looks like a light bulb.

In Figure 5 we see that a small bulb, S, the size of a flashbulb is evacuated and sealed in the center of a larger bulb, i.e. (light bulb size), also evacuated. The neck of the larger bulb, N, is coated with tinfoil and connected by one wire to an AC power source. The device is so fundamental to Tesla's theories that we must use his description in its entirety. Tesla states:

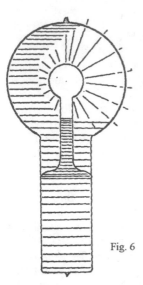

Fig. 6

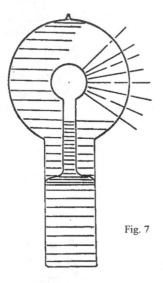

Fig. 7

"The construction shown in Fig. 5 was chosen in order to remove from the brush any conducting body which might possibly affect it. The bulb consists in this ease of a lamp globe L, which has a neck N, provided with a tube B, and small sphere S, sealed to it, so that two entirely different independent compartments are formed, as indicated in the drawing. When the bulb is in use, the neck N, is provided with a tinfoil coating, which is connected to the generator and acts inductively upon the moderately rarefied end highly conducting gas enclosed in the rack. From there the current passes through the tube B, into the small sphere S, to act by induction upon the gas contained in the globe L.

It is of advantage to make the tube T very thick, the hole through it very small, and to blow the sphere S very thin. It is of the greatest importance the sphere S be placed in the centre of the globe L. Figures 6, 7, and 8 indicate different forms, or stages, of the brush. Figure 5 shows the brush as it first appears in a bulb provided with a conducting terminal: but in such a bulb it very soon disappears often after a few minutes. I will confine myself to the description of the phenomenon as seen in a bulb without conducting electrode. It is observed under the following conditions:

When the globe L (Figs. 5 and 6) is exhausted to a very high degree, generally the bulb is not excited upon connecting the wire W (Fig 5) or the tinfoil coating of the bulb (Fig 5) to the terminal of the induction coil. To excite it, it is usually sufficient to grasp the globe L with the hand. Intense phosphorescence then spreads at first over the globe, but soon gives place to a white, misty light. Shortly afterward one may notice that the luminosity is unevenly distributed in the globe, and after passing the current for some Lime, the bulb appears as in Fig. 7. From this stage the phenomenon will gradually pass to chat indicated in Fig. B, after some minutes, hours, days or weeks, according as the bulb is worked. Warming the bulb or increasing the potential hastens the transit.

When the brush assumes the form indicated in Fig. 8, it may be brought to a state of extreme sensitivity to electrostatic and magnetic influence. The bulb hanging straight down from a wire, and all objects being remote from it, the approach of the observer at a few paces from the bulb will cause the brush to fly to the opposite side, and if he walks around the bulb, it will always keep on the opposite side. It may begin to spin around the terminal long before it reaches that sensitive stage. When it begins to turn around principally, but also before, it is affected by a magnet, and at a certain stage it is susceptible to magnetic influence to an astonishing degree. A small permanent magnet, with its polcs at a distance of no more than two centimeters, will affect it visibly at a distance of two meters, slowing down or accelerating the rotation according to how it is held relatively to the brush. I think I have observed that at the stage when if is most sensitive to magnetic, it is not most sensitive to

110

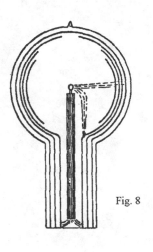

Fig. 8

electrostatic influence. My explanation is that the electrostatic attraction between the brush and the glass of the bulb, which retards the rotation, grows much quicker than the magnetic influence when the intensity of the stream is increased.

When the bulb hangs with the globe L down, the rotation is always clockwise. In the southern hemisphere, it would occur in the opposite direction and on the equator, the brush should not turn at all. The rotation may be reversed by a magnet kept at some distance. The brush rotates best, seemingly, when it is at right angles to the lines of force of the earth. It very likely rotates, when at this maximum speed, in synchronism with the alternations, say 10,000 times a second. The rotation can be slowed down or accelerated by the approach or receding of the observer, or any conducting body, but it cannot be reversed by putting the bulb in any position. If the potential or frequency is varied, while in a highly sensitive state, its sensitivity rapidly diminishes. Changing either of these slightly will generally stop the rotation. The sensitivity is likewise affected by the variations of temperature. To attain great sensitivity it is necessary to have the small sphere S in the centre of the globe L, as otherwise the electrostatic action of the glass of the globe will tend to stop the rotation. The sphere S should be small and of uniform thickness; any dissymmetry of course has the effect to diminish the sensitivity.

The fact that the brush rotates in a definite direction in a permanent magnetic field seems to show that in alternating currents of very high frequency the positive and negative impulses are not equal, but that one always preponderates over the other.

Of course, this rotation in one direction may be due to the action of two elements of the same current upon each other, or to the action of the field produced by one of the elements upon the other, as in a series motor, without necessarily one impulse being stronger than the other. The fact that the brush turns, as far as I could observe, in any position, would speak for this view. In such case it would turn at any point of the earth's surface. But, on the of her hand, it is then hard to explain why a permanent magnet should reverse the rotation, and one must assume the preponderance of impulses of one kind. As to the causes of the formation of the brush or stream, I think it is

due to the electrostatic action of the globe and the dissymmetry of the parts. If the small bulb S end the globe L, were perfect concentric spheres, and the glass throughout of the same thickness and quality, I think the brush would not form, as the tendency to pass would be equal on all sides. That the formation of the stream is due to an irregularity is apparent from the fact that it has the tendency to remain in one position, and rotation occurs most generally only when it is brought out of this position by electrostatic or magnetic influence. When in an extremely sensitive state it rests in one position, most curious experiments may be performed with it. For instance, the experimenter may, by selecting proper position, approach the hand at a certain considerable distance to the bulb, and he may cause the brush to pass off by merely stiffening the muscles of the arm. When it begins to rotate slowly, and the hands are held at proper distance, it is impossible to make even the slightest motion without producing a visible effect upon the brush. A metal plate connected to the other terminal of the coil affects it at a great distance, slowing down the rotation often to one turn second.

I am firmly convinced that such a brush will, when we learn how to produce it properly, prove a valuable aid in the investigation of the nature of the forces acting in an electrostatic or magnetic field. If there is any motion, which is measurable going on in the space, such a brush ought to reveal it. It is, so to speak, a beam of light, frictionless, devoid of inertia.

I think that it may find practical applications in telegraphy with such a brush it would be possible to send dispatches across the Atlantic, for instance, with any speed since its sensitivity may be so great that the slightest changes will affect it. If it were possible to make the stream more intense and very narrow, its deflections could be easily photographed."

I am convinced that this remarkable invention, which never found any practical use, is Tesla's model of how the earth would behave if properly charged, or how the sun would behave if ·influenced by the charge on earth. Tesla never tells us what he may have found when he used two or more rotating brush bulbs as a model of the sun and planetary interactions. In its simplest interpretation, the central small bulb S represents the magnetosphere sheath that surrounds the earth, and the globe L represents the conducting ionosphere, and the intervening space is the atmosphere. But some seven more years were to pass before Tesla could test this model on a planetary scale.

In this period, 1892, Tesla worked feverishly to comprehend the vast scheme of nature that he was trying to unravel and, hopefully, to control. Besides overwork he was also using his own body mercilessly as a guinea pig in his experiments. One of the effects (which we now recognize as the result of repeated electro shock therapy) is a loss of memory for recent events. Tesla writes of such an experience:

"I will tell of an extraordinary experience, which may be of interest to students of psychology. I had produced a striking phenomenon with my ground transmitter and was endeavoring to ascertain its true significance in relation to the currents propagated through the earth. It seemed a hopeless undertaking and for more than a year I worked unremittingly but in vain. This profound study so entirely absorbed me that I became forgetful of everything else, even of my undermined health. At last, as I was on the point of breaking down, Nature applied the preservative, inducing lethal sleep.

Regaining my senses, I realized with consternation that I was unable to visualize scenes from my life except those of infancy, the very first ones that had entered my consciousness. Curiously enough, these appeared before my vision with startling distinctness and afforded me welcome relief. Night after night when retiring, I would chink of them and more and more of my previous existence was revealed. The image of my mother was always the principal figure in the spectacle that slowly unfolded, and a consuming desire to see her again gradually took possession of me. This feeling grew so strong that I resolved to drop all work and satisfy my longing. But I found it too hard to break away from the laboratory, and several months elapsed during which I succeeded in reviving all the impressions of my past life up to the spring of 1892.

In the next picture that came out of the mist of oblivion, I saw myself at the Hotel de la Pair in Paris just coming to from one of my peculiar sleeping spells caused by prolonged exertion of the brain. Imagine the pain and distress I felt when it flashed upon my mind that a dispatch was handed to me at that very moment hearing the sad news that my mother was dying.

It was especially remarkable that all during this period of partially obliterated memory I was fully alive in everything touching on the subject of my research. I could recall the smallest details and the least insignificant observations in my experiments and   recite pages of texts and complex mathematical formulae." [9]

Tesla foresaw this event in late 1891 while working in New York City in February 1892 he gave his famous London Lecture, and from there went to Paris to give a lecture to the French Society of Electrical engineers. As he returned from the second lecture to his hotel, he was handed a telegram informing him of his mother's coming death. He caught the next train home to Gospic and arrived their lust in time to see her off this plane of existence. This was the fifth great shock in Tesla's life. If was all the more tragic in that it came at the very pinnacle of fame, fortune, and self-satisfaction with his work. At age 35 he seemed to have everything, but when he lost his mother, the only person he loved, he collapsed. Overwork, anxiety, self-experimentation, travel, his mother's death – all combined into another one of his grave illnesses, which kept him in bed for weeks.   During this period of enforced recuperation Tesla's review of his life program continued. He did not like; what he saw on his mind's screen. He had been too indulgent with himself. He had allowed himself to be idolized and lionized. He was creeping into commercialism beyond his immediate and real needs. He had not penetrated nature's secret - he had only, like Edison, proliferated gadgets and toys. He resolved to let the world go by, to be true to his calling, to lead a more spartan and stoic life - and to do it alone.

While convalescing, Tesla returned to the mountains where he had spent his nineteenth year in solitude. He tells how his inspiration came to put may of his ideas together: [10]

"I sought shelter from an approaching storm. The sky became overhung with heavy clouds, but somehow the rain was delayed until, all of a sudden, there was a lightning flash, and a few minutes after, a deluge. This observation set me thinking. It was manifest that the two phenomena were closely related as cause and effect, and a little reflection led me to the conclusion that the electrical energy involved in the precipitation of the water was inconsiderable, the function of lightning being much like that of a sensitive trigger. Here was a stupendous possibility of achievement, If we could produce electrical effects of the required quality, this whole planet and the conditions of existence on it could be transformed. The sun raises the water on the oceans, and winds drive it to distant regions where it remains in a state of most delicate balance. If it were in our power to upset it when and wherever desired, this mighty life-sustaining stream could be, at will, controlled. We could irrigate arid deserts, create lakes and rivers, and provide motive power in unlimited amounts.

This would be the most efficient way of harnessing the sun to the uses of man. The consummation depended on our ability to develop electric forces of the order of those in Nature. It seemed a hopeless undertaking; but I made up my mind to try it, and immediately on my return to the United States in the summer of 1892, work was begun."

By his 36th birthday, he had moved out of the opulent Waldorf-Astoria and into the smaller Hotel Gerlach on 27th Street. With his new resolve, he plunged again into a merciless schedule as the new superman who would control the secrets of Nature.

SECTION THREE

Titanic Forces Released On Earth By Tesla
1892 - 1899
Age 36 to 43 Years

While Tesla made a serious attempt to lead a more pure life of monasticism, celibacy, and science, the world would not leave him alone. On the one hand, he had determined during his recent visit to his homeland in the Balkans that he would try to invent the "trigger" that could control the weather forces of Nature. Toward this goal he made invention after invention which was to culminate (1899) in the super-giant Tesla Coil, which he named "The Magnifying Transmitter"· His philosophy at this time (1893) is best summarized by a statement he made in 1919:

"My belief is firm in a law of compensation. The true rewards are ever in proportion to the labor and sacrifices made. This is one of the reasons why I feel certain that of all my inventions, the Magnifying Transmitter will prove most valuable and important to future generations. I am prompted to this prediction not so much by thoughts of the commercial and industrial revolution which it will surely bring about, but of the humanitarian consequences of the many achievements it makes possible. Considerations of mere utility weigh little in the balance against the higher benefits of civilization. We are confronted with portentous problems, which cannot be solved just by providence for our material existence, however abundantly. On the contrary, progress in this direction is fraught with hazards and perils not less menacing than those born from want and suffering. If we were to release the energy of atoms or discover some other way of developing cheap and unlimited power on any one point of the globe, this accomplishment instead of being a blessing, bring disaster to mankind in giving rise to dissension and anarchy which would ultimately result in the enthronement of the hated regime of force.

The greatest good will come from technical improvements tending to unification and harmony, and my wireless transmitter is preeminently such. By its means, the human voice and likeness will be reproduced everywhere and factories driven thousands of miles from waterfalls furnishing the power; aerial machines will be propelled around the earth without stop, and the sun's energy controlled to create lakes and rivers for motive purposes end transformation of arid deserts into fertile land. Its introduction for telegraphic, telephonic and similar uses will automatically cut out the statics and all other interferences which at present impose narrow limits to the application of the wireless."

114

These were his visions and goals for humanity. He had only to implement them with his inventions, which of course required materials, men and money. The U.S. was undergoing another depression and money was hard to get. The Westinghouse Company was having financial difficulties, and had to be reorganized. Its new management insisted that the company could not survive if it continued to pay Tesla the royalty rate of one dollar per horsepower, and they put pressure on George Westinghouse to break the contract with Tesla. O'Neill reports this meeting between the two friends, Westinghouse and Tesla, in detail [2]. The conclusion was that Tesla tore up the contract just to help the man who had once helped him. In this gesture, Tesla gave up millions of dollars in future income and thereby bound himself to a life where he was forever handicapped by limited funds for his work. But Tesla's gesture was thoroughly in the spirit of his new resolves and his belief in a law of compensation. Other pressures came from Thomas A. Edison and his companies. As the success of the Tesla alternating current increased, the fortunes of the Edison investments decreased. Edison, it is believed, was instrumental in having New York State introduce electrocution as the form of capital punishment at Sing Sing Prison in Ossining some 30 miles up the Hudson River from New York City [3]. The means used for electrocution was Tesla's alternating current! Of course, this was the worst possible kind of publicity for Tesla's invention.

In order to counteract this bad publicity, Tesla had to do something that would prove the human safety of the AC system. Fortunately, the Columbia World Exposition was being planned to open in 1893 in Chicago. Westinghouse got the contract to light it with AC, the first world's fair to be so lighted. Then a large exhibition was planned by Westinghouse which would feature the new Tesla inventions. Tesla himself decided to use this event to publicize the safety of the alternating current. He stood on the high tension end of one of his coils that passed over a million volts through his body so that he was enveloped in a sheet of electrical flame. Of course, he had done this experiment privately in his laboratory many times for his friends, but this was the first time he had done it for publicity. The Edison campaign backfired, and Tesla's genius for the theatrical and for safety prevailed in the public mind. But more than this effect, the management of the Edison General Electric Company saw the handwriting on the wall, and quietly licensed the Tesla patents from Westinghouse. But the battle of the electrical giants had one more round to go.

In 1890, the Morgan financial group had started a company to try to develop the electric power potential of Niagara Falls. If was called the Cataract Construction Company, and its president was Edward Dean Adams. An international competition was held for the best design for the electrical power plant. The international commission was chaired by Lord Kelvin. The commission found that none of the plans offered were feasible. So the Cataract Company asked for plans and bids from companies in order to get the work under way. It was an incredible personal triumph for Tesla when his concept and plan were adopted by the commission. In October, 1893, two companies were awarded the contracts to electrify Niagara Falls: Westinghouse won the contract to build the AC power plant at the Falls; and General Electric, using the licensed Tesla patents, was awarded the contract to build the transmission lines and distribution systems to Buffalo, New York, 22 miles away. Tesla had set huge ideas and forces in motion years ago, and now the tide of industrial civilization was lifting them higher and higher to the thundering crest of Niagara Falls' worldwide reputation.

Meanwhile, back at the laboratory Tesla pushed on toward his goal. He had many problems to solve. Foremost was to build an electrical generator that could duplicate the trigger effect of lightning bolts in releasing rain from clouds. We can reconstruct how Tesla's mind was working in these days from scattered comments he made over the following forty years. He formulated a concept of the electrical system of the earth globe floating and

spinning in space on his mental computer screen modeled after the
rotating brush bulb. He had determined that one charged body influenced another charged body by electrostatic force oscillations between them. He set up the hypothesis that if this were true of the planets, then an electrostatic force "piston" effect on one planet should be transmissible to another planet. The missing knowledge in this theory was: "Does the planet earth have an electrical charge?"

As simple as this question is, no one had ever been able to answer it scientifically. If the earth had a charge, then Tesla knew that he could "pump" this charge with electrostatic forces and begin to create the effects he envisioned: namely, weather control, illumination of the skies at night, electric power transmission through the earth, and so forth.

In order to determine whether the earth had a charge, he must build a machine that would perturb the earth and its possible mass of electricity to find out what it's natural period of oscillation would be. In other words, he would oscillate the earth very much like one oscillated a clock pendulum, and after it started, determine its natural period of swing. He knew by mathematical calculations that this should be around eight cycles per second. Not knowing whether he would also have to find a mechanical resonance of the earth -- by striking it like a bell -- or if he would have to find an electrical resonance by oscillating it like a coil -- he set about developing both types of machines. One machine was a mechanical vibrator which would tap the earth - and then he would listen for the sonic resonant tone of response. The other machine was a huge Tesla electrical oscillator coil that would tap the earth electrically, and he would then listen for the resonant electrical tone of response. He proceeded to work on both of these machines.

With respect to the high voltage Tesla coil, he made improvement upon improvement on it in his laboratory at 33 South Fifth Avenue until he was able to produce 4,000,000 volts potential. This was the limit he could go within the confines of a small building within New York City. To go to higher voltages he would have to go out into open country, which he did in 1899. In the meantime, he turned from getting higher voltages, to get finer and finer tuning of his coils. The reason he had to do this was that in order to measure the resonant frequency of the earth, he would have to tune his coil to within a fraction (1/10) of a cycle. He tested the sharpness of "tuning principles" with his coils by building pairs of coils -one being a transmitter and the other a receiver of electrical energy. He would then send out one of his workmen in the vicinity of his laboratory with a receiver coil, and Tesla would send electrical energy to it from a transmitter coil inside his laboratory. In these "tuning" experiments, Tesla was laying the foundation for all future "wireless" or "radio" technology. He planned to make a test of all of his new tuning refinements in the spring of 1895 by sending a boat up the Hudson River north of New York City. With his laboratory as a fixed platform for radio transmission, and the boat as a mobile platform for radio reception: he would make his final tests for circuit tuning and earth electrical resonance.

Unfortunately, his plans were frustrated and delayed. On the night of March 13, 1895, a fire broke our in the basement of 33 South Fifth Avenue and swept through the entire structure, including Tesla's laboratory. All of his hundreds of invention models, plans, notes, laboratory data, tools, photographs- all, were destroyed. Not only was all of Tesla's intellectual capital destroyed on that night, but he had no funds with which to start anew. It was a black, black day for Tesla, blacker than the smoldering ruins of his laboratory.

Even as the disaster of his laboratory loss was still ashes in his heart, the power from Niagara Falls began to flow in August, 1895 -- Tesla's greatest triumph to date. The builders and backers of this biggest of all electrical power planes on the planet were highly pleased with the success of the Tesla polyphase system. One of them, Mr. Adams, president of the Cataract Construction Company, gave Tesla his full sympathy for the laboratory loss, and offered to finance all of his future work. Here Tesla stood at the crossroads of his entire future. If he accepted Adams's offer, he would have the full backing of the powerful J.P. Morgan financial empire. Adams proposed to finance Tesla immediately with $100,000 for a small percentage of shares in Tesla's laboratory. More financing would follow, and in the hands of good financiers, Tesla's money problems would be over, and he could concentrate purely on scientific research. Tesla was in no position to bargain, but he did not want partners, to be controlled, or to be owned. No one knows why he refused the generous overall financial plan, but he accepted only $40,000 to get him going again [4]. For the second time in his life, he waived the opportunity to be rich. His monastic ideal was firmly in command.

It was not long before he had a new laboratory at Houston Street near Mulberry Street. He plunged into his work with renewed vigor to make up for a lost year of research. His work was prodigious, and to the uninitiated reader he seems to have been going in all directions at once as he turned out new steam engines, perfected radio transmitters and receivers, founded the art of electro-therapy, discovered X-rays, and laid the foundations far the entire art of teleautomatics, or remote-controlled robots. In addition, he discovered different kinds of mechanical vibrators that were useful both for personal massage, and could cause "controlled" earthquakes. But now that we know what his goal was, we can see that each of these pioneering areas of invention were simply new tools toward his end goal.

He developed mechanical vibrators because he wanted to know what the resonance vibration (mechanical) was of the earth. Since he had calculated that this could be either in the 8Hz region, or in the region of 0.00015 Hz, he built devices that could tap the earth at these frequencies. But he was prudent and did a lot of research to make sure that these mechanical vibrators were safe -- for himself, for others, and for the planet. He found that the 0.00015 Hz waves were too long to affect a human. However, in the 8 Hz range he found extraordinary effects. He built a vibrating platform driven by a magnetic drive which he could vary over the vibratory range of 6 Hz to 15 Hz. He would stand on this platform and record the effects on himself. He found that in the low range, 6 Hz, one would develop a feeling of pleasant dizziness, some mild nausea, and a profound pelvic relaxation that could lead to sudden massive diarrhea. Today we would call this a cholinergic effect on the parasympathetic nervous system. One day his friend Samuel Clemens, better known as Mark Twain, was experimenting with the sensations induced by this vibrator. Clemens was feeling so good that he persisted in staying on the machine after Tesla has warned him that he had had enough. Clemens giggled and said he was having the best sensation of his life, and remember, he was the one who had said, "A woman is just a woman, but a cigar is a good smoke." Suddenly, he turned green, leaped off the vibrator, and headed directly into the nearby water-closet: he had just been triggered into massive colic and diarrhea. Thus Tesla explored the effects of extra low frequency (ELF) mechanical waves, electric waves, and magnetic waves on the human body to check for possible hazardous effects, as well as beneficial effects.

Tesla, in exploring the effects of alternating electric currents running through his body, founded the art and science of scientific electrotherapy [5]. Thus he was able to learn which frequencies, voltages, and currents of alternating current were beneficial to man. In working with very high voltages in vacuum tubes, Tesla observed as early as 1892 that there were "invisible radiations" emanating from them. He was pursuing this subject vigorously at the time his laboratory burned on March 13, 1895 and had laid out the whole science of X-rays including shadowgraph pictures on photographic plates. When Prof. Wilhelm Konrad Roentgen announced his discovery of X-rays in December of 1895, Tesla was immediately able to reproduce his results. However, Tesla did not publish his findings until March 11,1896.[6] It is interesting to note that although Tesla worked for years near the X-rays produced by his high voltage vacuum tubes, he was sufficiently aware of their danger so that he did not ever suffer any X-ray burns as did so many of his contemporaries. In the course of the same studies, he discovered cosmic ray particles as early as 1893.[7] These were rediscovered 30 years later by Robert A. Milliken. Thus, in his wide ranging hazard studies, Tesla clearly learned the biological effects of extra low frequencies, mechanical and electrical, alternating currents from the top of the hearing range (20,000 Hz) and upwards, light radiation, energetic radiation (X-rays), and energetic particles—electrons, protons, etc. Today we can only laud his caution and concern far living things as he carefully explored the bio-effects of the new energies he was generating.

In addition to the bio-effects, Tesla was also concerned about the effects of these new energies on the dielectrics in his condensers, the insulation in his coils, and the effects of the ozone and nitric oxides thus produced on men and materials. All of these effects he pursued in minute detail, ever seeking to find new materials that would withstand the high voltages he planned and the high frequencies and energetic particles he produced. One of his concerns was for the ability of his structures to withstand the vibratory stresses of the extra low frequency he planned to use. So he methodically vibrated the structures he built with his mechanical generators. One day in his new Houston Street laboratory, he attached a small (hand-sized) vibrator to a steel post of his building. Unknown to Tesla, within a few minutes after the vibrator started, in the surrounding neighborhood windows began to shatter, structural elements of buildings suddenly split, the ground trembled -- all the effects of a minor earthquake! People rushed to the local Mulberry street Police Station to complain. The police, knowing of Tesla's presence, rushed to his laboratory. There, as they entered, they saw Tesla swinging a sledgehammer at a small piece of iron attached to a pillar. As Tesla dislodged the Iron piece, the trembling and rumbling of the building immediately ceased. Tesla had suddenly realized that his vibrator had hit the resonant frequency of his building -- and that it would soon collapse [8]. From such studies he learned to build his coils, towers, and condensers to withstand the dangerous resonances inherent in such extra low frequencies.

We earlier mentioned Tesla's plans to test his finely tuned radio transmitters and receivers just before his lab burned down. In the spring of 1897 Tesla carried out the final tests that gave birth to modern radio. He mounted a radio transmitter and receiver in his laboratory at 46 East Houston Street, and a like pair aboard a ship that slowly moved up the Hudson River in the direction of Indian Point opposite Bear Mountain just south of the famed West Point Military Academy.

The ship and the laboratory continually broadcast messages back and forth over various distances. The tests were a complete success. He announced the results on July 9, 1897 in the ELECTRICAL REVIEW -- on his 41st birthday. So complete was Tesla's command of the art and science of Radio transmission, that after years of litigation, the U.S. Supreme Court ruled in October of 1943 that Tesla had established clear priority over all other claimants in the world for the invention of radio.

But the perfection of radio was only a small part of the tools that Tesla needed for his grand experiment. His highly tuned radios would be used to make a radio map of the earth, and to precisely clock distances and velocity over the globe. In addition, he knew that he would have large complex equipment to operate and control during his experiments. Since he worked alone -- usually with only one assistant -- he would need some means of doing fine tuning, running controls, and reading data points at a distance from where he stood in the experiments. To accomplish this he created the art of teleautomatics. This is the science of remote control of devices and apparatus by means of radio commands. Today this art lies behind the control of gun firing systems, torpedoes, aircraft, rockets, submarines, drone airplanes, trains, subways, etc., etc. Yet Tesla mastered this entire art in a few years. In September 1898 Tesla hired Madison Square Garden in New York City through the generosity of his friend John Hays Hammond. Here he displayed a steel boat several feet long floating in a large tank. Anyone present in the hall could call out a command to have the boat go forward or backward, turn right or left, go in a circle, etc. The boat promptly maneuvered as directed. The impression created on the public was truly wondrous. It seemed to them that the boat had a mind of its own—an intelligent robot. What in fact happened was that as Tesla stood at a simple control board, he heard the command given, and through his controls relayed the proper command via radio signals to the boat. The excitement created by this invention can best be shared by reading the editorial in the staid *Electrical Review of New York* from November 9th 1898,

"A NEW MARVEL:

As we go to press an invention of Tesla's is announced which must produce a profound impression all over the world. Tesla has already identified himself with a number of most remarkable scientific advances, and great things may still be confidently expected as the fruit of his earnest and persistent labors, but it is difficult for us to see how he could ever produce a more beautiful result than he now makes known through a United states patent issued this week.

To direct and control to the minutest detail, by a subtle agent, the operations of a mechanism however complicated and ponderous, to change its speed and direction at will, to make it perform an unlimited number of movements, without any tangible connection and from great distance, is indeed a closing days of this century of wonders. When Bell transmitted the human voice over a wire so that the faintest of its modulations could be recognized, it was marvelous triumph; when Edison showed his fascinating invention of the phonograph, this, too, was justly looked upon as a wonder; when Tesla first showed the phenomena of the rotating magnetic field, or when he presented the magical effect of a tube of glass brilliantly lighted in his hand, the world stood astonished; when Roentgen announced the epoch-making discovery of the rays bearing his name, the scientific world was thrilled as never before; but we believe that the beauty and importance of the invention Tesla has just announced, in its ultimate developments, will be such as to place it among the most potent factors in the advance and civilization of mankind. The fact that the invention has been thoroughly and practically developed makes its immediate application sure.

In this issue we begin the publication of the clear and exhaustive wording of Tesla's own description of this invention, which will be concluded next week."

Now, as the end of 1898 approached Tesla had invented dozens of new tools, and carried out thousands of measurements in hundreds of experiments in preparation of his grand experiment. Money was always a problem. But his enthusiasm had been conveyed to many of his rich friends, and their respect for his genius made it easy for them to donate the money he needed.

Tesla could only carry out his next experiments in an area where large amounts of alternating current were available. The number one candidate of course, was Niagara Falls, N.Y., but here the cold weather, and long winter snowfall limited the working time to only a few summer months. However, there was a large electric installation in Colorado used to power Telluride Mining Company near Colorado Springs. An old friend, Leonard E. Curtis, was in charge, and when he heard of Tesla's requirements, he invited him to work there. The climate was ideal for year-round research, and the electric power available was adequate, so Tesla decided to go to Colorado Springs. Besides, he would get away from the press who were always hot on his heels looking for a story. He arrived in Colorado Springs in early May 1899. Before him were the tasks of getting land on which to build a laboratory, build a magnifying transmitter from scratch, and test and calibrate all his measuring instruments. This was a Herculean order which even a modern engineering firm could not undertake with a cost plus contract and expect to deliver in less than a few years. Tesla arrived with a few of his laboratory workers, and his engineering associate, Fritz Lowenstein. From Tesla's Colorado Springs notebook, drawings and 436 pages of notes, calculations, photographs, we read that he is conducting definitive experiments by the 11th of June 1899—and by July 3, 1899 has made his first great breakthrough. The genius of Tesla and his workers in setting up one of the great experiments of all time in such a short time is one of the eighth wonders of the world— even to this day. And now let us recount what Tesla discovered.

Even though we have Tesla's notebook before us, we know that he never wrote down novel discoveries until he could file for a patent. He had a life-long habit of keeping all the vital elements of an invention filed secretly in his vast mental computer archive. Therefore, in this reconstruction of his Colorado experiments we include elements that are not recorded, but have become subsequent solid scientific knowledge.

We note that one of the first patents that Tesla filed with the U.S. Patent Office after his return from Colorado Springs is "Means for Increasing the intensity of Electrical Oscillations". (Patent No. 685,012 dated October 22, 1901, filed March 21, 1900.) This patent shows (see Fig. 9) that Tesla had discovered the superconductivity effect. This means that when electrical current conduction occurs at temperatures approaching absolute zero, the resistance to electron flow drops virtually to zero, and electron conductivity becomes maximal. Now this effect of superconductivity made higher resonances possible with the greatest art that Tesla could command. (Room temperature superconductors are now finally being discovered 100 years afterwards. – Ed. note) Let us explain.

No. 685,012.   Patented Oct. 22, 1901.

N. TESLA.

MEANS FOR INCREASING THE INTENSITY OF ELECTRICAL OSCILLATIONS.
(Application filed Mar. 21, 1900. Renewed July 8, 1901.)

(No Model.)

Fig. 9

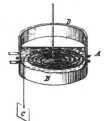

Previously we illustrated how electrical energy (electrons) can be likened to sloshing water between two glasses at right angles to each other one glass representing the condenser with electric force; and the other glass representing the coil with magnetic force. Now imagine that the force of the water transferred from glass to glass gets higher and higher. Obviously, if we used ordinary drinking glasses the water would

Witnesses:

Raphael Netter
Benjamin Miller

Nikola Tesla, Inventor

by Kerr, Page & Cooper Attys.

splash out due to the low walls of the glass, and we would have losses of water (i.e., electricity). Now to conserve the water with use of high pressures and force, we would need glasses with higher and higher walls. These could eventually become long tubes, perhaps hundreds of feet long. We now realize that with such long tubes we could not slosh the water by hand, back and forth, with the rubes at right angles. We would need some high speed apparatus to do the transfer operation. But having done this we would find that the capillary attraction between the water and the glass nails would slow down the transfer from one tube to the other, end this is comparable to a resistance increase in the electric force current, and a self-inductance increase (impedance increase) in the magnetic force. This would prevent resonance attainment—or the maximum swing of force outwards to alternatively maximize first the electrical potential, then the magnetic field potential. The higher the water swings in each tube the greater the "potential" output of the water analog we use to portray the Tesla Coil. By placing the electrical oscillations at the superconducting temperature, Tesla removed the <u>resistance</u> to electron flow. This is equivalent to removing all the capillary attraction drag exerted on the water flow by the tube well.

Tesla further discovered that the greatest resonance is achieved in such a superconductive state if the coils –the primary and the secondary are round in flat pancake spiral geometry. This is illustrated in Fig. 8 from Tesla's patent drawing. A is the primary coil. B is the flat wound secondary coil with the free end up and unlabelled (going to capacitor ball), and the lower end C, grounded in the earth. D is the rank which holds the artificial refrigerant such as liquid air, or liquid helium.

Now the primary A was connected through a condenser to the alternating current generator from the Telluride Mining Company. Each cycle of current surge in the primary induced a current in the secondary. The secondary is so wound, with it condensers tuned for maximum resonance, that it resonates at 20,700 Hz, and the voltage is likewise stepped up to attain potentials of millions of volts. Now Tesla had achieved far the first time on the planet earth by means of superconductivity two freely oscillating circuits each with maximum resonance, the primary and the secondary. We need no longer concern ourselves with the primary – it is a given constant in the system. We now focus only on the secondary coil.

Tesla measured the length of the wire that goes into the winding of the secondary so that it is exactly one quarter of a wave length long. Thus for 20,700 Hz the secondary wire is 3,838 meters in length (Tesla gives these figures in the Colorado notebook as 2.25 miles.) Now this is another secret discovered by Tesla in his radio tuning experiments. If the coil wire length is exactly one quarter of the wave length of its oscillating frequency, <u>all the energy</u> (electrons) will slosh to one end of the wire (remember our water tube model) leaving the other end of the wire with zero energy (no electrons). With this knowledge Tesla could direct all of the energy into the earth with each cycle, and have virtually no radiation loss in the air; or he could direct all of the electrical energy of the secondary coil to the free end in the air with each cycle, in which ease he would get big sparks and radiation into the air. In the first case he could transmit power through the earth. In the second case he could transmit radio energy and information through the air.

He further arranged to have a large ball at the upper (Zero) end of the quarter wave length long secondary coil which served as a condenser in the atmosphere. We have now described in essence the machine which Tesla called a magnifying transmitter, the TMT. It was huge. The primary coil formed a huge tub eighty-five feet across. The secondary coil in pancake form (we do not have its dimensions) was placed inside of the primary as in Fig. 8. The mast holding up the huge, shiny copper ball was several hundred feet high. Tesla was now Prometheus, prepared to steal electrical fire from the Colorado sky. How did he go about this unprecedented adventure?

First, he turned on the generator switch that energized the primary coil so that it was freely oscillating at 900 cycles. This induced a current in the secondary which was tuned to 10,700 Hz. Now the secondary oscillations per cycle must be clearly understood. Tesla had placed high in the air at the upper end of the secondary wire, a large ball. When the secondary (0°-90°) surged all of its electrons into the earth, they were directed toward the large iron core in the center of the earth – a huge magnet. When the electric potential hit this magnet it caused transient iron-cooling with a large release of energy. This iron-liberated electron energy was sloshed back to the ball in the atmosphere on the return swing (900-2700). This charged up the ball to a maximum and then the electron energy was released back into the coil (270°-0°-90°) and earth to complete the full cycle. Now the oscillation had gained more energy than the primary pumped into it because of two effects:

Let it be noted that two of Tesla's early patents (No. 396,121 Thermomagnetic Motor, Jan. 15, 1889; and No. 428,057 Pyromagnetic Electric Generator, May 13, 1890) converted heat directly into electricity. Today in modern industrial metal casting techniques, hot castings are quickly and safely cooled without cracking by applying large electric potentials. In his magnifying transmitter, Tesla used that principle to draw electrical energy from the hot iron core of the earth and feeding it into the oscillator to get higher and higher potentials, In addition, by placing the condenser ball in the atmosphere he drew on the free charge in the air during the part of the cycle (0°- 90°) when the maximum potential on the lower end of the secondary wire was in the earth, and its upper end was at zero potential.

Now he further increased the potential by now only having maximum resonance within the secondary coil-condenser circuit itself, but by getting this entire oscillator to get in perfect harmony (or resonance) with the absolute pitch of the earth itself.

This is how this was accomplished. Tesla had to establish whether the earth had electrical charge, or not. With his sophisticated instruments he probed the earth, and found that it was alive with charge, and that this charge had a natural heartbeat (of the earth) at 8 Hz [11]. Furthermore he found that the earth had an electrical charge in the atmosphere up to the ionosphere which increased about 100 volts per meter, thus placing the earth/ionosphere potential difference at hundreds of millions of volts. His findings clearly showed that the earth by itself was resonant at 6 Hz (6.67Hz precisely). He now saw the earth as the inner conductor plate of a spherical capacitor, with the ionosphere as the outer conductor plate [of a spherical capacitor (same as a condenser)], and that the atmosphere was sandwiched between, as the dielectric (or charge bearing gas). This condenser structure formed a resonating cavity with a peak at 8 Hz (7.83Hz). Thus Tesla desired to get the electrical beat of the magnifying transmitter secondary coil in resonance with these two resonant cavities of the earth. He did this by pulsing (or chopping) the 20,700 Hz wave into segments with his magnetic interrupter so that the pulses hit the earth 6 times per second. In this way, he was hitting the earth with weak pulses of energy so that the earth began to ring like a bell. He discovered two new things in this test. First, the electrical 6 Hz pulses tapped the earth with a sound which traveled to the opposite pole (the antipode) of the earth and bounced back in 108 minutes (This is about twice as fast than sound would travel in water if traveling the diameter of the earth and back. – Ed. note). By keeping up this tapping he discovered the sonic stationary wave of the earth. Secondly, he found that the electromagnetic wave also reached electrical resonance in the earth, and that it traveled faster than the speed of light through the earth – or 471,240 kilometers per second. The standard speed of light in a vacuum is 299,000 kilometers per second [12]. Tesla had found that at electrical resonance with the earth, electromagnetic waves had become Supraluminal and Superconducting! In the supraluminal sense, they moved faster than the speed of light! Modern science has never tested this basic discovery.[1]

Later on, Tesla made the statement that if he tapped the earth in this fashion sonically for just two weeks, it would split the earth in two like a sliced apple! [13] Remembering his experience in New York, he was careful not to trigger any earthquakes.

The electrical resonances at 8 Hz made it possible to electrostatically oscillate the earth/ionosphere capacitor and create many startling effects. By a method he never disclosed, Tesla was able to condense (or attract) mist and water from clear air. This was the first step toward his plan of weather control. By putting two identical alternating current frequencies into the air and carefully phase shifting them as in his rotating magnetic field (Fig. 3), he found that he could precisely steer a high potential, high energy beam at any velocity he desired around the planet. [14] By this means he planned to influence jet streams and trade winds in order to move clouds around the planet and deliver water where needed. With his ability to produce potentials as high as 20,000,000 volts (measured), he could create lightning bolts over a hundred feet long. He believed that with improved technology of the magnifying transmitter it was possible to produce 100,000,000 volts potentials.[15]

By the end of 1899 Tesla had achieved all of his scientific goals. He now knew that he could transmit human information (voice and pictures) by wireless anywhere on the globe— simply and efficiently. He knew that he could transport electrical power anywhere on the earth by using earth as a conductor. He lit a bank of 200 light bulbs (10 kilowatts) by this wireless means at a distance of 26 miles just to spot check his theory.[16]

He knew that he could control weather by either electrically collecting clouds for gathering or dispersing their moisture), or by moving clouds along predetermined paths and releasing their contents with an electrical trigger to produce rain.

He knew that with his sonic lapping of the earth—he called this the science of Telegeodynamics —he could release and adjust the pent-up tectonic forces of the earth and thereby prevent serious earthquakes.

He also knew that these extra low frequencies (ELF) both mechanical and electromagnetic, had powerful effects on living things and must be used with caution until they were fully understood.[17]

But most spectacular of all, he felt that he had achieved a test of his theory that all planets and the sun were electrically charged, and that by perturbing this charge with his TMT electrostatic forces, the planets could communicate with each other. To this end he wrote a most fascinating article in Colliers for 1901 entitled, "Talking With The Planets"[18]. He further believed that by the same means man could tap on the breast of Mother Sun and release her energy toward the earth as needed— magnetic as well as light. In his 43rd year, Tesla, in spite of every possible
obstacle, succeeded in penetrating deeper into Nature's mysteries than any man before, and had even grasped in his hand, the Promethean electrical fire from the sun—even if but for a moment.

He dismantled his Colorado equivalent and laboratory in January, 1900 and returned to New York City. The working knowledge of his secrets was not to be rediscovered until the Soviets silently announced their success on July 4, 1976.

When Tesla returned to New York in early 1900, he made up his mind to try to start a worldwide broadcasting service. But true to his usual solo pattern, he did not want to go

---

[1] The blue glow of nuclear reactor cooling ponds is caused by *Cherenkov* radiation which is a well-known physics phenomena where the particles exceed the speed of light in that medium. The formula for the speed of EM phase velocity is $v = 1/\sqrt{LC}$ which can be faster than light speed in a vacuum. See Dr. H. Milnes, "Faster than Light?" *Radio Elec.*, Jan. 1983. –Ed. note

public in any way by trying to finance his operation with many stockholders, or even partners. Instead he went to the most powerful financier in the world, J.P. Morgan, who had been impressed with the brilliant success of the Tesla polyphase system in powering the budding industrial world of the United States. Morgan agreed to finance him on a private basis so that Tesla could be his own boss. Tesla decided to build the improved commercial version of the Colorado Springs Magnifying Transmitter 65 miles east of New York City an long Island. This was the first step toward developing a "World Telegraphy System" whose aim was to bring all people on the planet earth into one shared information brain.

As the structure went up in the midst of a 175 acre trace of land, rumor was rife in the little farming community. No one had ever seen such a strange mushroom-shaped copper-capped 185 foot high tower on this planet. It was designed by one of the most famous architects of the day, Stanford White. The construction was closely guarded as some 50 workers toiled. Great curiosity centered on a mysterious stairway that went deep into the ground in he center of a huge concrete platform that served as the base for the tower. The *New York Times* for March 17, 1901 makes these curious observations:

> "Mr. Scherff, the private secretary of the inventor, told an inquirer that the stairway going into the ground led to a small drainage passage which was to keep the ground below the tower dry. But villagers who watched the construction day by day tell a different story.
> They declare char it leads to a well-like excavation as deep as the tower is high, with walls of masonwork and a circular stairway leading to the bottom. From there, they say, tunnels have been built in all directions until the entire ground below the little plain on which the tower is raised has been honeycombed with subterranean passages. They tell with awe how Mr. Tesla, on his weekly visits to Wardenclyffe, spends as much time in the underground passages as he does in the power plant on the surface."

So secretly did work proceed on this Magnifying Transmitter complex that in reading newspapers and journals about it twenty years later we find the mystery still persisted. The tower and structures were completed in 1902, just as Tesla ran out of money. Tesla was unable to finance the construction and installation of the electrical equipment. There were many simple ways that Tesla could have financed the completion of his grand project. Why he avoided success in the completion of the world Telegraphy System has never been explained satisfactorily, by Tesla or anyone else although many speculations were offered. Tesla, in all subsequent interviews, never wavered in his faith that the design was a practical possibility. In spite of this faith and a burning desire to finish his lifelong quest of taming the electrical forces of the earth, he never made a practical move to bring the effort to conclusion. And even more strangely, no one else on the planet ever tried to duplicate his design—at least, not until the mysterious Russian signals appeared all over the planet in early July, 1976.

SECTION FOUR

Tesla's Wisdom: The Legacy of his Planetary Thinking
1900 – 1943

Tesla spent the first half of his life penetrating the secret of planetary ecology. Having opened the Pandora's box of planet earth, he was overwhelmed by the responsibility this knowledge placed on him. Although he quickly shut the lid on the box, the awesome knowledge he now possessed haunted him for the remainder of his life. We find in his speeches, conversations with friends, and in his writings, a total preoccupation with the problems of war and peace. Since planet earth is suddenly faced with the knowledge of the positive and negative potential effects coming from the use of the Tesla Magnifying Transmitter, it would be well to get a grasp of the future possibilities we face by reviewing Tesla's pioneering thoughts.

By June of 1900, Tesla had framed the problem facing humanity in a masterful statement published in THE CENTURY his good friend Robert Underwood Johnson a magazine ably edited by Tesla contemplated life in its broadest sense. As he surveyed the earth, sun, the moon, the planets, and the far reaches of space, he could only state that man has no way of knowing from where he comes, why he is on earth, or where he is going. Tesla also saw so vividly that all observable processes in Nature -- living or non-living -- follow a rhythmic cycle (Figure 2) of birth (0°), growth to physical maturity (90°), decline from peak physical powers (180°, a resurging of mental and spiritual powers (270°), and a weakening of such powers with death (360°)

The cycles are similar for plants and animals, the seasons, for the earth, the sun, and the stars. Someday, the sun's fuel would be expended, and earth would become a block of ice and, eventually, barren like the moon. But in this cycle there is always rebirth -- and most importantly with the rise and cyclical fall -- ever the onward movement to an unknown destiny. Being a scientist, Tesla knew that he could not deal with unknown past origins or with the unknown future, but he could deal with the present; and for him the present with characterized by the "onward movement."

In this philosophical perspective, Tesla recognized that whatever life was, in essence it was the embodiment of some mysterious unknown principle in the flesh. Earthly life is then a process in which this unknown principle faces the challenge of mastering the flesh -- the earth component of life. And, of course, Tesla's own life was a living proof of this thesis as he tried to master the powers of the flesh, and bent his powers of soul and mind to communicate with the laws of Nature -- for him, the ultimate dialogue. Then, looking at earthly life from its place of flesh, of mass, he abstracted a few rules for the management of human life. As the mass of each individual and the collective mass of the entire biosphere hurtles on through space, it acquired extra force from the velocity imparted to the mass, or, in sparse language:

Human Energy = ½ Mass Times Velocity Squared
or
$$E = \frac{1}{2} M V^2$$

But he considered the velocity squared, V^2, to be made up of two parts, a velocity V, which we can compute; for example, the velocity of the spin of the earth, about 1,000 miles per hour; and the velocity of movement through space, the sum of our orbital velocity around the sun plus our velocity around the galaxy plus the velocity of the galaxy around something else, [2] and so on to some unknown end. The latter unknown velocity, V', he called a "hypothetical velocity" since it could not be quantified in his day. The equation he used, $E = \frac{1}{2} MV^2$ is the ordinary equation from mechanics for the kinetic energy of a moving mass. Einstein went beyond this simple formulation and said that any mass has potential energy, ultimately, in the amount where V^2 is replaced by the velocity of light, C, squared, or C^2 thus:

$$E = MC^2$$

where the ½ term is dropped.

Now analyzing the dynamics of this equation for energy, Tesla points out that when a mass is accelerated at a given velocity there are two main forces acting on it. One is the force which holds it back, the resistance to motion: the retardant force. The other is the force which moves it ahead: the advanced force. The balance of these two forces will determine how fast the mass will move, and this in turn determines its energy content. The energy content can be converted by various means to do work. Ultimately, Tesla said, humans are here on earth to work -- to convert their inherent mass-energy into work energy which will further personal evolution, and the collective evolution of earth life.

But he also pointed out that in addition to modulating the retardant forces and the advanced forces, one could increase the mass of humanity by increasing its numbers on the biosphere in order to increase its energy content. The human mass can show increased energy content, he said, by conserving itself hygienically by clean food, environment, and proper exercises; by "the promotion of marriage, by the conscientious attention to the child, and generally stated, by the observance of all the many precepts and laws of religion and hygiene [3]. Tesla was most insistent on a high moral guidance of living processes. He expanded on this theme in many of his writings. But he noted that the key obstacle to increasing the human mass was the limitation of the food supply. Here Tesla made another practical and brilliant contribution. He pointed out the well-known fact that the soil needed soluble nitrogen to have a high yield of food. It was Tesla who discovered that high frequency, high voltage electrical discharges were the cheapest method of getting the nitrogen from the air to combine with oxygen and hydrogen to produce nitrogen oxides (NOX) and ammonia (NH3) -- the starting point of all soil fertilizers. Today hundreds of millions of tons of such fertilizers are used on earth every year to increase agricultural production. However, after 75 years of such nitrogen fertilizer use, man has begun to learn that it too presents a hazard to existence. The nitrogen oxides evaporate from the soil and rises to the upper layers of the stratosphere where they have a strange chemical action. This layer of the stratosphere contains a thin layer of ozone, sufficient to act as a shield against deadly ultraviolet rays reaching the surface of the earth. If these deadly rays reached earth, all life would be extinguished eventually beginning with a dramatic increase in cancer in all living things. It is now known that the nitrogen oxides have a peculiar catalytic effect in destroying ozone -- each N02 molecule can destroy about 10,000 ozone molecules! There is strong evidence that the ozone layer is now slowly being depleted and thinned. This is due to the catalytic breakdown of ozone by many agents other than the nitrogen oxides. But the TMT, wisely used, can selectively repair damage to the ozone layer, and regulate the amount of ozone in the shield. Tesla was aware of this potential use of the TMT. The Soviets have created this beneficial by-product of their global warfare system, but have failed to tell the world about it. This will be explained later.

Tesla went on to analyze the retardant forces acting on the onward movement of man. He compared these to frictional effects which can be cured by the proper lubrication. The greatest of these retardant forces, says Tesla, is ignorance. And the greatest manifestation of ignorance, he points out, in "civilization" is organized warfare. The best lubricant for ignorance is education which becomes an "eraser" of ignorance. But the problems of war, Tesla realizes, are ultra-complex. First, he recognizes that an orderly society needs to have the governing influences of law, and a means to enforce laws, i.e., police. He analyzes brilliantly how war feeds on ignorance, its appeal to bravery and sacrifice, and the lure of medals and other rewards and honors that come from battle. In his incisive way he proposes a solution unthought of in 1900: eliminate men as combatants of war! In the process of education, he says, we should escalate war to the level of a sophisticated game in which man has no role except to produce teleautomatons, i.e., robots. Since he was the one who created this science he could also brilliantly advise as to how to use it. But these robots would have all the intelligence and skill of thinking man. They would exist in the form of "thinking" aircraft and rocket spacecraft -- maneuvering with respect to each other's intentions without interference by man. But they would only outwit, or destroy each other—not man or the planet earth. When this was achieved technically, man would be cheated out of getting a personal thrill out of war, and this barbarous practice would be replaced by true, creative peaceful processes. Today as some 4,000 satellites clutter space, and we read press reports of the U.S. and the Soviet Union having developed satellites that can carry on warlike acts between themselves, Tesla's vision approaches reality.

What Tesla believed in strongly, up to 1930, was that war could be abolished by pushing the tools of war to such sophistication that they could operate independently of man. To this end he developed from the TMT a machine that could put up an invisible curtain of defense around any country, which was not penetrable by aircraft or missiles. However, he later modified his thinking when he realized that all automation of war weaponry was in the direction of "overkill." That Tesla himself did not actively develop the many teleautomatons he designed torpedoes, submarines, aircraft, etc., is proven by a curious set of historical events.

One of the private benefactors of Tesla beginning in the 1890s was John Hays Hammond, the great mining engineer who became the U.S. Ambassador to the Court of St. James in England. Hammond lived in Gloucester, Massachusetts on the seacoast facing the reef of Norman's Woe. His great friend and neighbor was Colonel William House, who in turn was the personal political adviser of President Woodrow Wilson, the U.S. President from 1914-1918, and 1918-1922. John Hays Hammond had a son, John Hays Hammond, Jr., or, as my friend Jack Hammond. When Jack was a student at the Sheffield Scientific School of Yale University he met Tesla through his father, Jack told me that this meeting changed his life – he wanted to follow in Tesla's footsteps for a career. During many conversations between Col. House, John Hammond, Tesla, and Jack before World War I started, they discussed the inevitable coming of this war, and how unprepared America was. When Wilson was elected President in 1914, Col. House was in a key position to inform the President of Tesla's ideas with respect to peace and war. President Wilson wanted peace, and kept America out of the war as long as possible. But he was also responsible for making sure that America was prepared for war. Col. House persuaded President Wilson to start a secret program of putting Tesla's teleautomatics into practical use by the U.S. Navy. Tesla declined to participate in this program to build war machines. But he allowed young Jack, who was only 25 years old to go ahead with his ideas [4]. Jack entered into contractual arrangements with the U.S. Navy for several millions of dollars, and perfected and applied the Tesla teleautomatic patents to battleships, torpedoes, gun-fire control, submarines and aircraft. Jack received several hundred U.S. Patents as result of his work. When the U.S. entered World War I in 1917, it was armed with the best of Tesla's ideas for automation of warfare.

When Tesla saw World War II coming as early as 1935, he secretly offered his "Wall of Force" defense system to Great Britain, Canada, and the United States. His aim was not to get rich, because he was now thoroughly adjusted to his monastic life, but only to prevent war, and secondly to save the best values in Western civilization against the satanic doctrine of Adolf Hitler. But the military advisers of these great powers from 1935 to 1939 were blind to Tesla's advanced technology. As we now sadly know, it was not until 1976 that the Soviets were able to recapture the advanced thinking and technology of Tesla. And now, as in 1914, and 1938, the world teeters on the brink of another world conflagration, but this time armed with nuclear weapons, military computers, and mind-enslaving global magnetic warfare methods.

But in 1900 Tesla was still optimistically trying to chart the long course man had to work through to find peace. In addition to working out methods to increase human mass-energy as outlined, and his vast contributions to lubricating the retardant forces with teleautomatic science and technology, he contemplated methods of increasing the advancing forces in man's energy equation.

His analysis is superb. He points out that all of the energy on earth available to man coal, oil, wood, waterfalls, tides, wind, solar radiation, etc., all come from the sun. The best way to increase the energy contribution of the advancing forces on man is to capture more of the motive power of the sun, thereby amplifying man's natural energy endowment. This amplification would increase his work output, acting on himself, the environment, and in evolution. The onward movement of man would be accelerated. He points out in no uncertain terms how barbarous it is of man to burn up millions of year's inheritance of wood, coal and oil in a few generations. Man must find renewable, non-depletable sources of energy. He then produced a long list of inventions to bring about this result. He invented an advanced type of steam engine to more efficiently utilize coal; he invented and perfected the best high pressure steam turbines ever known; he devised apparatus to capture the vast geothermal energy of the earth. He invented solar energy devices that converted sunlight by day into electricity, and the sun's radio waves by night into electricity. He invented devices to convert heat directly into electricity – thermoelectric machines. He devised methods to convert magnetism, or gravity directly into motive power. And all the electrical output of these inventions was to be transported wherever needed on the planet by a grid of Tesla's Magnifying Transmitter Towers.

But he reserved his highest skill and greatest dedication to finding ways to extract energy from what he called the ambient medium – literally the thin air and/or vacuum around us. Today we would call this goal the extraction of energy from zero-point vacuum fluctuations. This process also contemplates penetrating the wall of light, the velocity-of-light barrier, to extract energy from the domain of tachyons, particles exceeding the speed of light, at supraluminal velocities. Theoretically, this energy transfer across the wall of light occurs through mini-black holes, and mini-white holes, a theory not yet proven, but under active discussion by scientists today.

Tesla outlined the five elements that must be mastered to extract energy from the ambient medium. But his approach did not contemplate extracting energy from mini-holes (something of the order of 10^{-39} centimeters diameter), but from the vacuum regions between the planets. He planned by modulating the electrostatic forces around the earth, and thereby perturbing the sun, to draw energy from the sun to earth. His five elements were outlined by him.

First, he needed a mechanical modulator that would tap the earth to bring out full mechanical resonance. This would produce a sonic wave bouncing from pole to anti-pole every 108 minutes. He solved this problem by means of a unique steam engine which he exhibited as early as 1893 at the Chicago World's Fair.

Second, he needed to develop an air compressor that would liquify the air in sufficient amounts to run his huge TMT secondary coil at superconducting temperatures. He worked for a long time at this problem, but it was solved by Dr. Carl Linde working independently, and Tesla was the first to give him recognition for priority.

Third, he had to perfect the TMT in order to utilize the upper stratosphere and the certain layers of the ionosphere as a conductor of electricity. In Figure 5 we see the bulb model of the earth/ionosphere condenser. The earth is represented by the central bulb, S; the ionosphere is represented by the outer glass globe, L. The intervening space between S and L

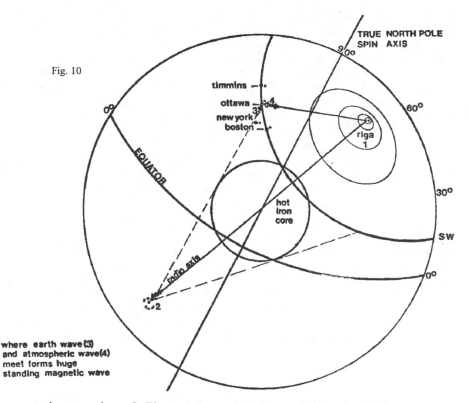

Fig. 10

where earth wave (3)
and atmospheric wave(4)
meet forms huge
standing magnetic wave

represents the atmosphere. In Figure 6, the earth, S, is energized by the TMT and light (radio waves) resonating between the earth, S, and the ionosphere, L, in different patterns as shown also in Figures 7 and 8. Figure 8 shows that a beam of light can be formed in this model. Figure 10 shows more clearly how such a beam is formed which we have been calling a "standing wave". In Figure 10, we have two velocities of electromagnetic waves interacting: velocity AC-1 emanates from Riga (1) as ionosphere electromagnetic waves moving at the velocity of light, c, and goes to Point (4). Velocity AC-2 goes from Riga (1) directly into the earth through the hot iron core where it picks up energy and a velocity greater than c, the tachyonic level, to reach the antipode at (2). At (2), much of the energy dissipates into space, and some of it is reflected by scattering around the iron core in the cone shape to (3) and thus the standing wave, SW, is formed by velocity-phase interference patterns. These phase relations work very much like the rotating magnetic field phasing of Figure 2 to produce a <u>magnetic</u> standing wave. This magnetic standing wave can be moved, steered or made stationary (at any rate desired) by adding another wave to the system, emanating from Gomel. When the wave from Riga, and the one from Gomel are properly phase-locked, the wave, SW, can slowly sweep around its great circle route, or can be made to stand still over cities like Timmins, Ottawa, New York, Peking, etc. The Soviets have used the region between Timmins, Ontario and Ottawa, Ontario, very heavily as a human experimental territory because they had a large spy network in this area ready to report the local effects back to Moscow by secure diplomatic pouch. This spy ring was broken up by the Royal Canadian Mounted Police (RCMP) in February, 1978, and 26 KGB agents operating out of the Russian Embassy in Ottawa were expelled from the country. [7] As we have already amply documented from Tesla's models, his Colorado Springs experiments, and the Soviet experiments, this third element of ionospheric conduction has been mastered. The Soviets perfected Tesla's art by having a large satellite in polar orbit with a nuclear power plant of 100KW aboard to help control the standing wave pattern. On January 24, 1978, this nuclear powered Soviet satellite crashed in the Great Slave Lake region of Canada. [8] It is not known yet whether the U.S. government shot it down with one of its Killer Satellites. Only time will reveal the depths of the secret struggle going on between the Soviet Union and the United States for mastery of the skies over planet earth. Thus the third element involves an electric oscillator (the TMT) that will resonate the electrostatic forces operating between the earth conductor and the ionospheric conductor.

The fourth element requires all of Tesla's art of teleautomatics, including computers and satellites to precisely tune the earth to resonance to produce the desired effects.

The fifth element requires the earth tuning of the fourth element to be extended to bringing the sun into resonance with the earth. [9] This immediately sets up interplanetary communications of much energy between the two bodies. The trick is to control the flow of energy from earth to sun in perfect balance for life functions. This is not easy, and too little is known about what can happen.

The ideal full wavelength for an 8 Hz wave is approximately 23,500 miles long (comparable with the circumference of the earth – Ed. note). A one-quarter wavelength for 8 Hz is 5875 miles, and this is 1.48 earth radii long. A one-quarter wave length signal for 8 Hz generated on the earth would resonate as shown in Figure 11. The Table and Figure 11 also show that the inner belt of protons can be used to resonate frequencies from 6.66 Hz to 10.80 Hz. This is the filter mechanism most likely used by the Soviets, in addition to the satellite orbiting at this same altitude.

TABLE

Inner Belt Proton Resonance as a Function of Wavelength and Distance as Earth Radii.

Hertz	Wavelength as Circumference	Radius of Circumference =Earth Radii		¼ Wavelength = Earth Radii	
6.66	27,928 Mi.	4444 Mi	1.12	6928 Mi.	1.76
8.00	23,500 Mi	3740 Mi	0.93	5875 Mi	1.48
10.80	17,222 Mi	2741 Mi	0.69	4305 Mi	1.09

Since the solar wind produces particles that have an average kinetic energy eight times that of the magnetic energy density, we have a constant source to energize the inner belt protons in order to maintain resonance, at a center frequency of 8 Hz. It is now beginning to be realized that such a mechanism does indeed affect the earth's magnetic field, which in turn affects the weather. In the early 1960's, Dr. Robert Uffen pulled together a lot of data collected since the 1700's to show that the magnetic flux variations from the sun, known as the 11.1 year sun spot cycle, had a controlling effect on the fluctuations of the earth's magnetic field. But his most important finding was that long term world climatic changes could be correlated with changes in the earth's magnetic field. This basic discovery was quickly followed by a report from Columbia University, which showed that when the intensity of the earth's magnetic field decreases the average world temperature also decreases, as well as the reverse effect.

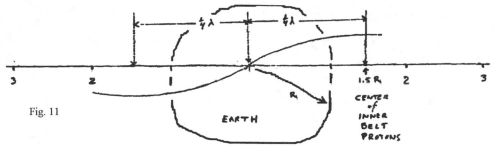

Fig. 11

Dr. J.W. King of the Appleton Laboratory in the United Kingdom extended these findings in 1974. He found that over the northern hemisphere of our planet, the magnetic field strength contour lines matched the atmospheric pressure contour lines. The centers of low atmospheric pressure coincide with the centers of highest magnetic intensity. Furthermore, Dr. King showed that atmospheric pressure patterns take the form of a dumb-bell shape with a low in each end, and this is matched by a similar contour for the magnetic field. Dr. King's paper published in NATURE, January 18, 1974, founded the new science of magnetometeorology linking geomagnetism with meteorology.

We have tried to show, in simplified and abbreviated form, how an 8 Hz magnetic wave is a common universal factor connecting protons in the brain with protons in the sun, and how the earth and the solar wind are organized as a filter mechanism for the 8 Hz magnetic frequency which is so fundamental to a balanced healthy state for living things. We have also indicated the mechanism whereby very small, plus or minus, deviations from an 8 Hz center frequency can induce pathological states in people, and most likely in all living things on this planet. [22] It is also very clear that in carrying out their experiments with the Tesla Magnifying Transmitter, the Soviets are exploiting only the pathological-inducing effects, as we shall show in the succeeding chapters.

But there is one effect that nobody dares mention, except one man. This is the possibility of impressing human thoughts, neutral, good, or bad, upon the ELF waves as the carrier system. There is no easy explanation of this phenomenon, but I will give a personal experience that may convey the essential idea. In 1961, I was with the late Aldous Huxley, and his wife Laura. From them I learned that Laura had a gift for making non-contact motions with her hands over sick people, and in some cases had helped the patient. One case in particular intrigued me, where Laura had cured a patient of paroxysmal ventricular tachycardia (a sudden and dangerous form of heart racing) I arranged to have Laura and her "now cured" heart patient monitored electronically while non-contact hand passes were carried out by Laura. To my great surprise, I found out that when Laura's hands passed about four inches over the skin of the patient, that both she and the patient came into resonance by producing enormous ELF waves in their separate EEG's at a peak power centered on the 8 Hz frequency. This test was repeated over and over again until I was satisfied that this was a genuine EM induction effect. Others have repeated my work since then and confirmed it.

Since healing has been observed with a process of 8 Hz EM wave induction, it was logical to assume that <u>precise information</u> transfer had occurred in order to correct the organic defect in the patient. This also meant that potentially one could transfer other kinds of information if one knew how to impress thoughts upon an 8 Hz carrier wave. The question now arises as to whether the Soviets have taken this step, and added this mental modulation possibility to their ELF, TMT experiments.[23] My friend, Thomas Bearden, (Lt. Col.

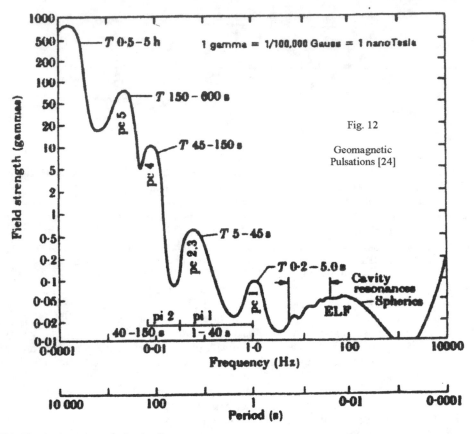

Fig. 12

Geomagnetic
Pulsations [24]

USA, Ret.) says, yes, emphatically.

Thus we now begin to understand what Tesla's visionary mind was seeing as he contemplated the TMT as a means of controlling the solar system for the benefit of earth and man. We shudder, as today we contemplate the awesome power wielded by the military machine of the Soviet Union. It is fearsome enough to contemplate Soviet scientists' manipulation of the earth's weather patterns, and men's mind -- in secret disregard for humanity's needs. But to play Jovian games with the sun's energies and thereby imperil the entire solar system is unquestionably the ultimate insanity of "civilized" man! We can now understand why Tesla kept his knowledge of the sun secret from men. He knew that they could not be trusted to wisely use this power. But now that the secret is out, and is being used for anti-humanitarian and evil ends, all of mankind must rise in protest and stop this barbarous abomination against man and God.

When the original manuscript of this book was written in 1978-1979, I did not know that the U.S.A. had started to build TMT's in reaction to the Soviet TMT's. Now (1985) we know that the U.S.A. has operational TMT's directed against the Soviet Union. Thus all humanity and all forms of life are helplessly being bombarded by ELF signals designed to weaken and damage them. (The HAARP Project has recently been connected to such Tesla technology as indicated on the latest Public Broadcast Station show on Tesla, available from www.pbs.org/tesla , but the cold war EMI battle cited here hopefully may have subsided by now. – Ed. note)

SECTION FIVE

Titantic Forces Unleashed by the USSR (1976 – 1986) and by the USA (1980 – 1986)

The first definitive analysis of the new Soviet TMT technology appeared in the authoritative journal Aviation Week & Space Technology, May 2, 1977, "Soviets Push for Beam Weapon," [1] and I quote:

"The Soviet Union is developing a charged-particle beam device designed to destroy U.S. intercontinental and submarine-launched ballistic missile nuclear warheads. Development tests are being conducted at a facility in Soviet Central Asia. The Soviets also are exploring another facet of beam weapons technology and preparing to test a spaceborne hydrogen fluoride high-energy laser designed for a satellite killer role. U.S. officials have coined the term directed-energy weapons and high-energy lasers.

A charged-particle beam weapon focuses and projects atomic particles at the speed of light which could be directed from ground-based sites into space to intercept and neutralize reentry vehicles, according to U.S. officials. Both the USSR and the U.S. also are investigating the concept of placing charged-particle beam devices on spacecraft to intercept missile warheads in space. This method would avoid problems with propagating the beam through the earth's atmosphere. Because of a controversy within the U.S. intelligence community the details of Soviet directed-energy weapons have not been made available to the President or to the National Security Council.

Recent events have persuaded a number of U.S. analysts that directed-energy weapons are nearing prototype testing in the Soviet Union. They include:

1) Detection of large amounts of gaseous hydrogen with traces of tritium in the upper atmosphere. The USAF TRW Block 647 defense support system early warning satellite with scanning radiation detectors and infrared sensors has been used to determine that on seven occasions since November, 1975, tests that may be related to development of a charged-particle beam device have been carried out in a facility at Semipalatinsk.

2) Test of a new, far more powerful fusion-pulse magnetohydrodynamic generator to provide power for a charged-particle beam system at Azgir in Kazakhstan near the Caspian Sea. The experiment took place late last year in an underground chamber in an area of natural salt dome formations in the desert near Azgir and was monitored by the TRW early warning satellite stationed over the Indian Ocean.

3) Point-by-point verification by a team of U.S. physicists 126 and engineers working under USAF sponsorship that the Soviets had achieved a level of success in each of seven areas of high-energy physics necessary to develop a beam weapon.

4) Recent revelations by Soviet physicist Leonid I. Rudakov [2] during a tour last summer of U.S. fusion laboratories that the USSR can convert electron beam energy to compress fusionable material to release maximum fusion energy. Much of the data outlined by Rudakov during his visit to the Lawrence Livermore

Laboratory has since been labeled top secret by the Defense Department and the Energy Research and Development Administration.

5) Pattern of activity in the USSR, including deployment of large over-the-horizon radars in northern Russia to detect and track U.S. ICBM reentry vehicles, development and deployment of precision mechanical/phased-array anti-ballistic missile radar and massive efforts aimed at civil defense. The Semipalatinsk facility where beam weapons tests are taking place has been under observation by the U.S. for about 10 years. The central building at the facility is believed by some officials to contain a collective accelerator, electron injector and power stores. The building is 200 ft. wide and 700 ft. long, with walls of reinforced concrete 10-foot thick, the entire facility, with its associated support equipment is estimated to have cost $500 million. The test site is at the southern edge of the Semipalatinsk nuclear test area, and it is separated from other test facilities. It is surrounded by a series of security fences. The total amount invested by the USSR in the test project for the 10 years' work there is estimated at $3 billion by U.S. analysts. The U.S. used high-resolution photographic reconnaissance satellites to watch as the Soviet technicians had four holes dug through solid granite formations not far from the main large building at the facility. Mine heads were constructed over each opening, and frames were built over the holes. As tons of rocks were removed, a large underground chamber was built deep inside the rock formation. In a nearby building, huge extremely thick steel gores were manufactured. The building has since been removed. These steel segments were parts of a large sphere estimated to be about 18 meters (57.8 ft.) in diameter. Enough gores for two complete spheres were constructed. U.S. officials believe the spheres are needed to capture and store energy from nuclear-driven explosives or pulse-power generators. [3] The steel gores are believed by some officials to be among the earliest clues as to what might be taking place at the facility. The components were moved to the nearby mine heads and lowered into the chamber. One of the major problems in gaining acceptance of the concept within the U.S. scientific community was to convince high-energy physics experts that the Russians might be using nuclear explosive generators as a power source to drive accelerators capable of producing high intensity proton beams of killing potential. U.S. officials, scientists and engineers queried said that the technologies that can be applied to produce a beam weapon include:

1) Explosive or pulsed power generation through either fission or fusion to achieve peak pulses of power.

2) Giant capacitors capable of storing extremely high levels of power for fractions of a second.

3) Electron injectors capable of generating high-energy pulse streams of electrons at high velocities. This is critical to producing some types of beam weapons.

4) Collective accelerator to generate electron pulse streams or hot gas plasma necessary to accelerate other subatomic at high velocities.

5) Flux compression to convert energy from explosive generators to energy to produce the electron beam.

6) Switching necessary to store the energy from the generators in large capacitors.

7) Development of pressurized lines needed to transfer the pulses from the generators to power stores. The lines must be cryogenically cooled because of the extreme power levels involved.

For several years, Air Force Maj. Gen. George J. Keegan, who until his recent retirement headed USAF's intelligence activities has been trying to convince the

Central Intelligence Agency and a number of top U.S. high-energy physicists that the Soviets are development a charged-particle beam weapon for use in an antiballistic missile role.

It was anticipated by Gen. Keegan and his advisers that the USSR would be forced to vent gaseous hydrogen from the experiments at Semipalatinsk and that early warning satellites could detect it. Liquid hydrogen in large amounts is believed by some officials to be utilized to cushion the nuclear explosive generator sphere and for cryogenic pumping of large drift tubes nearly a kilometer in length through which the beams are propagated for underground testing. In both cases, large amounts of gaseous hydrogen are formed and released into the atmosphere, probably carrying large amounts of nuclear debris or radioactive tritium that can be exploded at altitude and dispersed to avoid harming the people below, according to some U.S. scientists. "Explosions of such gaseous hydrogen discharges are now being detected with regularity from Soviet experiments," a U.S. official said, "and scientific studies of the gas releases and explosions have confirmed their source as being near the Semipalatinsk facility."

In recent public pronouncements, Gen Keegan has taken the CIA to task for having rejected Air Force Intelligence information about Soviet beam weapon development. He also has spoken bitterly about a number of top U.S. physicists who refuse to accept even the possibility that the Soviets are involved in beam weapon development. Most of the physicists who would not accept the data were older members of the scientific community who had been involved in research and development from the early days of a project called "Seesaw" The U.S. attempted unsuccessfully to develop a charged-particle beam device under the project code named Seesaw. It was funded by the Defense Department's Advanced Research Projects Agency but abandoned after several years.

A number of influential U.S. physicists sought to discredit General Keegan's evidence about Soviet beam development. The general attitude was that, if the U.S. could not successfully produce the technology to have a beam weapon, the Russians certainly could not. "It was the original not-invented-here attitude," one of the U.S. physicists said.

There were about 20 hypotheses advanced by these physicists and the CIA's Nuclear Intelligence Board as to what the facility at Semipalatinsk was being used for by the USSR. One theory was that it was a supersonic ramjet test site and another was that it was a nuclear reactor test site for commercial applications. That was based on the layout, which resembled some reactors in the USSR "There is now no doubt that' there is dumping of energy taking place at the site with burning of large hydrogen flames," one official said [5] "What bothered the Nuclear Intelligence Board at first was that it was hard to imagine that some seven technologies critical to the weapons concept could be perfected there within the time frame presented and not be detected by us."

It is obvious that the splashing of radio interference all over the planet originating in the Soviet Union has military scientists on edge as they try to second guess Soviet intentions and technology. It is not too surprising that the U.S. military analysts would "guess" that the Soviet Union were developing a "directed-beam" weapon, for this idea has a long history. The invention of the directed-beam goes back to the year 1900, when Nikola Tesla invented his "magnifying transmitter" which developed such high voltages that it equaled lightning bolt effects found in nature. However, he did not reveal to the world the development of his "directed-beam" until his 78th birthday. Joseph W. Alsop, Jr., writing in the *New York Herald Tribune* on Wednesday, July 11, 1934, says under the headline:

> Beam to Kill Army at 200 Miles, Tesla's Claim on 78th Birthday.
> "Dr. Nikola Tesla, inventory of polyphase electrical current, pioneer in high frequency transmission, predecessor of Marconi with the wireless, celebrated his 78th birthday yesterday by announcing his invention of a beam of force somewhat similar to the death ray of scientific romance.
>
> It is capable, he believes, of destroying an army 200 miles away. It can bring down an airplane like a duck on the wing, and it can penetrate all but the most enormous thicknesses of armor plate. Since it must be generated at stationary power.plants by machines which involve four electrical devices of the most revolutionary sort, Dr. Tesla considers it almost wholly a defensive weapon. In peace times, he says, the beam will also be used to transmit immense voltages of power over distances limited only by the curvature of the earth.
>
> He came to the idea of a beam of force, he said, because of his belief that no weapon has ever been found that is not as successful offensively as defensively. The perfect weapon of defense, he felt, would be a frontier wall, impenetrable and extending up to the limits of the atmosphere of the earth. Such a wall, he believes, is provided by his beam of force. It is produced by a combination of four electrical methods or apparatuses. First and most important is a mechanism for producing rays and other energy manifestations in free air. Hitherto vacuum tubes have always been necessary. Second is an apparatus for producing unheard-of quantities of electrical current and for controlling it when produced. The current is necessary as power for the first mechanism. Without this, no rays of sufficient strength could be produced. The third is a method of intensifying and amplifying the second process, and the fourth is a method of producing 'tremendous electrical repellent force.'
>
> 'These four inventions in combination enable man to let loose in free air forces beyond conception,' Dr. Tesla remarked mildly. By scientific application we can project destructive energy in thread-like beams as far as a telescope can discern an object. The range of the beams is only limited by the curvature of the earth. Should you launch an attack in an area covered by these beams, should you say, send in 10,000 planes or an army of a million, the planes would be brought down instantly and the army destroyed. The plane is thus absolutely eliminated as a weapon; it is confined to commerce. And a country's whole frontier can be protected by one of the plants producing these beams every 200 miles. Nor should they be much more costly than an ordinary power plant.
>
> The beam of force itself, as Dr. Tesla described it, is a concentrated current it need be no thicker than a pencil, of microscopic particles moving at several hundred times the speed of artillery projectiles. The machine into which Dr. Tesla combines his four devices is, in reality, a sort of electrical gun. He illustrated the sort of thing that the particles will be, by recalling an incident that occurred often enough when he was experimenting with a cathode tube. Then, sometimes, a particle larger than an electron, but still very tiny, would break off from the cathode, pass out of the tube and hit him. He said that he could feel a sharp, stinging pain where it entered his body, and again at the place where it passed out. The particles in the beam of force, ammunition which the operators of the generating machine will have to supply will travel far faster than such particles as broke off from the cathode, and they will travel

in concentrations, he said. (Tesla probably discovered "charge clusters" that penetrated his body; which were acting as an anode at ground potential. Ken Shoulders rediscovered fifty years later that millions of electrons can coagulate into clusters and penetrate any object, even steel. He also patented the process as a new source of energy #5,018,180 and #5,148.461. – Ed. note)

As Dr. Tesla explained it, the tremendous speed of the particles will give them their destruction-dealing qualities. All but the thickest armored surfaces confronting them would be melted through in an instant by the heat generated in the concussion. Dr. Tesla declared that the two most important of the four devices involved in his force beam generator, the mechanism for producing rays in free air, and the mechanism for producing great quantities of electrical current, had both been constructed and demonstrated by actual experiments. The two intensifying and amplifying apparatuses are not yet in existence but he displayed the most perfect confidence that when they are, they will work as he expects them to do."

In a letter to His Majesty's government of Britain, dated August 28, 1936, Tesla offered the secrets of his "beam of force" weapon. During the ensuing correspondence with the Director of Mechanization, the War Office, London, S.W. 1 [84/T/3458 (M.G.O.4.b.)], there is a letter from Tesla dated October 26, 1937 in which he states:

"My discoveries and inventions for securing complete immunity from any form of attack constitute the most revolutionary technical advance in history and will affect profoundly the future of humanity. They will save the lives of millions of people and prevent destruction of property of inestimable value in all countries. They may also be the means of preserving and strengthening the greatest empire on earth."

Tesla in this letter pressed for an early decision on the part of His Majesty's government, and then said, "I am indifferent now as to whether these terms are accepted or not, but venture to point out in all deference that if England does not take advantage of the present opportunity, some foreign power might later exact a price so great as to strain to the utmost the financial resources of Great Britain and cripple it seriously."

I might add that Tesla estimated in 1937 that the first beam of force plant could be built for 10 million pounds sterling. Who knows, but God, what the history of World War II might have been, had the British, in 1937, developed Tesla's beam-of-force defense system? Today, forty years later, there is a possibility that the scientists of the Soviet Union have solved the secrets of Tesla's beam-of-force system, now called the directed-beam system. The full implications of the possible Soviet breakthrough are best summarized in the editorial in Aviation Week for May 2. 1977 [8]:

"The Soviet Union has achieved a technical break-through in high-energy physics application that may soon provide it with a directed-energy beam weapon capable of neutralizing the entire United States ballistic missile force and checkmating this country's strategic doctrine. The hard proof of eight successful Soviet tests of directed-energy beam weapon technology gives new and overriding urgency to bring these developments into the public domain and rip the veil of intelligence secrecy so that this whole matter of vital national urgency and survival will finally be brought to the attention of the President of these United States, the Congress and the citizens of this republic whose future is at risk. In all of the previous four years that these Soviet developments have been known to the official intelligence community, they have been stifled by a conspiracy of skepticism and silence and never once penetrated to the highest decision-making councils of this country. The incredible story of how the Soviets leap-frogged a generation of high-energy physics technology and developed a workable experimental model of a directed-energy beam weapon now has been largely verified by the successive Soviet tests at Semipalatinsk and Azgir and the brilliant work of a small group of extremely young physicists in this country. The fact that this country still has a chance of avoiding a crippling technological surprise that could render its entire strategic missile force ineffective is due to the courageous, dogged and perceptive work of a handful of U.S. Air Force intelligence specialists who polarized around the leadership of Maj. Gen. George Keegan, Jr., recently retired chief of Air Force intelligence.

We do not suggest any formal conspiracy to suppress the mounting evidence of a massive Soviet research development, and industrial push aimed at the goal of an anti-ICBM directed-energy beam weapon. Rather it was a combination of smug American assurance that the Soviets were simply not capable of out-reaching us in any technological race and the intellectual arrogance of elderly scientists who through the ages have spent their twilight years proving that the next generation of breakthroughs is 'impossible.'

In modern times, we have the continuing example of Dr. Vannevar Bush, who thundered that the ICBM was a technical impossibility, and the assortment of scientists in the Eisenhower era who firmly believed that manned spaceflight should be abandoned because the human system could not survive its rigors. It was a similar group of high-energy physicists, some heavy with Nobel laurels, who encouraged the natural technical illiteracy of the Central Intelligence Agency to discount the steadily growing stream of Soviet developments and to lead the bitter intramural battles that suppressed the evidence from higher government councils for crucial years.

There is still considerable debate over the real significance of the Soviet tests at Semipalatinsk and Azgir and how long it will take the Soviets to translate their experimental developments into a usable weapon. But there is no longer much doubt among top-level U.S. high-energy physicists that it is feasible to develop a directed-energy beam device.

There is also an element in the Pentagon that can visualize the eventual Soviet deployment of the directed-energy beam weapon as the end game of an intricate chess exercise that began with the 1972 negotiation of the anti-ballistic missile treaty, which effectively stopped not only U.S. deployment of an anti-ICBM system but also most of its significant on-going research and development. The hypothesis for this chess game, which ends in the early 1980's with the triumphant Soviet shout of 'check and mate,' involves the U.S. finding its strategic deterrent ballistic missile force stripped of any defensive system, with the Soviets using their anti-ICBM directed-energy beam weapon to negate a U.S. retaliation and a strong civil defense shield to minimize damage from the few warheads that night penetrate.

The race to perfect directed-energy weapons is a reality. Despite initial skepticism, the U.S. scientific community now is pressuring for accelerated efforts in this area."

While the Western military mind was focusing its tunnel vision on hardware and its physical effects on material systems, the Soviets were creating subtle but profound effects on the entire planetary biosphere. We shall review a number of the various effects that the Soviets induced in living things in this chapter. The history of Soviet research in affecting living things with electromagnetic radiation has a long history, and the interested reader is referred to the recent book by Alexandr P. Dubrov, of the Academy of Sciences of the USSR, which has some 1228 references on this subject. [9]

Of immediate interest is the fact that the Soviets began to irradiate the U.S. Embassy in Moscow beginning in 1960, and for years afterward, with microwaves, or very high radio frequencies. [10] (It is an historic fact that the U.S. installed copper sheets on the walls of the embassy but a lot of the U.S. embassy workers were still stricken with cancer within a few years. – Ed. note) This practice has been a puzzle to U.S. experts over the years. The Soviets have never explained their intent, or their goals. It is my opinion that the Soviets have tried to deflect the interest of the American experts away from the ELF radio spectrum by directing their attention to the opposite end of the radio spectrum. [11]

It is small consolation to note that all the ideas and technologies for the TMT, including weather modification, were carried out by Nikola Tesla in 1900. Tesla almost received recognition for this work, and we quote from the New York Times, November 7, 1915. The headline states:

Tesla's Discovery: Nobel Prize Winner

"Transmission of Electrical Energy without Wires, which affects Present-day Problems." "To Illuminate the Ocean, Scientist says, Collisions will be Avoided, and Unlimited Water Drawn to Irrigate Deserts."

Nikola Tesla, who with Thomas A. Edison, is to share the Nobel Prize in Physics, according to a dispatch from London, said last evening that he had not yet been officially notified of the honor. His only information on the matter was the dispatch in the *New York Times*.

"I have concluded," he said, "that the honor has been conferred upon me in acknowledgement of a discovery announced a short time ago which concerns the transmission of electrical effects of unlimited intensity and power can be produced, so that not only can energy be transmitted for all practical purposes to any terrestrial distance, but even effects of cosmic magnitude may be created."

Mr. Tesla said the discovery had a direct and vital bearing on the problems now foremost in the public mind. For instance, he said, wireless telephony would be brought to perfection undreamed of through the application of this discovery. He added, "We will deprive the ocean of its terrors by illuminating the sky, thus avoiding collisions at sea and other disasters caused by darkness. We will draw unlimited quantities of water from the ocean and irrigate the deserts and other arid regions. In this way we will fertilize the soil and derive any amount of power from the sun. I also believe that ultimately all battles, if they should come, will be waged by electrical waves instead of explosives."

Alas for Tesla, the Nobel Prize was never formally awarded to him. But his ideas and experiments on weather modification were quite specific, and used the same magnifying transmitter power source, as would be used in over-the-horizon radar, or a directed-beam weapon. What is not generally known is that Tesla invented a device that performs the functions of a true over-the-horizon radar which he called "telegeodynamics". [14] In this invention he introduced controlled seismic tapping of the earth, and with proper receivers, he claimed to be able to keep track of every moving vehicle on the planet. But it would be more colorful to quote his own words in an interview on his 79th birthday from the *New York Times*, July 11, 1935.

> One of the subjects which he hoped, he said, will come to be recognized as his "greatest achievement in the field of engineering," was, he said, the perfection by him of "an apparatus by which mechanical energy can be transmitted to any part of the terrestrial globe."
>
> This apparatus, he said, will have at least four practical possibilities. It will give the world a new means of unfailing communication; it will provide a new and by far the safest means for guiding ships at sea and into port; it will furnish a certain divining rod for locating ore deposits of any kind under the surface of the earth; and finally, it will furnish scientists with a means for laying bare the physical conditions of the earth, and will enable them to determine all of the earth's physical constants.
>
> He called this discovery "tele-geodynamics", motion of earth-forces at a distance.

Further details about this invention are given in *The New York American* of the same date. The headline is:

TESLA'S 'CONTROLLED' EARTHQUAKE

> Nikola Tesla, father of radio and of the modern method of electric power transmission, observed his 79th birthday yesterday by drinking a quart of boiled milk and outlining the latest of his many startling discoveries.
>
> This is an apparatus by which energy can be transmitted to any part of the earth, with practical possibilities in the navigation of ships, discovery of ore deposits and determination of the physical properties of the earth's interior.
>
> His experiments in transmitting mechanical vibrations through the earth, called by him the art of telegeodynamics, were roughly described by the scientist as a sort of "controlled earthquake."
>
> The rhythmical vibrations pass through the earth with almost no loss of energy, he said, and predicted the system in time will be universally adopted, since it furnishes an "unfailing means of communication".[15]
>
> He asserted: "It becomes possible to convey mechanical effects to the greatest terrestrial distances and produce all kinds of unique effects of inestimable value to science, industry and the arts."
>
> The invention could be used with destructive effect in war, he said, by exploding bombs thousands of miles away which had been equipped with apparatus to receive the vibrations.[16]

Tesla once said to a reporter with respect to his telegeodynamic art of "controlled earthquakes" that with a small amount of power it would take about two weeks to two months of vibration of the earth to bring it to full resonance, "and it could be made to split in two -- like an apple."[17] Is there a possibility that the Soviets are experimenting with Tesla's method of "controlled earthquakes"? (As the cold war ended with the fall of the Berlin wall, the Soviet Union being dissolved, the new Russia has little funding for such activities since the U.S. has become its ally. However, the technological process still exists. – Ed. note)

REFERENCES

SECTION ONE

1. ELECTRICAL EXPERIMENTER, February 1919, pp. 696-697 and p. 743. "My Inventions - 1. My early life." by Nikola Tesla. This autobiography was published in five installments. O'Neill, John J.
2. PRODIGAL GENIUS, The Life of Nikola Tesla. Ives Washburn, Inc. New York, 1944.
3. ELECTRICAL EXPERIMENTER. Op. Cit.
4. ELECTRICAL EXPERIMENTER, April 1919, pp 864-865. "My Inventions: - My Later Endeavours. The Discovery of the Rotating Magnetic Field." by Nikola Tesla. My principal source for this information was the late John Rays Hammond, Jr. of Gloucester, Mass.
5. ELECTRICAL EXPERIMENTER, April 1919. Op. Cit. When a conducting metal is cooled below a certain critical temperature (usually near absolute zero, minus 273 C.), it loses all resistance to the flow of electricity. Therefore, once an electrical the electrical current continues to flow on and on without damping. Such "infinite" conduction will continue as long as the temperature remains below the critical temperature. This phenomenon is called superconductivity Brian D. Josephson received the Nobel prize in 1973 for a most important discovery connected with this phenomenon signal is initiated in such a super-cooled conductor, current continues to flow on and on without damping.
6. ELECTRICAL EXPERIMENTER. April 1919. Op. Cit.
7. Figure 3 is most important in that it is the foundation for understanding how the Soviets could transmit a magnet wave around the planet. One looks at the region between 270° and 0°, where AC-l and AC-2 cross. The crossing point represents the region of the two Soviet transmitters, one in Gomel, and the other in Riga. (Compare with Figure 10.) By adjusting the proper frequency phase differences between the two transmitters they will create a standing magnetic wave which follows a great circle route on the planet whose center is Rigs and Gomel, and whose circle course is some 45° from this center. See SECTION FOUR for a more detailed explanation.
8. PRODIGAL GENIUS. Op. Cit.

SECTION TWO

1. The Inventions, Researches, and Writings of Nikola Tesla. Ed. by Thomas Commerford Martin. Published by the ELECTRICAL ENGINEER, New York, 1894. 496 pp. Chapter Ill, pp. 9-25. "The Tesla Rotating Magnetic Field."
2. A kind of magnetic "back-pressure" that damps free-oscillation in a circuit.
3. The Maxwell Equations define the relation between electric fields, E and magnetic fields, H, where v denotes the electric charge, and j the current density; c is the velocity of light.

$$\nabla \cdot E = 4 \pi v$$
$$\nabla \cdot H = 0$$
$$\nabla \times E = -\frac{1}{c} \frac{\partial H}{\partial t}$$
$$\nabla \times H = (4 \pi j)/c + \frac{1}{c} \frac{\partial E}{\partial t}$$

4. The Inventions, Researches, and Writings of Nilrola Tesla. Op. Cit. Chapter XXVIII, pp. *294-378.* "On Light and other High Frequency Phenomena." February, 1893.
5. Ibid. Chapter XXVII, pp. 198-293. "Experiments with Alternate Currents of High Potential and High Frequency." A Lecture delivered before the Institution of Electrical Engineers, London. February, 1892.
6. This is a clue already known to Tesla as to the frequencies required to produce electrical plasmas, and ball lightning. These have been produced over Canada, and the Soviet Union by the Soviet ELF emissions.
7. At the Columbia World's Fair in *Chicago,* Tesla demonstrated before the public that he could pass lore than a million volts through his body with *safety.* This was in August 1893.
8. London Lecture, Op. Cit. supra Note 5 .
9. PRODIGAL GENIUS, Op. Cit.
10. ELECTRICAL EXPERIMENTER, June 1919, pp. 112-113. "My Inventions - 5. The Magnifying Transmitter." by Nikola Tesla.

SECTION THREE

1. ELECTRICAL EXPERIMENTER, June 1919,
2. PRODIGAL GENIUS, Op. Cit.
3. PRODIGAL GENIUS, Op. Cir.
4. PRODIGAL GENIUS, Op. Cit.
5. "High Frequency Oscillators for Electro-Therapeutic and other Purposes." Paper read by Nikola Tesla, September 13, 1898 at the Eighth Annual Meeting of the American Electro-Therapeutic Association, Buffalo, New York. THE ELECTRICAL ENGINEER, December 23, 1891,p.670."Massage with Currents of High Frequency" by N. Tesla.
6. ELECTRICAL REVIEW, March 11, 1896. "On Roentgen Rays" by Nikola Tesla.
7. PRODIGAL GENIUS, Op. Cit. Also, Letter to the Editor by N. Tesla.
8. PRODIGAL GENIUS, Op. Cit. Also, The New York Times, February 6, 1932.
9. COLORADO SPRINGS NOTES, 1899 - 1900 (Eng Pub. by NOLIT, Beograd, Yugoslavia, 1978 (English Edition), Nikola Tesla.
10. Ibid., p. 169. Tesla explicitly makes a note in his diary that he is omitting making notes of his researches on superconductivity as follows: "The following items, partly worked out, omitted for want of time: From 1-30 September 1899. Method of increasing magnifying factor of resonant circuits by coiling."
11. It was not until 1960 that Balser and Wagner (published in NATURE, Vol. 188, No. 4751, 1961) proved what Tesla had already proved in 1895. Namely, they showed that the earth-cavity was resonant at 8 Hz. More precise studies done by M.J. Rycroft showed (in Figure 6) that the ELF spectrum had its fundamental at 7.8 + 0.2 HE· RADIO SCIENCE, Journal of Research, NBS/USNC - URSI, Vol. 69D, No. 8, August 1965, pp. 1071-1081. "Resonances of the Earth-ionosphere Cavity observed at Cambridge, England." Neither of these authors mentions Tesls's researches of six decades earlier.
12. COLORADO SPRINGS NOTES, Op. Cit.
13. THE WORLD TODAY, Vol XXI, No. 8, February 1912, pp. 718-722. "Nikola Tesla, Dreamer" by Allan t. Benson.

14. ELECTRICAL EXPERIMENTER, May 1919, p. 21.   "The True Wireless" by Nikola Tesla.
15. In 1976, it was just these effects which the Soviets had replicated, and which were measured in Canada.
16. PRODIGAL GENIUS, Op. Cit.
17. Tesla mates careful notes in the COLORADO SPRINGS NOTES, Op. Cit., to study the ELF effects on both planes and animals.
18. TALKING WITH THE PLANETS, by Nikola Tesla. Colliers Illustrated Weekly, Vol. XXVI, No. 19, January 9, 1901.

SECTION FOUR

1. THE CENTURY Illustrated Monthly Magazine, June 1900, "The Problem of Increasing Human Energy," by Nikola Tesla.
2. This velocity has recently been measured and is of the order of 600 km pet second, or 372.82 per second, or expressed as 1,342,159 miles per hour.
3. THE CENTURY, Op. Cit.
4. Personal communication from John Hays Hammond, Jr. to the author.
5. Letters from Tesla to the governments named, now in the Tesla Museum, Belgrad, Yugoslavia.
6. Tesla's unpublished description of his Telegeodynamic Oscillator - method and means.
7. News items about the expulsion of the KGB from Canada in early February 1978, These references have been misplaced by author (April 24, 1979).
8. The New York Times, January 29, 1978, "Plunge from Orbit," Section 4. The Week in Review.
9. RADIO ELECTRONICS, March 1977, p. 6. Scientists were able for the first time to determine the structure of the sun's magnetic field from data returned by Pioneer II satellite. The sun's magnetic field is roughly spherical, and envelops the entire solar system - some 3,700,000,000 x 2 miles in diameter.
10. Kamiya, Joe, "Conscious Control of Brain Waves," PSYCHOLOGY TODAY, April 1, 1968, pp. 57-60.
11. Wiener, Norbert, CYBERNETICS, Wiley, New York, 1948.
12. Bremner, Frederick J·. V· Begnignus, and F. Moritz. NEUROPSYCHOLOGIA, Vol. 10, 1972, pp. 307-312.
13. Akasofu, Syun-ichi and Sydney Chapman, SOLAR TERRESTRIAL PHYSICS. The International Series of Monographs on Physics. Oxford University Press, Ely House, London W.1, 1972. Figure 11 is adapted from Figure 6.44 on page 439.
14. Rycroft, M.J. "Resonances of the Earth-Ionosphere Cavity observed at Cambridge, England," RADIO SCIENCE, A Journal of Research, NBS/USNC-URSI, Vol. 69D, No. 8, August 1965. Figure 12 is adapted from Figure 6 of this article.
15. Puharich, Andrija, PROTOCOMMUNICATION. A lecture given at the Twentieth International Conference, August 27, 1971, St. Paul de Vence, France. Published in THE PROCEEDINGS of the Parapsychology Foundation, 29 West 57th Street, New York, 10022. 1972.
16. I do not have the exact references to this work at hand, but they can be found by reference to the work of Dr. Enrici Clementi of IBM, San Jose, California.
17. The velocity of the first orbital in the Bohr hydrogen atom is 1359 miles/sec. This yields a factor of 4.678.

18. I am indebted to Hugh Harleston, Jr. for help and advice in helping to solve this part of the problem.
19. Akasofu, Op. Cit. Figure 13 is adapted from Figure 3.3 on page 115.
20. Akasofu, Op. Cit. Figure 14 is adapted from Figure 6.15 on page 388.
21. Akasofu, Op. Cit. Page 321 and 389
22. See "Ozone Paper," where the author proved this theory in developing a new and successful treatment for cancer. (Reprinted as "How Transdermal Electrotherapy Led to Highly Efficient Water Electrolysis With Anomalous Organic Molecule Formation and a Spinoff that Successfully Treated Neoplasms in Mice," *Energetic Processes, Volume I,* Xlibris Pub., 2002, p. 238, www.xlibris.com – Ed. note)
23. "Soviet and Czechoslovakian Parapsychology Research," DEFENSE INTELLIGENCE AGENCY. Prepared by the U.S. Army Medical Intelligence and Information Agency, Office of the Surgeon General, DST-1810S-387-75, DZA TASK PT-1810-12-75. September 1975.
24. Galejis, J., Terrestrial Propagation of Long Electromagnetic Waves, Pergammon Press, New York, 1972

SECTION FIVE

1. This authoritative article was written by Military Editor, Clarence A. Robinson, Jr. Editor Robert Hotz in an editorial on this article, entitled "Beam Weapon Threat", states: (p. 11) " We do not suggest any formal conspiracy to suppress the mounting evidence of a massive Soviet research, development and industrial push aimed at the goals of an anti-ICBM directed-energy beam weapon. Rather it was a combination of smug American assurance that the Soviets were simply not capable of out-reaching us in any technological race, and the intellectual arrogance of elderly scientists who through the ages have spent their twilight years proving that the next generation of breakthroughs is impossible".
2. Leonid I. Rudakov, Kurchatov Institute of Atomic Energy, Moscow, is listed as one of the key developers of high-current acceleration technology in the Soviet Union. (AIR FORCE MAGAZINE, September 1977, page 126).
3. Col. Bearden describes in his writings how these steel gores (spheres) can be used in a psychotronic warfare.
4. Project Seesaw was an attempt to repeat Nikola Tesla's particle beam accelerator weapon.
5. It is interesting to note that on the day following his discovery of the stationary waves in the earth, Tesla in his notes of July 5, 1899 makes plans to produce large amounts of hydrogen. COLORADO SPRINGS NOTES, Op. Cit.
6. The limitation of the energy travel, in a line of sight, by the curvature of the earth, tells us that this particle beam, as described in Appendix A, is he prototype of the types being developed in the USSR, and most recently in USA.
7. This letter is in the author's files.
8. Op. Cit. under Reference 1, Supra
9. Dubrov, A.P. THE GEOMAGNETIC FIELD AND LIFE, Plenum Press, New York and London, 1978.
10. There is a vast literature on this subject, much too large to quote here. The interested reader can find all of the pertinent and up to date literature in : Paul Brodeur: "The Zapping of America: Microwaves, their deadly risk and the cover up". W.W. Norton and Co., New York, 1977.

11. Unpublished documents in author's files.

12. The function of this satellite is to keep the Soviet electromagnetic radiation pulsing (at ELF frequencies) aimed at the inner proton belt in order to maintain peak power through resonance. Einaudi, F., and Wait, J.R., "Analysis of the excitation of the earth-ionosphere waveguide by a satellite-borne antenna." Parts II and CANADIAN J. PHYSICS, 49: 11, No. 4, 1971.

13. Author's note: As of this writing (February 1978), the winter of 1977-78 has turned out to be the worst since 1883 in the northeast US. (Later note: April 1979). The winter of 1978-1979 has turned out to be the worst on record for the Midwest US.

14. Tesla's own unpublished description of this art.

15. It is well to point out here that in the event of a nuclear all-out war, all radio communication "signal" would be lost in the "static and noise" resulting from excessive atmospheric ionization. It may well be that the Soviets are also developing Tesla's Telegeodynamic system for communication in the event of an all out nuclear war.

16. There is real concern among intelligence officials with whom I have talked in the US, Canada and from countries of Western Europe that the Soviets already have this capability.

17. Benson, Allan L.: "Nikola Tesla, Dreamer," THE WORLD TODAY, Vol.XXI, No.8, February 1912, p.722.

Andrija Puharich was an author, inventor, and a medical doctor. His patents include #3,586,791, #3,629,521, #3,497,637, #2,995,633, #3,170,993, and #4,394,230. His books include *Uri, Beyond Telepathy, The Sacred Mushroom*, and *The Iceland Papers*. An interesting example of the battles he fought and lost is contained in the patent Court of Appeals case, No. 22286, Puharich vs. Brenner (*US Patent Quarterly*, V.162, p.136, June 25, 1969).

7 Worldwide Wireless Power Prospects

Kurt Van Voorhies, Ph.D., P.E.
Adapted from *Proceedings of IECEC, 1991*

ABSTRACT

Worldwide wireless power began as a concept with the pioneering work of Nikola Tesla about 100 years ago. His principal approach is summarized. The viability of such a system must still be demonstrated and many questions remain. Potentially, a wireless system can transfer power more efficiently and flexibly, especially to and from remote regions. The principal elements of worldwide wireless power transfer include: 1) the source: an oscillator/transmitter, 2) the path: the cavity bounded by the earth and the ionosphere and 3) the receiver: a means of extracting power from the path. The system transfers and stores energy via the resonance modes of the cavity. The key challenges facing demonstration of technical feasibility are in finding an efficient means of coupling power into and out of the earth-ionosphere cavity, and in devising a feasible receiver that is both small and efficient. Along with demonstrating technical feasibility, new research must consider safety, environmental impact, susceptibility to weather, and effects on weather.

INTRODUCTION

Nikola Tesla pioneered the concept of worldwide wireless power transfer about 100 years ago, beginning with work on high voltage, high frequency single electrode lighting systems, and following with development of the Tesla Coil, The Magnifying Transmitter, and the single electrode x-ray tube. The Tesla Wireless system and concepts leading thereto are documented in Tesla's notes [1,2] patents [3,4], lectures [4-8] and published articles [4, 5, 9-11] and described by Tesla's biographers [12,13] and others [14, 15]. Following the death of Tesla in 1943, the concept lay dormant until referenced by Wait in 1974 [16,17] in conjunction with extremely low frequency communications, followed by Marincic's illuminating review in 1982 [18] and subsequent technical analysis by Corum and Corum [19-24], Golka [25,26] replicated the oscillator used in Tesla's Colorado Springs experiments for studying ball lightning and plasma containment for nuclear fusion. Corum and Corum [27-31] have also replicated Tesla's ball lightning experiments but with smaller scale equipment. However, Tesla's worldwide wireless power concept remains unverified.

PRINCIPLES OF WORLDWIDE WIRELESS POWER TRANSFER

Consider the earth as a large spherical capacitor or cavity resonator, comprising the *terra firma* as the inner conductor, the lower atmosphere as the insulating dielectric, and the upper atmosphere (electrosphere) and ionosphere as the outer conductor. Power is coupled into the cavity via either direct conduction/displacement, or radiation, with high power RF oscillators or transmitters tuned to the cavity's resonant frequency. A remove receiver, also tuned to this resonant frequency, then extracts this power wirelessly. The propagation loss in the earth-ionosphere cavity increases with frequency but, at the fundamental frequency, is about 11% less than the equivalent loss on a 200KV power line. The wireless concept described here differs from that used in microwave wireless power transmission in that the latter beams power along a line of sight path, normally from outer space to earth [32]

PROMISES OF WORLDWIDE WIRELESS POWER TRANSFER

The benefits of wireless power transfer have not changed since originally described by Tesla in 1900 [9] and 1904 [10]. A cheap, efficient means of distributing energy would revolutionize development and improve access to new energy sources. Energy could be coupled into the cavity at the source, eliminating the need for the costly and time-consuming process of constructing and maintaining power transmission lines. The system would enable better utilization of remote sources of energy and would facilitate power transfer to remote users worldwide. While Tesla primarily proposed supplying power for lighting in conjunction with his high frequency single electrode lighting systems, he also envisioned "...energy of a waterfall made available for supplying light, heat and motive power anywhere – on sea, or land or high in the air..."[10]. Of course, the economic viability of such a system depends upon either 1) a technical means for controlling/measuring the supply and use of wireless power around the world, or 2) a very low cost source energy.

Nikola Tesla

Nikola Tesla was a prolific inventor best known for the AC induction motor and AC polyphase distribution system which are the basis for our present AC power system. His other inventions include the Tesla coil, high frequency generators, the Tesla Magnifying Transmitter, key elements of radio, single electrode high frequency, the single electrode x-ray tube, a viscous turbine, and remote control. Following his developments in low frequency AC machines and power distribution systems , Tesla experimented with single electrode, high frequency, high voltage lamps utilizing rarefied gases, the forerunner of present fluorescent lights. Initially he utilized patented high frequency alternators with 384 poles to produce the necessary 20 KHz power, but subsequently invented the disruptive discharge high voltage transformer, a.k.a. Tesla Coil, in 1891 [33].

In a Tesla Coil, low frequency AC power is amplified in voltage with a conventional transformer. The output of this transformer feeds the Tesla Coils' resonant LC primary circuit through a spark gap. The spark gap creates a broad spectrum of energy, components which resonate the primary and secondary circuits of the Tesla coil. The secondary of the Tesla Coil is tuned to be electrically ¼ wavelength long, with one terminal grounded, and acts as a "slow wave" device to resonantly amplify the voltage further.

Tesla found that the high frequency output from the Tesla coil could readily power lights and motors utilizing a single wire with a ground return. Tesla presented these results in this lecture to the IEE in London in 1892[7]. Following the work of Kelvin and Crookes, Tesla also noted that slightly rarefied gases were excellent conductors, leading him to propose a system for " ...transmitting intelligence or perhaps power, to any distance through the earth or environing medium". [34] In February 1893, at his lecture on high frequency currents before the Franklin Institute of Philadelphia (repeated in March in St Louis.) Tesla proposed to determine the capacitance of the earth and the period of oscillations resulting from a disturbance of the earth's charge . After subsequent patented improvements to the Tesla Coil, Tesla patented the single wire power distribution system in March., 1897, [35] and patented the wireless power distribution 6 month later [36,37]. In the wireless system , the single wire conductor was replaced by a conductive path through a slightly rarefied gas coupled to bodies of large surface area, or open capacitors, connected to the high tension terminals of the transmitter and receiver, thus forming an open resonator circuit between the body and the earth. In his patent, Tesla claimed the use of the conductive layers in the upper atmosphere as the conductive path.

In the 1892 lecture in London, Tesla noted that " It is quite possible, however, that such 'no wire' motors, as they might be called, could be operated by conduction through the rarefied air at considerable distances. Alternate currents, especially of high frequencies, pass with astonishing freedom trough even slightly rarefied gases. The upper strata of the air are of difficulties of a merely mechanical nature. There is no doubt that with the enormous potential is obtainable by the use of high frequencies and oil insulation, luminous discharges might be passed through many miles of rarefied air, and that by thus directing the energy of many hundreds of thousands of horsepower, motors or lamps might be operated at considerable distances from stationary sources. But such schemes are mentioned merely as possibilities. We shall have no need to transmit powers in this way. We shall have no need to transmit powers at all. Ere many generations pass, our machinery will be driven by a power obtainable at any point of the universe... "[38] Tesla demonstrated plasma conduction in a glass tube with rarefied air surrounding a central axial platinum electrode, he observed that the wire was heated only at the ends, and not in the middle. He also observed that the pressure at which the gas becomes conducting is directly related to the applied voltage.

Colorado Springs Laboratory

Tesla moved to Colorado Springs in May 1899, after reaching the limits of his New York Laboratory with Tesla Coils operating at 4 million volts. The dry, electrostatic filled air at the 2000 m facility in Colorado Springs facilitated his developments. His primary and secondary coils were 51 ft. in diameter, and it was here that he developed the concept of an extra coil placed in series with the secondary but with loose inductive coupling so as to enable large resonant amplification of voltage. In addition to the development and improvement of the high power Tesla coil, Tesla concentrated on the development of sensitive receivers necessary for detecting communication signals. On July 3, 1899, using these devices, Tesla monitored the progression of a passing thunderstorm, observing electrical standing waves which he attributed to the storm's disturbance of the earth's electrical charge and a corresponding propagation of this disturbance around the conductive globe. Tesla also experimented with his single electrode x-ray tubes. The oscillator reportedly operated at frequencies between 45KHz and 150 KHz, at voltages between 12 MV and 18MV, and with secondary currents as high as 1100A [1,12,12,39].

Wardenclyffe Laboratory

Funded principally by J.P. Morgan, Tesla proceeded with the construction of a system of "World Telegraphy" at Wardenclyffe on Long Island upon his return from Colorado Springs in 1900 [12,13]. While he intended to use the facility publicly for communications, Tesla's secret aim was to implement wireless power transfer. The facility featured at 187 ft. wooden tower designed to support a 68 ft. diameter copper hemisphere, which was not completed because of Tesla's difficulty in obtaining funding following Marconi's success in demonstrating transoceanic wireless communication with much simpler equipment (albeit using Tesla's patents in the process) The transmitted was to have operated at 30 MV, which Tesla claimed was sufficient for worldwide power distribution; however, the transmitter was designed to handle up to 100MV. Aside from its toroidal elevated capacitor, patent 1,119,732 [40] filed in 1902 shows the Wardenclyffe configuration of the transmitter, which incorporated the 'extra coil" from the Colorado Springs experiments.

Tesla's Concept of Worldwide Wireless Power Transfer

Tesla outlined the requirements for wireless power distribution in patent 787,412, describing the earth as "....behaving like a perfectly smooth or polished conductor of inappreciable resistance with capacity and self-induction uniformly distributed along the axis of symmetry of wave propagation"[41]. He described reflections of signals from antipodes, the points on the globe diametrically opposite from the transmitter, as being similar to those from the end of a conducting wire, thus creating stationary waves on the conductive surface. He provided three requirements for resonance: 1) the earth's diameter should be equal to an odd number of quarter wavelengths, 2) the frequency should be less than 20 KHz to minimize Hertzian radiation; and 3) most critical, the wave train should continue for a minimal period, which he estimates to be 1/12 second, and which represents the period of time for a wave to propagate from and return to the source at a mean speed of 471,240 Km/sec. Tesla conceived the wave as propagating through the earth along a straight line path, the effect on the outside surface being that of concentric rings expanding to the equator and then contracting until reaching the opposite pole. Tesla also applied a fluid analogy to the earth and the water level representing the earth's state of charge at any given point. While his earlier work emphasized ground currents as the mechanism for transferring power, he later indicated that he had conclusively demonstrated that "...with two terminals maintained at an elevation of not more than thirty thousand to thirty five thousand feet above sea level, and with an electrical pressure of fifteen to twenty million volts, the energy of thousand of horse-power can be transmitted over distances which may be hundreds, and, if necessary, thousands of miles. In am hopeful , however that I may be able to reduce very considerably the elevation of the terminals now required..."[42].

Summary of Tesla's Proof of Concept

Tesla claimed to have observed the effects of the Colorado Springs transmitter at a distance of up to 600 miles. An advertising brochure for the World Telegraphy system claims the transmission of power around the globe in sufficient quantity to light incandescent lamps (50watts). Others report that a bank of 200 watt lamps, 50 watts each, were lit at a distance of 26 miles [12, 13]. The article in Century magazine shows photographs of an isolated extra coil powering and incandescent lamp as evidence of "...electrical vibrations transmitted to it through the ground from the oscillator..." [43]. However, this extra coil was most likely within the inductive field of primary transmitter, with the ground serving as a return path.

Rationale for a Renewed Interest in Wireless Power Transfer

Given Tesla's firm and unending belief in the feasibility of wireless power transfer, yet his inability, after considerable expenditure of time and money, to conclusively demonstrate its viability, the reader may question why there is a renewed interest in demonstrating the feasibility of wireless power transfer. Aside from the benefits outlined initially, the best reason probably lies in both 1) the legacy of Tesla himself, and 2) the fact that because of insufficient funding, Tesla was never able to teat a facility that had been developed strictly for power transfer, and thus hi wireless power transfer concept remains to be proven.

The legacy of Tesla speaks for itself in terms of his many and varied significant inventions, his insightful pioneering understanding of physics and electrical engineering, his tremendous drive and creative energy enabling him to constructively, work long hours on a protracted basis guided by a keen sense of vision, his ability to visualize and test concepts in his mind enabling him to achieve good results with little trail and error, and his genuine concern for improving the condition of humanity. The breath of his accomplishments at Colorado Springs with less than 8 months exemplifies these. The Colorado Springs experiments focused primarily in the development of wireless communications, i.e., radio rather than wireless power transfer. As indicated by Marincic [18], 56% of his time was spent on developing the Tesla Coil, 21% on receivers for small signals, 16% on measuring the capacity of the vertical antenna, and 6% on miscellaneous other research, including fireballs. Wireless power transmission experiments were limited to small distances.

While Tesla shared much with the world in the form of his patents, publications, lectures, he was also a very secretive person, and never fully documented his intended configuration for the wireless power system, even though he was confident there would be a workable solution. He believed that that his Magnifying Transmitter (Tesla Coil w/extra coil designed to excite the earth) would ultimately be recognized as his greatest invention [11], and felt that there would be no problem in wireless disturbing the earth's energy. He also believed the universe to be so full of energy that, ultimately, wireless distribution would not be necessary. Modern day researchers attempting to follow his path, must also be part detective. Tesla's belief and confidence in wireless power transfer is clear, however, so too was Edison's belief in magnetic ore separation, which, like Tesla's experience with Wardenclyffe, left him in deep financial debt. [44]

Recent Developments

In recent years, there has been a renewed interest in Tesla's work on high voltage, high frequency phenomena. Beginning in 1968, R. Golka formed Project Tesla to measure, under Air Force Contract , aircraft susceptibility to lighting discharge and to repeat Telsa's ball lighting experiments for application to laser fusion. In the process, he replicated Tesla's Colorado Springs transmitter and succeeded in operating it at twice Tesla's original power levels [25,26]. In 1986, Golka and Grotz proposed the application of this device to artificially resonating the earth-ionosphere waveguide [45].

Cheney reports on wireless power projects that had been planned and some partially implemented circa 1977-1980 in Canada, Central Minnesota and Southern California. [13]

Wait indicated how Tesla's early wireless experiments were the forerunner of modern developments in ELF. He observed that Tesla's fluid analogy for the process is faulty in its assumption that all of the signal energy would propagate through the fluid medium, i.e. the earth. Also faulty was Tesla's notion that energy propages to the antipode via the center of the earth, although it is not known if Tesla had viewed this as a conceptual model as opposed to a physical model as presently interpreted.

Marincic, in his annotations of Tesla's Colorado Springs Notes [1,2] and his excellent review of Tesla's wireless work [18] applies results from recent ELF experimental data to show that the transfer of power via ELF radiation would be extremely inefficient. He indicates that for a typical gridded ELF antenna, 106 m. total length, that the antenna operating efficiency would be only 0.026% and for both receiving and transmitting antennas, the total efficiency would be (0.026%), not to mention the path loses, which are as low as 0.25 dB/Mm at 10Hz and 0.8dB/Mm at 50Hz. For a fixed size antenna, efficiency increases with operating frequency, but so do path losses, so that for long distance power transfer, the overall efficiency of a radiation-based system will be low.

Corum and Corum [27-31] also replicated some of Tesla's Colorado Springs fireball experiments but with much smaller scale equipment. This work extended to a critical engineering evaluation of Tesla's wireless power concept.[20-23], showing how the current moment in the tower of Tesla's transmitter could be used to excite the Shummann resonances in the earth-ionosphere cavity. They also hypothesized that Tesla intended to use his single electrode x-ray to both ionize a current path to the sphere of elevated capacitance and to rectify the RF energy enabling the sphere to be electrostatically charged at RF rates [20,21] The sphere would then be discharged to ground, either naturally or via a second x-ray device, at a Schumman resonance frequency. Corum and Corum have also verified that Tesla's electrical measurements such as the attenuation constant, phase velocity, cavity resonant frequency and Q are consistent with modern measurements [23] and that the loses due to glow discharge around the transmitter would be small [21].

J. F. Corum patented a toroidal helical antenna [46,47] one of whose applications could be a waveguide probe for either ELF communications or wireless power transfer. This antenna is physically small while reportedly possessing good radiating efficiencies with vertical polarization. Since the propagating Schumman modes are primarily vertically polarized, a vertically polarized antenna would have a distinct advantage over the horizontally polarized example presented by Marincic. However, in applying Corum's design formula to the 8Hz example presented in his patent, one finds that an antenna with a 6 Km major radius (0.0002 free space wavelengths) would require a virtual continuum of 43, 200 semicircular loops each 600 m in diameter, with a total conductor length equivalent to half the circumference of the earth.

The Q of the earth-ionosphere cavity is generally reported to be about 6-8 but Corum and Spaniol [48] indicate that a low Q cavity does not necessarily limit the practicability of wireless power. However, Sutton and Spaniol [49] found that the previously measured Q values were limited by instrument noise and using modern equipment they measured levels as high as 1000, which they say were also confirmed by others. [50].

In 1986-1988, Nash, Smith, Craven and Corum of WVU utilized a ¼ wave coaxial resonator to develop a high frequency "Tesla Coil" and proposed coupling this device to a Tesla single electrode x-ray tube to generate ionizing radiation with possible application to wireless power transfer [53].

THE KEY ELEMENTS OF WORLDWIDE WIRELESS POWER TRANSFER

The key elements of worldwide wireless power transfer consist of:
1. source/transmitter
2. path
3. receiver
4. system considerations
5. environmental impact
6. economic viability

Each of these will now be explained in more detail, along with their subgroups.

Source/Transmitter

The source/transmitter, consisting of Tesla's Magnifying Transmitter is the most highly developed elements of the system, as evidenced by the standard terminology of "wireless power transmission". In this paper, the term "transfer" emphasizes the importance of other system elements as well. The Tesla Coil is remarkable efficient power processing element, and Corum and Corum have shown that Tesla's Colorado Springs Transmitter operated at power levels high by even modern standards, with peak average power levels some four orders or magnitude higher that those of the Stanford Linear Accelerator. [21]

Earth-Electrosphere/Ionosphere Cavity with Dielectric Atmosphere

The path comprises the earth (ground) and the atmosphere. The ground is a good conductor at lower frequencies, conductivity decreasing with frequency due to the skin effect. The lower atmosphere is normally a good insulator. At higher altitudes the air becomes conductive due to ionization casued by cosmic rays. The conductive layer, termed the electrosphere, [54] provides an electrostatic shield and an equipotential surface due to its high conductivity relative to the ambient currents. Lord Kelvin, in 1860 [55] originally postulated the existence of such a conductive layer based upon the fact that rarefied gases act as good conductors, and he thus postulated that this conductive layer together with the earth and intervening insulating atmosphere forms a capacitor. The potential of the electrosphere is about 300KV. The ionosphere, located above the electrosphere, is caused by ionizing solar radiation, different ionospheric layers (D,E,F) being attributed to different components of the radiation. The ionosphere is that part of the earth's atmosphere which reflects radio waves [54,56] . The properties of the path are normally measured under conditions (voltage, current, frequency) quite different from those expected for wireless power transfer, and this should be considered before drawing conclusions on the suitability of the path for such purposes. Also, the effects of weather on conductivity and the effects of magnetic storms must be considered.

Spherical Cavity Modes

The spherical cavity between the ground and the ionosphere resonates at specific modes as predicted by Schumman [57,58] and discussed by Wait[59] and Galejs[60]. The transverse electric field mode (TE) is cutoff below 1.5 KHz, so for the ELF frequencies normally considered for wireless power transfer, the cavity will only support transverse magnetic TM waves, [61]. The first seven Schumann resonances are naturally excited by lightning and this fact has been used to track lightning strikes around the globe. [61-67]. The polarization and ellipticity of the waves vary diurnally. Waves propagating in the cavity are attenuated with distance due to the finite conductiveness of the conductive and dialectric layers, and the attenuation increase exponentially with frequency, increasing from 0.25 dB/Mm at 10 Hz to 20 dB/Mm at 1 KHz. (compared with 1.15 dB/Mm for a conventional 200KV power line [24]. Tesla has indicated that very little power is required to maintain a state of resonance in the cavity [21].

Waveguide Coupling

The key issue in wireless power transfer is how to couple power into and out of the cavity with minimal, or at least acceptable loss. Corum and Corum have indicated that Tesla more likely created the necessary current moments to excite the cavity by electrostatically charging an isolated capacitance at RF rates via a single electrode x-ray tube and then suddenly discharging this capacitance at a resonant frequency of the cavity [20-21]. They reported that the currents measured by Tesla would have been sufficient to generate relatively weak ELF global field strengths . Tesla noted that the discharge tended to pass upward away from ground, which he attributed to either electrostatic repulsion, or convection of the heated air. However, with such an electrically short tower, radiation into the cavity at cavity resonant frequencies would not be suffiently efficient for technical or commercial viability. And while a resonating cavity would have purely reactive fields, and hence zero point radiation resistance together with non-stationary fields would be required for power transfer within the cavity. A radiative coupling approach appears to be infeasible for reasons stated above by Marincic.

Transmission Line Coupling

A second method for coupling power into the cavity would be via direct conduction/displacement with the conductive surfaces of the waveguide, which appears to be Tesla's original concept dating back to 1892. Several mechanisms could be considered as follows: 1) Recall that, in 1900, he proposed using balloons at 30-35 thousand feet of elevation. Conceivably the power could be conducted to these via an ionization path, created by a single electrode x-ray tube driven by the transmitter. 2) The conducting path formed by ionizing radiation might be used to couple directly into the electro sphere without the elevated conductive sphere. 3) An approach might also be borrowed from those used in present ionospheric modifications experiments [68]. 4) Perhaps with the extremely high operating voltages that Tesla had proposed, the displacement coupling with the atmospheric conduction path would be direct, as apparent from an artist's rendition of wireless power distribution from Tesla's Wardenclyffe facility [69]. Tesla originally indicated that the atmosphere could be made conductive at lower elevations with either high voltage or high frequency so this should be studied further. . With such a direct coupling approach, the power transfer mechanism would then be a spherical "transmission line", rather than a spherical wave guide.

Ground Currents

The ground currents in Tesla's Colorado Springs experiments were reported to have caused sparks within the ground, and to have shocked horses through their metal shoes within ½ mile from his transmitter. [70]. As an aside, ground currents were separately exploited for communications during WW I, when conversations over the then prevalent single wire telephone systems were susceptible to enemy interception by differentially amplifying the signals extracted from two separate and displaced ground plates. The phenomenon of magnetospheric plasma whistler waves was first noticed with these receivers, but was not identified until later [71].

Power Loss

Power loss can occur in all elements of the path, which have finite conductivity: the ground, the dielectric lower atmosphere, and the conductive upper atmosphere. Elaborate and extensive ground planes are often constructed with antenna systems in order to minimize resistive power loss to the ground. Since the ground is an intrinsic conductive element, losses are inevitable, but can be reduced by operating at lower frequencies and/or establishing distributed area contacts at the transmitter and receiver sites. The poor conductivity of the Colorado Springs soil appears to have caused Tesla some difficulty[1]. At Wardenclyffe, Tesla was planning to use saltwater filled with viaducts under the transmitter to establish a good ground connection. Similar to the ground, atmospheric losses can be reduced by operating at lower frequencies. This appears to conflict with Tesla's notion that gases conduct better at high frequencies, but could be explained by higher dialectric losses. One important feature to the wireless system is the possibility of storing power in the resonating fields within the earth-ionosphere cavity, however, the feasibility of doing this will be dependent upon the Q of the cavity and upon the relative amount of excess power being stored therein. As Tesla had indicated, the power losses are reduced with higher operating voltage since power would then be distributed at lower current levels. Precipitation can dramatically change the conductivity of the atmosphere, and the effects of this on power coupling need to be considered further.

Receiver

The receiver is the least understood element of the system, and one that is most crucial to the system's success. For system using a radiative coupling mechanism, an antenna's efficiency and size both benefit from higher operating frequencies which, as noted above, increased the system's path losses. A transmission line approach would require conductive/displacement coupling into the electrosphere, which requires invention and development.

Tesla expressed confidence in being able to extract power for both individual and home use as well as for powering ground and air transportation vehicles, as illustrated in an artist's rendition [69]. He indicated in patent 649,621: "Obviously the receiving coils, transformer, or other apparatus may be movable – at, for instance, when they are carried by a vessel floating in the air or by a ship at sea. In the former case the connection of one terminal of the receiving apparatus to the ground might not be permanent, but might be intermittently or inductively established without departing from the spirit of my invention. IT is to be noted here that the phenomenon here involved in the transmission of electrical energy is one of true conduction and is not to be confounded with the phenomenon of electrical radiation which have heretofore been observed and which from the very nature and mode of propagation would render practically impossible the transmission of any appreciable amount of energy to such distances as are of practical importance [36].

Tesla separately described the utilization of energy from ionized air, in connection with his description of the art of telautomatics; "Most generally I employed receiving circuits in the form of loops, including condensers, because the discharges of my high-tension transmitter, ionized the air in the hall so that even a very small aerial would draw electricity from the surrounding atmosphere for hours. Just to give an idea, I found for instance, that a bulb 12 inches in diameter, highly exhausted, and with one single terminal to which a short wire was attached, would deliver well on to one thousand successive flashes before all charge of the air in the laboratory was neutralized…" [72]

Systems Considerations

A wireless system would entail a multiplicity of transmitters and receivers each coupling into a common propagation and storage cavity, each requiring proper phasing and balance.

Safety

A wireless power system would expose the entire biosphere to ELF fields of varying intensity. The 78 Hz Seafarer/Sanguine/ELF submarine communication system provoked health concerns, as do high-tension power lines. The fields of wireless and wire-based power transmission systems need to be compared for equivalent power levels. There is much speculation about the adverse effects of magnetic fields on health. However, recent reports from PACE indicate that ELF energy at the lower Schumman resonance frequencies constitute a natural biological clock [71]. The first four Schumman resonances frequencies are within the range of brain wave activity. The fundamental mode is coincident with the theta wave spectrum, which ranges from 4 to 8 Hz, and is attributed to a normally unconscious state with enhanced mental imagery and a high level of creativity.[72] The next three Schumman modes are coincident with the beta wave spectrum which ranges from 13 to 26 Hz, and is associated with the normal conscious state.

Environmental Impact

Operating at high voltages and surrounded by a glow discharge, the transmitter could be a source of pollutants, including ozone, NO and nitric acid, as reported by Tesla during his experiments and steps would have to be taken to mitigate any such hazards if they exist.

Electromagnetic Interference (EMI) and Radio Frequency Interference (RFI)

The operating frequencies of a wireless system could be expected to be low enough so as to not interfer with present communications of electronic systems. The FCC does not make frequency allocations below 9Khz and Tesla had predicted the operating frequency to be below 20 Khz. Circuit interrupters in conventional Tesla coils could be expected to create a significant amount of wide-band EMI; however, modern transmitters could be expected to utilize more advanced switching devices which, together with shielding, could minimize radiated EMI/RFI. The glow discharge surrounding the high transmitter could also be a source of EMI/RFI.

Weather Modification

Since the potential of the electrosphere is about 300 KV relative to the earth, and the wireless system as proposed by Tesla was designed to operate at 30-100MV, there is a significant potential for electrically disturbing the atmosphere. It is not know whether this would be beneficial or harmful. Vonnegut [75] has suggested that the destructive effects of tornadoes may result from atmospheric electrical effects; however, Wilkins [76] concluded from laboratory model vortex experiments that the electrical effects were the effect, rather than the cause, of tornadoes.

Economic Viability

Given technical feasibility and safety, the wireless power transfer system must still be economically viable in order to succeed. Multiple transmitter could conceivably be phased to control the location of antinodes form which power could be extracted, however, this could be at best, a short term solution, unless wireless is constrained to a relatively few large scale facilities that will be expensive and technically difficult to construct. The worldwide regulation and control of wireless power distribution will be difficult if physically constrained to operate at selected resonant frequencies.

CONCLUSION

Times have changed since Tesla's initial investigations of wireless power. Tesla originally envisioned a distributed network of relatively low level suppliers and users of wireless power, and thought it would benefit remote users the most, although he also envisioned large scale power distribution. Our power needs have dramatically increased over the past 100 years, as have their complexity. Tesla expressed great confidence in the viability of wireless power distribution, yet was unable to see its fruition after nearly 50 years of effort. The fulfillment of his vision was undoubtedly impeded by limitation on funds and resources. Tesla demonstrated that the earth can be electrically resonated. The key challenge to feasible worldwide wireless power distribution is whether a means can be found for efficiently coupling power into and out of the cavity formed by the earth, the atmosphere, and the electrosphere/ionosphere. Radiative coupling does not appear to be viable . A conductive approach is proposed which is consistent with Tesla's original wireless concepts; this requires, however, further invention and development. The receiver is the element requiring the most development to make wireless power transfer feasible.

REFERENCES

1. Tesla, N Colorado Springs Notes 1899-1900 with commentaries by A Marincic , 1978 Nolti. Yugoslavia.
2. Ratzlaff, J.T., and Jost, F.A. Dr. Nikola Tesla (I English Serbo-Croatian Diary Comparisons, II Serbo-Croatian Diary Commentary by A. Marincic III Tesla/Sherff Colorado Springs Correspondence 1899-1900), 1979 Tesla Book Co., Millbrae, CA.
3. Ratzlaff, J.T. Dr Nikola Tesla, Complete Patents, 1983 Tesla Book Co. Millbare CA
4. Nikola Tesla Museum Nikola Tesla Lectures, Patents, Articles, 1956, Nolit, Beograd, 1973, Health Research, Mokelume Hill Ca.
5. Martin, T.C. The inventions Researches and Writings of Nikola Tesla , 1894, The Electrical Engineer , New York, 1986, Angriff Press, Hollywood Ca.
6. Tesla, N. Lecture delivered before the American Institute of Electrical Engineers at Columbia College, NY May 20, 1891 Reference [5] pp145-197 Reference 4 pp L-15-L47
7. Tesla, N. Lecture delivered before the Institution of Electrical Engineers in London February 1892, Reference 5 pp;.198-293. Reference 4 pp. L-48.
8. Tesla, N, Lecture delivered before the Franklin Institute Philadelphia Pa February 1893 and before the National Electric Light Association St. Louis March 1893, Reference 5 pp 294-373 Reference 4 pp L-107-L-155
9. Tesla, N "The Problem of Increasing Human Energy" The Century Illustrated Monthly Magazine, June 1900 also in Reference 4 pp A-109-A-152
10. Tesla, N. "The Transmission of Electric Energy without Wires", Electrical World and Engineering, March 5, 1904, also in Reference [4] pp. A-153-A-161.
11. Tesla, N., "My Inventions: The Autobiography of Nikola Tesla,", (introduction by B. Johnston), 1919, Hart Brothers, Williston, Vermont, 1982.
12. O'Neal, J.J, "Prodigal Genius" The Life of Nikola Tesla. David McKay Co., New York, 1955, also Angriff Press, Hollywood, Ca.
13. Cheney, M, "Tesla: Man out of Time, Dell New York, 1981.
14. Friedlander, G.D., "Tesla Eccentric Genius", IEEE Spectrum June 1972 pp.26-29
15. Trinkas, G. Tesla: The lost Inventions, High Voltage Press, Portland Oregon, 1988
16. Wait, J.R. "Historical Background and Introduction to the Special Issue on Extremely Low Frequency (ELF) Communication" IEEE Transactions on Communications. Vol. COM_22 No. 4, April 1974, pp.353-354.
17. Wait, J.R. "Propagation of ELF Electromagnetic Waves and Project Sanguine/Seafarer, IEEE J. Oceanic Engr. Vol OE-2 No.2 April 1977, pp. 161-172
18. Marincic, A.S. "Nikola Tesla and the Wireless Transmission of Energy" IEEE Trans.on Power Apparattus and Systems Vol. PAS-101, No.10 October 1982, 4054-4068

19. Corum, J.F., and Corum K.L., "A Technical Analysis of the Extra Coil as a slow wave Helical Resonator". Proceedings of the Second International Tesla Symposium, Colorado Springs, Colorado, 1986.
20. Corum, J.F. and Corum, K.L. "A Technical Analysis of the Extra Coil as a Slow Wave Helical Resonator", Proceedings of the Second International Tesla Symposium, Colorado Springs, Colorado, 1986.
21. Corum, JF. And Corum K.L., "Critical Speculations Concerning Tesla's Inventions and Applications of Single Electrode X-Ray Directed Discharges for Power Processing, Terrestrial Resonances and Particle Beam weapons." Proceedings of the Second International Tesla Symposium, Colorado Springs, Colorado, 1986.
22. Corum, J.F. and Aidinejad A, "The transient Propagation of ELF Pulses in the Earth-ionosphere Cavity". Proceedings of the Second International Tesla Symposium, Colorado Springs, Colorado, 1986.
23. Corum J.F. and Corum K.L., "A Physical Interpretation of the Colorado Springs Data" Proceedings of the Second International Tesla Symposium, Colorado Springs, Colorado, 1986.
24. Corum J. F. and Smith, J.E., "Distribution of Electrical Power by Means of Terrestrial Cavity Resonator Modes" Proposal submitted to Planetary Association for Clean Energy Inc, December 1986.
25. Golka R.K. " Long Arc Simulated Lightning Attachment Testing using a 150 Kwh. Tesla Coil (unknown publication status),
26. Golka, R.K. "Project Tesla" Radio, Electronics, February 1981, 48-49, also see Reference [13] pp 282-284
27. Corum, J.F., and Corum K.L. "Laboratory Generation of Electric Fire Balls" (unknown publication status).
28. Corum J.F. & Corum K.L. "The laboratory Production of Electric Fire Balls" (unknown publication status).
29. Corum, J. F & Corum, K.L "Production of Electric Fire Balls" (unknown publication status).
30. Corum, J.F., Edwards J.D. and Corum K.L, "Further experiments with Electric Fire balls (unknown publication status).
31. Michrowski, A., "Laboratory Generation of Electric Fireballs" Planetary Association for Clean Energy Newsletter" Vol.6, No.1 July 1990, pp.21-22
32. Glasser, P.E. Solar Power from Satellites" Physics Today, February 1977, pp.30-37, summarized in Reference [13] pp. 284-285
33. Tesla, N. Patent 462-418, Method and Apparatus for Electrical Conversion and Distribution " Application filed on February 14, 1891, Reference [3] p. 211, Reference [4], p; P-221
34. Reference [8]; Reference [5] p. 349
35. Tesla, N. Patent 593,138, "Electrical Transformer" Application filed on March 20, 1897, Reference [3] p. 301 Reference [4] p. P-252
36. Tesla N. Patent 649, 621, " Apparatus for Transmission of Electrical Energy". Application filed on September 2, 1897, Reference [3] p. 311
37. Tesla N. Patent 645,576 "System of Transmission of Electrical Energy" Application filed on September 2, 1897 Reference [3] p. 311.
38. Reference [7] Reference [5] p. 235
39. Tesla N. "Possibilities of Electro-Static Generators" Scientific American March 1934, 115, 132-134, 163-165 April 1934, 205.

40. Tesla N. Patent 1,119,732 "Apparatus for Transmitting Electrical Energy", application filed on September 2, 1897, Reference [3] p. 397, Reference [4] p. P-331
41. Tesla N Patent 787,412, " Art of Transmitting Electrical Energy through the Natural Media" Application field on May 16, 1900 Reference [3] p. 435 Reference [4] p. P-357
42. Reference [9] Reference [4] p. A-150
43. Reference [9], Reference [4] p. A-123
44. Peterson, M "Thomas Edison Failure" Inventions and Technology, Vol, 6 NO. 3 Winter 1991, pp.8-14
45. Golka R.K and Grotz, Toby "Proposal: Project Tesla The demonstration of Artificial Resonating the Earth's Ionosphere Waveguide, a precursor for the wireless transmission of vast amounts of electrical power using the Earth's Schumman's Cavity" October 28, 1986
46. Corum J.F. Patent 4,622,558 "Toroidal Antenna" Application filed on November 7, 1985
47. Corum J.F. Patent 4,751,515, "Electromagnetic Structure and Method" Application field on July 23, 1986
48. Corum J.F. Corum K.L and Spaniol C. "Concerning Cavity Q" Proceedings of the International Tesla Symposium 1988, summarized in Reference [49]
49. Sutton, J.F. and Spaniol C. "A Measurement of the Magnetic Earth-Ionosphere Cavity Resonances in the 3-30 Hz range", Presented at the International Tesla Symposium 1988.
50. Personal correspondence
51. Michrowski, A. The Planetary Association for Clean Energy Newsletter Vol. 5 Nos 3 &4 December 1987, p. 6
52. Reference [31] p.3
53. Nash, M Smith J.E, and Craven R P,M. "A Quarter-Wave Coaxial Cavity as a Power Processing Plant" p. 285 in Michrowski A New Energy Technology, Planetary Association for Clean Energy. 1990.
54. Chalmers, J.A., Atmospheric Electricity, Pergamon Press, N.Y. pp. 13, 33-35
55. Kelvin, Lord "Atmospheric Electricity" Royal Institute Lectures, Pap or Elec and Mag. Pp.208-226, summarized in Reference [54]
56. Davies, K, Ionospheric Radio Propagation, US Dept of Commerce, NBS, Monograph, 80, 1965
57. Schumman W.O. "On the radiation free selfoscillations of a conducting sphere which is surrounded by an air layer and an ionospheric shell" (in German) Z Naturfosch, 72, 1952, 145-154, summarized in Reference [59,60,61]
58. Schumman W.O. "On the damping of electromagnetic selfoscillations of the system earth-air-ionosphere (in German), Z. Naturforsch , 72, 1952, 250-252 summarized in references [59,60,61]
59. Wait, J.R. Electromagnetic Waves in Statified Media, Pergamon Press, New York 1970
60. Galejs, J. Terrestrial Propagations of Long Electromagnetic Waves, Pergamon Press, New York, 1972
61. Sentman D.D. "Magnetic elliptical polarization of Schumman resonances" Radio Science, Vol. 22, No. 4, July-August 1987, pp.595-606
62. Sentman D.D. "PC Monitors Lightning Worldwide" Computers in Science. Premier, 1987, page 25-34

63. Coroniti S. and Huges, J. Planetary Electrodynamics Vol. 2 Gordon and Breach, New York, 1969, summarized in Reference [61]

64. Jones, D.L. and Kemp, D.T. "Experimental and Theoretical observations on the transient excitation of Schumann resonances" Journal of Atmospheric and Terrestrial Physics, Vol 32, 1970 pp 1095-1108

65. Kemp, D.T. "The global location of large lightning discharges from single station observations of ELF disturbances in the Earth Ionosphere cavity" Journal of the Atmospheric and Terrestrial Physics Vol.33, 1981, pp. 919-927.

66. Jones, D.L.,"Extremely low frequency (ELF) Ionospheric Radio Propagation Studies Using Natural Sources" IEEE trans on Communications, Vol Com-22 No 4, April 1974, pp. 477-484.

67. Mitchell, V. B. "Schumman resonance – some properties of discrete events" Journal of Atmospheric and Terrestrial Physics, Vol 38, 1976, pp. 77-78

68. Eastlund, B.J and Ramo, S. Patent 4,712,155 "Method and Apparatus of Creating an Artificial Electron Cyclotron Heating Region of Plasma" Application filed on January 28, 1985

69. Reference [11] p. 89

70. Reference:[13] p. 138.

71. Stix, T.H. "Waves in Plasmas: Highlights from the past and present" Phys. Fluids B 2 (8) August 1990, 1729-1743

72. Reference:[11] p.107.

73. Reference: [31] p. 4.

74. Allen W. G. Overlords and Olympians, Health Research, 1974, Mokelume Hill, California, p. 12

75. Vonnegut, B. "Electrical theory of Tornadoes" J Geophys, Res. 65, 1950-203-212 summarized in Reference [54]

76. Wilkins, E.M. "The role of electrical phenomena associated with tornadoes" J. Geophys Res. 69, 1964, 2435-47 summarized in reference 54.

Kurt Van Voorhies holds patents #5,442,369, and #6,239,760 and can be reached at Vortekx, Inc., DeTour Village, MI, vortekx@sault.com

8 On the Transmission of Electricity Without Wires

Nikola Tesla
Reprinted from *Electrical World and Engineer*, March 5, 1904

It is impossible to resist your courteous request extended on an occasion of such moment in the life of your journal. Your letter has vivified the memory of our beginning friendship, of the first imperfect attempts and undeserved successes, of kindnesses and misunderstandings. It has brought painfully to my mind the greatness of early expectations, the quick flight of time, and alas! the smallness of realizations. The following lines which, but for your initiative, might not have been given to the world for a long time yet, are an offering in the friendly spirit of old, and my best wishes for your future success accompany them.

Towards the close of 1898 a systematic research, carried on for a number of years with the object of perfecting a method of transmission of electrical energy through the natural medium, led me to recognize three important necessities: First, to develop a transmitter of great power; second, to perfect means for individualizing and isolating the energy transmitted; and, third, to ascertain the laws of propagation of currents through the earth and the atmosphere. Various reasons, not the least of which was the help proffered by my friend Leonard E. Curtis and the Colorado Springs Electric Company, determined me to select for my experimental investigations the large plateau, two thousand meters above sea-level, in the vicinity of that delightful resort, which I reached late in May, 1899. I had not been there but a few days when I congratulated myself on the happy choice and I began the task, for which I had long trained myself, with a grateful sense and full of inspiring hope. The perfect purity of the air, the unequaled beauty of the sky, the imposing sight of a high mountain range, the quiet and restfulness of the place--all around contributed to make the conditions for scientific observations ideal. To this was added the exhilarating influence of a glorious climate and a singular sharpening of the senses. In those regions the organs undergo perceptible physical changes. The eyes assume an extraordinary limpidity, improving vision; the ears dry out and become more susceptible to sound. Objects can be clearly distinguished there at distances such that I prefer to have them told by someone else, and I have heard--this I can venture to vouch for--the claps of thunder seven and eight hundred kilometers away. I might have done better still, had it not been tedious to wait for the sounds to arrive, in definite intervals, as heralded precisely by an electrical indicating apparatus--nearly an hour before.

In the middle of June, while preparations for other work were going on, I arranged one of my receiving transformers with the view of determining in a novel manner, experimentally, the electric potential of the globe and studying its periodic and casual fluctuations. This formed part of a plan carefully mapped out in advance. A highly sensitive, self-restorative device, controlling a recording instrument, was included in the secondary circuit, while the primary was connected to the ground and an elevated terminal of adjustable capacity. The variations of potential gave rise to electric surgings in the primary; these generated secondary currents, which in turn affected the sensitive device and recorder in proportion to their intensity. The earth was found to be, literally, alive with electrical vibrations, and soon I was deeply absorbed in the interesting investigation. No better opportunities for such observations as I intended to make could be found anywhere. Colorado is a country famous for the natural displays of electric force. In that dry and rarefied atmosphere the sun's rays beat the objects with fierce intensity. I raised steam, to a dangerous pressure, in barrels filled with concentrated salt solution, and the tin-foil coatings of some of my elevated terminals shriveled up in the fiery blaze. An experimental high-tension transformer, carelessly exposed to the rays of the setting sun, had most of its insulating compound melted out and was rendered useless. Aided by the dryness and rarefaction of the air, the water evaporates as in a boiler, and static electricity is developed in abundance. Lightning discharges are, accordingly, very frequent and sometimes of inconceivable violence. On one occasion approximately twelve thousand discharges occurred in two hours, and all in a radius of certainly less than fifty kilometers from the laboratory. Many of them resembled gigantic trees of fire with the trunks up or down. I never saw fire balls, but as compensation for my disappointment I succeeded later in determining the mode of their formation and producing them artificially.

In the latter part of the same month I noticed several times that my instruments were affected stronger by discharges taking place at great distances than by those near by. This puzzled me very much. What was the cause? A number of observations proved that it could not be due to the differences in the intensity of the individual discharges, and I readily ascertained that the phenomenon was not the result of a varying relation between the periods of my receiving circuits and those of the terrestrial disturbances. One night, as I was walking home with an assistant, meditating over these experiences, I was suddenly staggered by a thought. Years ago, when I wrote a chapter of my lecture before the Franklin Institute and the National Electric Light Association, it had presented itself to me, but I dismissed it as absurd and impossible. I banished it again. Nevertheless, my instinct was aroused and somehow I felt that I was nearing a great revelation.

It was on the third of July--the date I shall never forget--when I obtained the first decisive experimental evidence of a truth of overwhelming importance for the advancement of humanity. A dense mass of strongly charged clouds gathered in the west and towards the evening a violent storm broke loose which, after spending much of its fury in the mountains, was driven away with great velocity over the plains. Heavy and long persisting arcs formed almost in regular time intervals. My observations were now greatly facilitated and rendered more accurate by the experiences already gained. I was able to handle my instruments quickly and I was prepared. The recording apparatus being properly adjusted, its indications became fainter and fainter with the increasing distance of the storm, until they ceased altogether. I was watching in eager expectation. Surely enough, in a little while the indications again began, grew stronger and stronger and, after passing through a maximum, gradually decreased and ceased once more. Many times, in regularly recurring intervals, the same actions were repeated until the storm which, as evident from simple computations, was moving with nearly constant speed, had retreated to a distance of about three hundred kilometers. Nor did these strange actions stop then, but continued to manifest themselves with undiminished force. Subsequently, similar observations were also made by my assistant, Mr. Fritz Lowenstein, and shortly afterward several admirable opportunities presented themselves which brought out, still more forcibly, and unmistakably, the true nature of the wonderful phenomenon. No doubt, whatever remained: I was observing stationary waves.

As the source of disturbances moved away the receiving circuit came successively upon their nodes and loops. Impossible as it seemed, this planet, despite its vast extent, behaved like a conductor of limited dimensions. The tremendous significance of this fact in the transmission of energy by my system had already become quite clear to me. Not only was it practicable to send telegraphic messages to any distance without wires, as I recognized long ago, but also to impress upon the entire globe the faint modulations of the human voice, far more still, to transmit power, in unlimited amounts, to any terrestrial distance and almost without loss.

With these stupendous possibilities in sight, and the experimental evidence before me that their realization was henceforth merely a question of expert knowledge, patience and skill, I attacked vigorously the development of my magnifying transmitter, now, however, not so much with the original intention of producing one of great power, as with the object of learning how to construct the best one. This is, essentially, a circuit of very high self-induction and small resistance which in its arrangement, mode of excitation and action, may be said to be the diametrical opposite of a transmitting circuit typical of telegraphy by Hertzian or electromagnetic radiations. It is difficult to form an adequate idea of the marvelous power of this unique appliance, by the aid of which the globe will be transformed. The electromagnetic radiations being reduced to an insignificant quantity, and proper conditions of resonance maintained, the circuit acts like an immense pendulum, storing indefinitely the energy of the primary exciting impulses and impressions upon the earth of the primary exciting impulses and impressions upon the earth and its conducting atmosphere uniform harmonic oscillations of intensities which, as actual tests have shown, may be pushed so far as to surpass those attained in the natural displays of static electricity.

Simultaneously with these endeavors, the means of individualization and isolation were gradually improved. Great importance was attached to this, for it was found that simple tuning was not sufficient to meet the vigorous practical requirements. The fundamental idea of employing a number of distinctive elements, co-operatively associated, for the purpose of isolating energy transmitted, I trace directly to my perusal of Spencer's clear and suggestive exposition of the human nerve mechanism. The influence of this principle on the transmission of intelligence, and electrical energy in general, cannot as yet be estimated, for the art is still in the embryonic stage; but many thousands of simultaneous telegraphic and telephonic messages, through one single conducting channel, natural or artificial, and without serious mutual interference, are certainly practicable, while millions are possible. On the other hand, any desired degree of individualization may be secured by the use of a great number of co-operative elements and arbitrary variation of their distinctive features and order of succession. For obvious reasons, the principle will also be valuable in the extension of the distance of transmission.

Progress though of necessity slow was steady and sure, for the objects aimed at were in a direction of my constant study and exercise. It is, therefore, not astonishing that before the end of 1899 I completed the task undertaken and reached the results which I have announced in my article in the Century Magazine of June, 1900, every word of which was carefully weighed.

Much has already been done towards making my system commercially available, in the transmission of energy in small amounts for specific purposes, as well as on an industrial scale. The results attained by me have made my scheme of intelligence transmission, for which the name of "World Telegraphy" has been suggested, easily realizable. It constitutes, I believe, in its principle of operation, means employed and capacities of application, a radical and fruitful departure from what has been done heretofore. I have no doubt that it will prove very efficient in enlightening the masses, particularly in still uncivilized countries and less accessible regions, and that it will add materially to general safety, comfort and convenience, and maintenance of peaceful relations. It involves the employment of a number of plants, all of which are capable of transmitting individualized signals to the uttermost confines of the earth. Each of them will be preferably located near some important center of civilization and the news it receives through any channel will be flashed to all points of the globe. A cheap and simple device, which might be carried in one's pocket, may then be set up somewhere on sea or land, and it will record the world's news or such special messages as may be intended for it. Thus the entire earth will be converted into a huge brain, as it were, capable of response in every one of its parts. Since a single plant of but one hundred horse-power can operate hundreds of millions of instruments, the system will have a virtually infinite working capacity, and it must needs immensely facilitate and cheapen the transmission of intelligence.

The first of these central plants would have been already completed had it not been for unforeseen delays which, fortunately, have nothing to do with its purely technical features. But this loss of time, while vexatious, may, after all, prove to be a blessing in disguise. The best design of which I know has been adopted, and the transmitter will emit a wave complex of total maximum activity of ten million horse-power, one per cent. of which is amply sufficient to "girdle the globe." This enormous rate of energy delivery. approximately twice that of the combined falls of Niagara, is obtainable only by the use of certain artifices, which I shall make known in due course.

For a large part of the work which I have done so far I am indebted to the noble generosity of Mr. J. Pierpont Morgan, which was all the more welcome and stimulating, as it was extended at a time when those, who have since promised most, were the greatest of doubters. I have also to thank my friend, Stanford White, for much unselfish and valuable assistance. This work is now far advanced, and though the results may be tardy, they are sure to come.

Meanwhile, the transmission of energy on an industrial scale is not being neglected. The Canadian Niagara Power Company have offered me a splendid inducement, and next to achieving success for the sake of the art, it will give me the greatest satisfaction to make their concession financially profitable to them. In this first power plant, which I have been designing for a long time, I propose to distribute ten thousand horse-power under a tension of one hundred million volts, which I am now able to produce and handle with safety.

This energy will be collected all over the globe preferably in small amounts, ranging from a fraction of one to a few horse-power. One its chief uses will be the illumination of isolated homes. I takes very little power to light a dwelling with vacuum tubes operated by high-frequency currents and in each instance a terminal a little above the roof will be sufficient. Another valuable application will be the driving of clocks and other such apparatus. These clocks will be exceedingly simple, will require absolutely no attention and will indicate rigorously correct time. The idea of impressing upon the earth American time is fascinating and very likely to become popular. There are innumerable devices of all kinds which are either now employed or can be supplied, and by operating them in this manner I may be able to offer a great convenience to whole world with a plant of no more than ten thousand horse-power. The introduction of this system will give opportunities for invention and manufacture such as have never presented themselves before.

Knowing the far-reaching importance of this first attempt and its effect upon future development, I shall proceed slowly and carefully. Experience has taught me not to assign a term to enterprises the consummation of which is not wholly dependent on my own abilities and exertions. But I am hopeful that these great realizations are not far off, and I know that when this first work is completed they will follow with mathematical certitude.

When the great truth accidentally revealed and experimentally confirmed is fully recognized, that this planet, with all its appalling immensity, is to electric currents virtually no more than a small metal ball and that by this fact many possibilities, each baffling imagination and of incalculable consequence, are rendered absolutely sure of accomplishment; when the first plant is inaugurated and it is shown that a telegraphic message, almost as secret and non-interferable as a thought, can be transmitted to any terrestrial distance, the sound of the human voice, with all its intonations and inflections, faithfully and instantly reproduced at any other point of the globe, the energy of a waterfall made available for supplying light, heat or motive power, anywhere-on sea, or land, or high in the air-humanity will be like an ant heap stirred up with a stick: See the excitement coming!

9   The True Meaning of Wireless Transmission of Power

Toby Grotz
Reprinted from *Tesla: A Journal of Modern Science*, 1997

Abstract

Many researchers have speculated on the meaning of the phrase "non-Hertzian waves" as used by Dr. Nikola Tesla.[1] Dr. Tesla first began to use this term in the mid 1890's in order to explain his proposed system for the wireless transmission of electrical power. In fact, it was not until the distinction between the method that Heinrich Hertz was using and the system Dr. Tesla had designed, that Dr. Tesla was able to receive the endorsement of the renowned physicist, Lord Kelvin.[1] To this day, however, there exists a confusion amongst researchers, experimentalists, popular authors and laymen as to the meaning of non-Hertzian waves and the method Dr. Tesla was promoting for the wireless transmission of power. In this paper, the terms pertinent to wireless transmission of power will be explained and the methods being used by present researchers in a recreation of the Tesla's 1899 Colorado Springs experiments will be defined.

Early Theories of Electromagnetic Propagation

In pre-World War I physics, scientists postulated a number of theories to explain the propagation of electromagnetic energy through the ether. There were three popular theories present in the literature of the late 1800's and early 1900's. They were:

1. Transmission through or along the Earth,
2. Propagation as a result of terrestrial resonances,
3. Coupling to the ionosphere using propagation through electrified gases.

We shall concern our examination at this time to the latter two theories as they were both used by Dr. Tesla at various times to explain his system of wireless transmission of power. It should be noted, however, that the first theory was supported by Fritz Lowenstein, the first vice-president of the Institute of Radio Engineers, a man who had the enviable experience of assisting Dr. Tesla during the Colorado Springs experiments of 1899. Lowenstein presented what came to be known as the "gliding wave" theory of electromagnetic radiation and propagation during a lecture before the IRE in 1915. (Fig. 1)

Dr. Tesla delivered lectures to the Franklin Institute at Philadelphia, in February, 1983, and to the National Electric Light Association in St. Louis, in March, 1983, concerning electromagnetic wave propagation. The theory presented in those lectures proposed that the Earth could be considered as a conducting sphere and that it could support a large electrical

---

[1] An honorary doctorate degree, was awarded Nikola Tesla in June, 1894 from Columbia College in the City of New York (Source: Columbia University Archives) – Ed. note.

charge. Dr. Tesla proposed to disturb the charge distribution on the surface of the Earth and record the period of the resulting oscillations as the charge returned to its state of equilibrium. The problem of a single charged sphere had been analyzed at that time by J.J. Thompson and A.G. Webster in a treatise entitled "The Spherical Oscillator." This was the beginning of an examination of what we may call the science of terrestrial resonances, culminating in the 1950's and 60's with the engineering of VLF radio systems and the research and discoveries of W.O. Schumann and J.R. Waite.

Figure 1

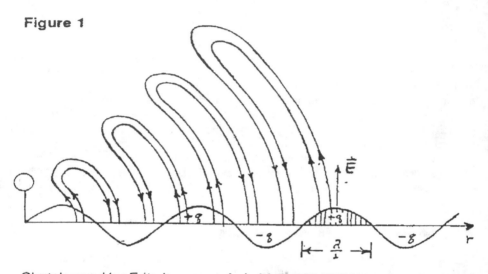

Sketch used by Fritz Lowenstein in his 1915 IRE lecture to explain the mechanism of radiation and propagation for radio waves. "...Q [is] the charge in the antenna and... q the electric charge of each half wave length gliding along the earth..." Even Zenneck was prepared to accept Lowenstein's explanation. Lowenstein believed that charge q was detached from the antenna and floated off along the ground as a "gliding wave."[5]

— Tesla Primer and Handbook, Chapter 2.

The second method of energy propagation proposed by Dr. Tesla was that of the propagation of electrical energy through electrified gases. Dr. Tesla experimented with the use of high frequency RF currents to examine the properties of gases over a wide range of pressures. It was determined by Dr. Tesla that air under a partial vacuum could conduct high frequency electrical currents as well or better than copper wires. If a transmitter could be elevated to a level where the air pressure was on the order of 75 to 130 millimeters in pressure and an excitation of megavolts was applied, it was theorized that; "...the air will serve as a conductor for the current produced, and the latter will be transmitted through the air with, it may be, even less resistance than through an ordinary copper wire". (Fig. 2)

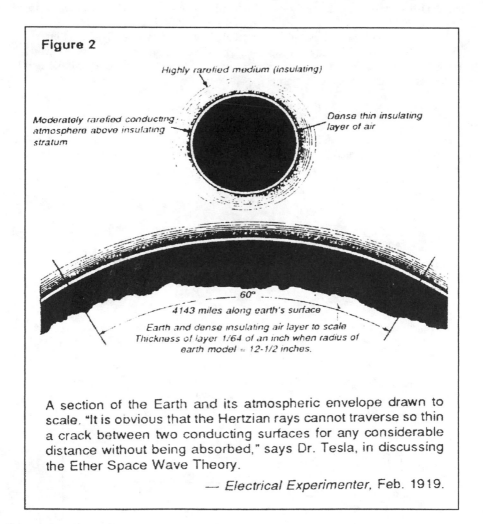

Figure 2

Highly rarefied medium (insulating)

Moderately rarefied conducting atmosphere above insulating stratum

Dense thin insulating layer of air

60°

4143 miles along earth's surface

Earth and dense insulating air layer to scale
Thickness of layer 1/64 of an inch when radius of
earth model = 12-1/2 inches.

A section of the Earth and its atmospheric envelope drawn to scale. "It is obvious that the Hertzian rays cannot traverse so thin a crack between two conducting surfaces for any considerable distance without being absorbed," says Dr. Tesla, in discussing the Ether Space Wave Theory.

— *Electrical Experimenter*, Feb. 1919.

Resonating Planet Earth

Dr. James T. Corum and Kenneth L. Corum, in Chapter 2 of their book, *A Tesla Primer*, point out a number of statements made by Dr. Tesla which indicate that he was using resonator fields and transmission line modes.

1. When he speaks of tuning his apparatus until Hertzian radiations have been eliminated, he is referring to using ELF vibrations: "...the Hertzian effect has gradually been reduced through the lowering of frequency." [3]

2. "...the energy received does not diminish with the square of the distance, as it should, since the Hertzian radiation propagates in a hemisphere."[3]

3. He apparently detected resonator or standing wave modes: "...my discovery of the wonderful law governing the movement of electricity through the globe...the projection of the

wavelengths (measured along the surface) on the earth's diameter or axis of symmetry...are all equal."[3]

4. "We are living on a conducting globe surrounded by a thin layer of insulating air, above which is a rarefied and conducting atmosphere...The Hertz waves represent energy which is radiated and unrecoverable. The current energy, on the other hand, is preserved and can be recovered, theoretically at least, in its entirety."[4]

As Dr. Corum points out, "The last sentence seems to indicate that Tesla's Colorado Springs experiments could be properly interpreted as characteristic of a wave-guide probe in a cavity resonator."[5] This was in fact what led Dr. Tesla to report a measurement which to this day is not understood and has led many to erroneously assume that he was dealing with faster than light velocities.

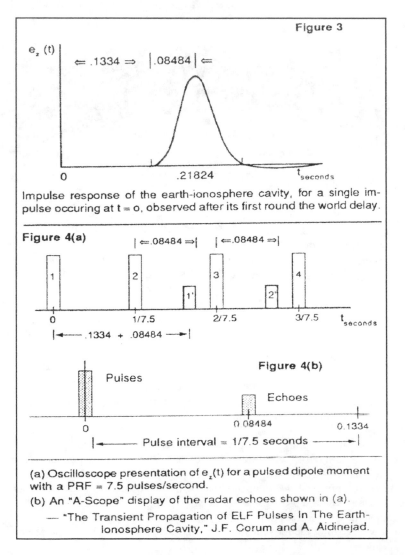

Figure 3

$e_z(t)$

$\Leftarrow .1334 \Rightarrow$ $\;|.08484| \Leftarrow$

0   .21824 $t_{seconds}$

Impulse response of the earth-ionosphere cavity, for a single impulse occuring at t = o, observed after its first round the world delay.

Figure 4(a)

$|\Leftarrow .08484 \Rightarrow|$ $|\Leftarrow .08484 \Rightarrow|$

1   2   1'   3   2'   4

0   1/7.5   2/7.5   3/7.5 $t_{seconds}$

$|\!\leftarrow\!\!-\!.1334 + .08484 \longrightarrow|$

Figure 4(b)

Pulses

Echoes

0   0.08484   0.1334

$|\!\leftarrow\!\!-$ Pulse interval = 1/7.5 seconds $\longrightarrow|$

(a) Oscilloscope presentation of $e_z(t)$ for a pulsed dipole moment with a PRF = 7.5 pulses/second.

(b) An "A-Scope" display of the radar echoes shown in (a).

— "The Transient Propagation of ELF Pulses In The Earth-Ionosphere Cavity," J.F. Corum and A. Aidinejad.

The Controversial Measurement

The mathematical models and experimental data used by Schumann and Waite to describe ELF transmission and propagation are complex and beyond the scope of this paper. Dr. James F. Corum, Kenneth L. Corum and Dr. A-Hamid Aidinejad have, however, in a series of papers presented at the 1984 Tesla Centennial Symposium and the 1986 International Tesla Symposium, applied the experimental values obtained by Dr. Tesla during his Colorado Springs experiments to the models and equations used by Schumann and Waite. The results of this exercise have proved that the Earth and the surrounding atmosphere can be used as a cavity resonator for the wireless transmission of electrical power. (Fig. 3)

Dr. Tesla reported that <u>0.08484 seconds was the time that a pulse emitted from his laboratory took to propagate to the opposite side of the planet and to return</u>. From this statement many have assumed that his transmissions exceeded the speed of light and many esoteric and fallacious theories and publications have been generated. As Corum and Aidinejad point out, in their 1986 paper, "The Transient Propagation of ELF Pulses in the Earth Ionosphere Cavity", this measurement represents the coherence time of the Earth cavity resonator system. This is also known to students of radar systems as a determination of the range dependent parameter. The accompanying diagrams from Corum's and Aidinejad's paper graphically illustrate the point. (Fig. 4 & Fig. 5)

Figure 4

COMPARISON OF PHYSICAL PARAMETERS

Physical Parameter	Accepted Experimental Values	Predicted from Tesla's Disclosures
1. Attenuation Constant (dB/Mm)	$.20 \leq \alpha \leq .30$	.26
2. Resonant Frequency (Hz)	$6.8 \leq f_o < 7.8$	6
3. Cavity Q	$3.8 \leq Q \leq 7.8$	$3.2 \leq Q \leq 6.4$
4. Coherence Time (sec.)	no data available	0.08484
5. Phase Velocity	$.71 \leq V_f \leq .83$	0.8
6. Cavity Mode Structure	$P_n (\cos \theta)$	"Projections of all the stationary nodes onto the earth's diameter are equal."
7. Cavity Thickness (Km)	$35 \leq h \leq 80$	{ "greater than 8 Km" "about 20 Km"

— "A Physical Interpretation of the Colorado Springs Data," Proceedings of the Tesla Centennial Symposium, 1984.

Figure 5

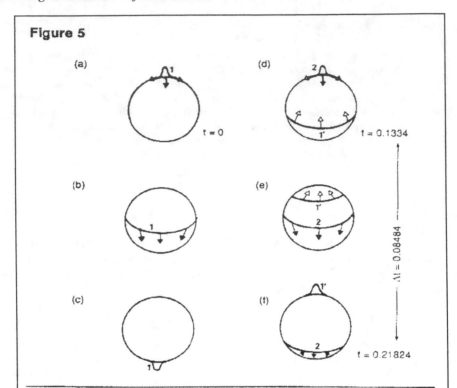

The sequence of events following the emission of pulses at the North Pole.

  (a) The first pulse emitted at t = 0.
  (b) The propagating disturbance 0.0444 seconds later.
  (c) The propagating pulse refocuses at the antipode at t = 0.08737 seconds.
  (d) The pulse energy propagates northward and is at the position shown, now called 1', at t = 1/7.5 seconds. At this same instant pulse #2 is emitted at the source.
  (e) An intermediate time after pulse #2 going southward has passed pulse #1' going northward.
  (f) At t = 0.21824 seconds, pulse #1' is refocused at the North Pole, having completed a terrestrial circuit. This is exactly 0.08484 seconds after pulse #2 was transmitted. Pulse #2 is shown on its southward journey.

One would, of course, employ a PRF equal to the resonant frequency of the cavity for a maximum response. It is tacitly assumed that the resonator system is linear.

—"The Transient Propagation of ELF Pulses In The Earth-Ionosphere Cavity," J .F. Corum and A. Aidinejad. Paper delivered to the 1986 International Tesla Symposium

We now turn to a description of the methods to be used to build, as Dr. Tesla did in 1899, a cavity resonator for the wireless transmission of electrical power.

PROJECT TESLA

The Wireless Transmission of Electrical Energy Using Schumann Resonance

It has been proven that electrical energy can be propagated around the world between the surface of the Earth and the ionosphere at extreme low frequencies in what is known as the Schumann Cavity. The Schumann cavity surrounds the Earth between ground level and extends upward to a maximum 80 kilometers. Experiments to date have shown that electromagnetic waves of extreme low frequencies in the range of 8 Hz, the fundamental Schumann Resonance frequency, propagate with little attenuation around the planet within the Schumann Cavity. Knowing that a resonant cavity can be excited and that power can be delivered to that cavity similar to the methods used in microwave ovens for home use, it should be possible to resonate and deliver power via the Schumann Cavity to any point on Earth. This will result in practical wireless transmission of electrical power.

Background

Although it was not until 1954-1959 when experimental measurements were made of the frequency that is propagated in the resonant cavity surrounding the Earth, recent analysis shows that it was Nikola Tesla who, in 1899, first noticed the existence of stationary waves in the Schumann cavity. Tesla's experimental measurements of the wave length and frequency involved closely match Schumann's theoretical calculations. Some of these observations were made in 1899 while Tesla was monitoring the electromagnetic radiations due to lightning discharges in a thunderstorm which passed over his Colorado Springs laboratory and then moved more than 200 miles eastward across the plains. In his Colorado Springs Notes, Tesla noted that these stationary waves "... can be produced with an oscillator," and added in parenthesis, "This is of immense importance."[6] The importance of his observations is due to the support they lend to the prime objective of the Colorado Springs laboratory. The intent of the experiments and the laboratory Tesla had constructed was to prove that wireless transmission of electrical power was possible.

Schumann Resonance is analogous to pushing a pendulum. The intent of Project Tesla is to create pulses or electrical disturbances that would travel in all directions around the Earth in the thin membrane of non- conductive air between the ground and the ionosphere. The pulses or waves would follow the surface of the Earth in all directions expanding outward to the maximum circumference of the Earth and contracting inward until meeting at a point opposite to that of the transmitter. This point is called the anti-pode. The traveling waves would be reflected back from the anti-pode to the transmitter to be reinforced and sent out again.

At the time of his measurements Tesla was experimenting with and researching methods for "...power transmission and transmission of intelligible messages to any point on the globe." Although Tesla was not able to commercially market a system to transmit power around the globe, modern scientific theory and mathematical calculations support his contention that the wireless propagation of electrical power is possible and a feasible alternative to the extensive and costly grid of electrical transmission lines used today for electrical power distribution.

The Need for a Wireless System of Energy Transmission

A great concern has been voiced in recent years over the extensive use of energy, the limited supply of resources, and the pollution of the environment from the use of present energy conversion systems. Electrical power accounts for much of the energy consumed. Much of this power is wasted during transmission from power plant generators to the consumer. The resistance of the wire used in the electrical grid distribution system causes a

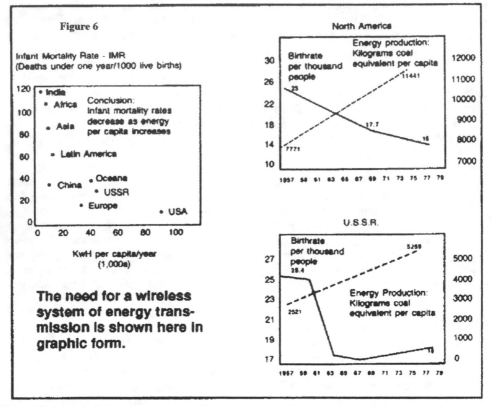

loss of 26-30% of the energy generated. This loss implies that our present system of electrical distribution is only 70-74% efficient (true only in the 1980's; today it is much worse. – Ed. note). A system of power distribution with little or no loss would conserve energy. It would reduce pollution and expenses resulting from the need to generate power to overcome and compensate for losses in the present grid system.

The proposed project would demonstrate a method of energy distribution calculated to be 90-94% efficient. An electrical distribution system, based on this method would eliminate the need for an inefficient, costly, and capital intensive grid of cables, towers, and substations. The system would reduce the cost of electrical energy used by the consumer and rid the landscape of wires, cables, and transmission towers.

There are areas of the world where the need for electrical power exists, yet there is no method for delivering power. Africa is in need of power to run pumps to tap into the vast resources of water under the Sahara Desert. Rural areas, such as those in China, require the electrical power necessary to bring them into the 20th century and to equal standing with western nations. As first proposed by Buckminster Fuller, wireless transmission of power would enable world wide distribution of off peak demand capacity. This concept is based on the fact that some nations, especially the United States, have the capacity to generate much more power than is needed. This situation is accentuated at night. The greatest amount of power used, the peak demand, is during the day. The extra power available during the night could be sold to the side of the planet where it is day time. Considering the huge capacity of power plants in the United States, this system would provide a saleable product which could do much to aid our balance of payments.

MARKET ANALYSIS

Of the 56 billion dollars spent for research by the the U.S government in 1987, 64% was for military purposes, only 8% was spent on energy related research. More efficient energy distribution systems and sources are needed by both developed and under developed nations. In regards to Project Tesla, the market for wireless power transmission systems is enormous. It has the potential to become a multi-billion dollar per year market.

Market Size

The increasing demand for electrical energy in industrial nations is well documented. If we include the demand of third world nations, pushed by their increasing rate of growth, we could expect an even faster rise in the demand for electrical power in the near future. In 1971, nine industrialized nations, (with 25 percent of the world's population), used 690 million kilowatts, 76 percent of all power generated. The rest of the world used only 218 million kilowatts. By comparison, China generated only 17 million kilowatts and India generated only 15 million kilowatts (less than two percent each) [7]

If a conservative assumption was made that the three-quarters of the world which is only using one-quarter of the current power production were to eventually consume as much as the first quarter, then an additional 908 million kilowatts will be needed. The demand for electrical power will continue to increase with the industrialization of the world.

Market Projections

The Energy Information Agency (EIA), based in Washington, D.C., reported the 1985 net generation of electric power to be 2,489 billion kilowatt hours. At a conservative sale price of $.04 per kilowatt hour that result in a yearly income of 100 billion dollars. The EIA also reported that the 1985 capacity according to generator name plates to be 656,118 million watts. This would result in a yearly output of 5,740 billion kilowatt hours at 100% utilization. What this means is that we use only about 40% of the power we can generate (an excess capability of 3,251 billion kilowatt hours). Allowing for down time and maintenance and the fact that the night time off peak load is available, it is possible that half of the excess power generation capability could be utilized. If 1,625 billion kilowatt hours were sold yearly at $.06/kilowatt, income would total 9.7 billion dollars.

Project Tesla: Objectives

The objectives of Project Tesla are divided into three areas of investigation:
1. Demonstration that the Schumann Cavity can be resonated with an

open air, vertical dipole antenna;
2. Measurement of power insertion losses.
3. Measurement of power retrieval losses, locally and at a distance.

Methods

A full size, 51 foot diameter, air core, radio frequency resonating coil and a unique 130 foot tower, insulated 30 feet above ground, have been constructed and are operational at an elevation of approximately 11,000 feet. This system was originally built by Robert Golka in 1973-1974 and used until 1982 by the United States Air Force at Wendover AFB in Wendover, Utah. The USAF used the coil for simulating natural lightning for testing and hardening fighter aircraft. The system has a capacity of over 600 kilowatts. The coil, which is the largest part of the system, has already been built, tested, and is operational. A location at a high altitude is initially advantageous for reducing atmospheric losses which work against an efficient coupling to the Schumann Cavity. The high frequency, high voltage output of the coil will be half wave rectified using a uniquely designed single electrode X-ray tube. The X-ray tube will be used to charge a 130 ft. tall, vertical tower which will function to provide a vertical current moment. The mast is topped by a metal sphere 30 inches in diameter. X-rays emitted from the tube will ionize the atmosphere between the Tesla coil and the tower. This will result in a low resistance path causing all discharges to flow from the coil to the tower. A circulating current of 1,000 amperes in the system will create an ionization and corona causing a large virtual electrical capacitance in the medium surrounding the sphere. The total charge around the tower will be in the range of between 200-600 coulombs. Discharging the tower 7-8 times per second through a fixed or rotary spark gap will create electrical disturbances, which will resonantly excite the Schumann Cavity, and propagate around the entire Earth.

The propagated wave front will be reflected from the antipode back to the transmitter site. The reflected wave will be reinforced and again radiated when it returns to the transmitter. As a result, an oscillation will be established and maintained in the Schumann Cavity. The loss of power in the cavity has been estimated to be about 6% per round trip. If the same amount of power is delivered to the cavity on each cycle of oscillation of the transmitter, there will be a net energy gain which will result in a net voltage, or amplitude increase. This will result in reactive energy storage in the cavity. As long as energy is delivered to the cavity, the process will continue until the energy is removed by heating, lightning discharges, or as is proposed by this project, loading by tuned circuits at distant locations for power distribution.

The resonating cavity field will be detected by stations both in the United States and overseas. These will be staffed by engineers and scientists who have agreed to participate in the experiment. Measurement of power insertion and retrieval losses will be made at the transmitter site and at distant receiving locations. Equipment constructed especially for measurement of low frequency electromagnetic waves will be employed to measure the effectiveness of using the Schumann Cavity as a means of electrical power distribution.

The detection equipment used by project personnel will consist of a pick up coil and industry standard low noise, high gain operational amplifiers and active band pass filters. In addition to project detection there will be a record of the experiment recorded by a network of monitoring stations that have been set up specifically to monitor electromagnetic activity in the Schumann Cavity.

Evaluation Procedure

The project will be evaluated by an analysis of the data provided by local and distant measurement stations. The output of the transmitter will produce a 7-8 Hz sine wave as a result of the discharges from the antenna. The recordings made by distant stations will be time synchronized to ensure that the data received is a result of the operation of the transmitter.

Power insertion and retrieval losses will be analyzed after the measurements taken during the transmission are recorded. Attenuation, field strength, and cavity Q will be calculated using the equations presented in Dr. Corum's papers. These papers are noted in the references. If recorded results indicate power can be efficiently coupled into or transmitted in the Schumann Cavity, a second phase of research involving power reception will be initiated.

Environmental Considerations

The extreme low frequencies (ELF), present in the environment have several origins. The time varying magnetic fields produced as a result of solar and lunar influences on ionospheric currents are on the order of 30 nanoteslas (nT).[2] The largest time varying fields are those generated by solar activity and thunderstorms. These magnetic fields reach a maximum of 0.5 microteslas (uT) The magnetic fields produced as a result of lightning discharges in the Schumann Cavity peak at 7, 14, 20 and 26 Hz.

The magnetic flux densities associated with these resonant frequencies vary from 0.25 to 3.6 picoteslas per root hertz (pT/Hz$^{1/2}$). Exposure to man made sources of ELF can be up to 1 billion (1000 million or 1 x 109) times stronger than that of naturally occurring fields. Household appliances operated at 60 Hz can produce fields as high as 2.5 milliTesla (mT). The field under a 765 kV, 60 Hz power line carrying 1 amp per phase is 15 uT. ELF antennae systems that are used for submarine communication produce fields of 20 uT. Video display terminals produce fields of 2 uT, 1,000,000 times the strength of the Schumann Resonance frequencies [9].

Project Tesla will use a 150 kW generator to excite the Schumann cavity. Calculations predict that the field strength due to this excitation at 7.8 Hz will be on the order of 46 picoteslas.

Future Objectives

The successful resonating of the Schumann Cavity and wireless transmission of power on a small scale resulting in proof of principle will require a second phase of engineering, the design of receiving stations. On completion of the second phase, the third and fourth phases of the project involving further tests and improvements and a large scale demonstration project will be pursued to prove commercial feasibility. Total cost from proof of principle to commercial prototype is expected to total $3 million.

---

[2] The unit of magnetic induction, formerly "Webers per meter squared" is now a "Tesla." It equals 10,000 gauss, which is a commonly used, smaller unit for magnets and "gaussmeters." – Ed. note

Project Tesla

It has been proven that electrical energy can be propagated around the world between the surface of the Earth and the ionosphere at extreme low frequencies in what is known as the Schumann Cavity.

Experiments to date have shown that electromagnetic waves of extreme low frequencies in the range of 8 Hz, the fundamental Schumann Resonance frequency, propagate with little attenuation around the planet within the Schumann Resonance Cavity.

The purpose of Project Tesla is to create pulses, or electrical disturbances, that would travel in all directions around the Earth and resonate in the thin membrane of nonconductive air between the ground and the ionosphere in the Schumann Cavity. The pulses, or waves, would follow the surface of the Earth expanding outward to the maximum circumference of the Earth, until meeting at a point opposite to that of the transmitter. This point is called the antipode. The traveling waves would be reflected from the antipode to the transmitter, be reinforced, and sent out again. This process, analogous to pushing a pendulum, would be repeated at 8.0 Hz, the resonant frequency of the Schumann Cavity. This is the basis for the wireless transmission of power. Tesla, Inc. has determined the exact method Nikola Tesla used to prove the feasibility of the wireless transmission of power during his experiments in Colorado Springs and the difference between that and the method he intended to use in the Wardenclyff Laboratory. This information will facilitate the successful wireless transmission of electrical power.

The need for a wireless system of energy transmission

There are areas of the world where the need for electrical power exists, yet there is no method for delivering power. Africa is in need of power to run pumps to tap into the vast resources of water under the Sahara Desert. Rural areas, such as those in China, require the electrical power necessary to bring them into the 20th Century and to equal standing with western nations. More than three billion people on this planet do not enjoy the access to electrical power that we in developed nations take for granted.

As first proposed by Buckminster Fuller, wireless transmission of power would enable world-wide distribution of off-peak demand capacity. This concept is based on the fact that some nations, especially the United States, have the capacity to generate much more power than is needed.

This situation is accentuated at night. The greatest amount of power use, the peak demand, is during the day. the extra power available during the night could be sold to the side of the planet where it is daytime. Considering the huge capacity of power plants in the United States, this system would provide a saleable product which could do much to aid balance of payments while raising the standard of living in Third World countries.

Environmental considerations

The extreme low frequencies (ELF), present in the environment have several origins. The time-varying magnetic fields produced as a result of solar and lunar influences on ionospheric currents are on the order of 30 nanoteslas. The largest time varying fields are those generated by solar activity and thunderstorms.

> ## *"So astounding are the facts in this connection, that it would seem as though the Creator, himself, had electrically designed this planet..."*
>
> — Nikola Tesla describing what is now known as Schumann Resonance (7.8Hz) in "The Transmission of Electrical Energy Without Wires As A Means Of Furthering World Peace," Electrical World and Engineer, Jan. 7, 1905, pp. 21-24.

REFERENCES

1. Tesla Said, Compiled by John T. Ratzlaff, Tesla Book Company, Millbrae, CA, 1984.

2. Dr. Nikola Tesla: Selected Patent Wrappers, compiled by John T. Ratzlaff, Tesla Book Company, 1980, Vol. I, Pg. 128.

3. "The Disturbing Influence of Solar Radiation on the Wireless Transmission of Energy", by Nikola Tesla, Electrical Review, July 6, 1912, PP. 34, 35.

4. "The Effect of Static on Wireless Transmission", by Nikola Tesla, Electrical Experimenter, January 1919, PP. 627, 658.

5. Tesla Primer and Handbook, Dr. James T. Corum and Kenneth L. Corum, unpublished. Corum and Associates, 8551 ST Rt 534, Windsor, Ohio 44099

6. Colorado Springs Notes, 1899 - 1900, Nikola Tesla, Nikola Tesla Museum, Beograd, Yugoslavia, 1978, Pg. 62.

7. Van Nostrands Scientific Encyclopedia, Fith Edition, Pg. 899.

8. "PC Monitors Lightning Worldwide", Davis D. Sentman, Computers in Science, Premiere Issue, 1987.

9. "Artificially Stimulated Resonance of the Earth's Schumann Cavity Waveguide", Toby Grotz, Proceedings of the Third International New Energy Technology Symposium/Exhibition, June 25th-28th, 1988, Hull, Quebec, Planetary Association for Clean Energy, 191 Promenade du Portage/600, Hull, Quebec J8X 2K6 Canada

ABOUT THE AUTHOR

Mr. Grotz, is an electrical engineer and has 15 years experience in the field of geophysics, aerospace and industrial research and design. While working for the Geophysical Services Division of Texas Instruments and at the University of Texas at Dallas, Mr. Grotz was introduced to and worked with the geophysical concepts which are of importance to the proposed project. As a Senior Engineer at Martin Marietta, Mr. Grotz designed and supervised the construction of industrial process control systems and designed and built devices and equipment for use in research and development and for testing space flight hardware. Mr. Grotz organized and chaired the 1984 Tesla Centennial Symposium and the 1986 International Tesla Symposium and was President of the International Tesla Society, a not for profit corporation formed as a result the first symposium. As Project Manager for Project Tesla, Mr. Grotz aided in the design and construction of a recreation of the equipment Nikola Tesla used for wireless transmission of power experiments in 1899 in Colorado Springs. Mr. Grotz received his B.S.E.E. from the University of Connecticut in 1973. He can be reached at 760 Prairie Ave., Craig, CO 81625, wireless @ rmi.net

10   Tesla's Self-Sustaining Electrical Generator

Oliver Nichelson
Adapted from *Proceedings of the Tesla Centennial Symposium*, 1984

Abstract

Before the discovery of the electron, the principle theory used to describe the electrical activity was that of the ether. At the turn of this century, the ether theory in use by science was a remnant of the concept common in western thought for several centuries. This situation favored the rise of atomic theory. This change in scientific paradigm requires a translation from 19th century terminology into 20th century language in order to understand Tesla's later research. Of particular interest is his magnifying transformer which claimed to produce resistanceless current.

The Historical Ether

Though science aims at giving accurate descriptions of the workings of nature, these descriptions change from historical period to historical period. In the same way that an object in one European country is called by a different name in an adjoining European country, so do the descriptions of nature change during different periods of man's history.

In the 19th century western science the broadest view of the physical world was that all objects were somehow each connected to one another through pre-material ether. Solid bodies were believed to be made from condensation of this ether. In this worldview, atoms and electrons did not exist as scientific realities.

Toward the end of the last century the atomic picture of the world emerged in steps. Solid bodies were explained by minute vortices in the ether – small whirlpools – forming lumps of matter. Lord Kelvin, the virtual spokesman of Victorian science, developed an ether vortex model of the electron in an effort to explain some of the properties of electricity. The electron as a discrete particle did not become a fact of science until Thompson discovered it in 1897.

The view of nature as a single entity formed out of the ether changed to the modern one of matter being made of collections of individual particles in 1905 [1]. In that year, Einstein presented his paper on Brownian motion explaining the movement of pollen particles on the surface of water in terms of discrete units of matter. From then until today, the atomic view has prevailed.

This difference between the 19th century description of nature and our presented description makes it difficult to have a complete picture of the work of the early electrical researchers. Today, Faraday, Maxwell and Tesla are recognized as valued contributors of the understanding of electricity, but their work was carried out before the electron- the fundamental carrier of electrical charge – was discovered. All of these scientist held a belief in an physical ether. Though Faraday's laws of induction are still accepted, and Maxwell's equations from electromagnetism are still used routinely, and Tesla's generators are still powering our lights, the 19th century physics that they learned and out of which their physics came, has been judged scientifically wrong.

The curious situation in which Faraday, Maxwell and Tesla can be seen be both right in their results but wrong in their beliefs about physics comes from an inability to translate the concepts of their historical periods into the language of our period. This lack of chronological translation, in contract to the spatial translation between European languages, is also an obstacle to understanding the physics of self-sustaining electrical ("free energy") generators based on the 19th century views.

In the last half of the 19th century, when researchers had to deal with the ether in practical engineering terms in order to guild their electrical devices, the concept of the ether, then, several centuries old, was a watered down theory. At that time, the ether was considered something like a thin gas that could be found everywhere. However, that was not a historically correct view of the ether.

The ether had been pictured traditionally as a non-material substance capable of condensing into ponderable matter. Gas, no matter how thin, is still ponderable matter; and because of that, could not qualify as the ether.

To find out what was meant historically by the concept of ether, an early writer on the subject can be cited. Robert Fludd, in 1659 described the "Ethericall...influences" as "far subtler condition than is the vehicle of visible light... so thin, so mobile, so penetrating, so lively, that they are able, and also do continually penetrate, and that without manifest obstacles or resistance, even unto the center or inward bosom of the earth where they generate metals of sundry kinds."[2]

Fludd quotes an even older source on the nature of ether, the writings of Plotinus (3rd Century AD) where the ether is described as being so fine "that it doth penetrate all bodies and... it maketh them not a jot bigger for all that because this inward spirit doth nourish and preserve all bodies." [3]

From these older descriptions of the ether, the following attributes can be seen missing from the late 19th century concept. First, the ether was held to be truly non-material – it does not make bodies "a jot bigger". If the ether were a gas, its addition to anything would be measurable. Second, the ether is a substance less material than "the vehicle of visible light", that is, something less than what today is known as a photon. Third, the ether was credited with generating metals and nourishing all bodies, clearly a distinct property not belonging to gases.

Whether or not the reality of the ether as put forth by these authors is accepted, it is historical fact that the tether Michelson and Moraly did not find in their experiments and that the modern atomists ridiculed so strongly when they came to scientific power in the early 20th century was never claimed to exist by people who first used the term. Taking a longer view of science, modern theorists fought a battle against an issue that never existed.

If, on the other hand, the ether is looked at in the earlier description of its properties, something can be learned about the operation of a least one type of self sustaining electrical generator. To do this, the ether concept has to be translated into an artifact of contemporary science.

The Modern Ether

The properties of having less mass than a massless photon, being able to interpenetrate a body but not add to it, and generating material bodies are encompassed in the modern view of the quantum wave nature of matter. In quantum theory, an object can be viewed as either made of particles or waves. It is not an idea everyone is comfortable with even now but one that is widely accepted and known to be verifiable by experiment. Transistors, tunnel diodes and even digital watches are a few of the real world objects operating on physical principles that are explained best by the quantum wave nature of matter.

If an object can be both a quantum wave and a particle, then it its wave state, it can be said to interpenetrate an object without making it "a jot bigger". Also, being a wave equivalent to a particle, the wave would not have the mass of a particle. It has amplitude instead. The quantum wave is also responsible for the generation of solid bodies. Present theory has it that a particle exists in its quantum wave state until a measurement is made, when the wave is then said, to collapse to form an object. The collapse of the quantum wave defines the state of the object, that is, it generates the particle.

The quantum wave state of nature very much resembles the 17th century picture of the ether.

With this conceptual parallel in mind, it is possible to understand better the work of Nikola Tesla, who held the ether theory as a scientific concept, who, also no the basis of this theory, build working electrical machines, and who is associated with the idea of an electrical generator which could maintain a current without an external prime mover.

Schooled during the 1860's, Nikola Tesla's understanding of physics was pre-atomic. In his biographical articles Tesla does not comment on the theoretical aspects of his education, but in his technical writings, he uses the term "the ether" in a positive sense and only in his later writings are found grudging references to atomic particles and electrons.

Tesla Magnifying Transformer

Tesla's most famous device was what he called a Magnifying Transformer, the principal tests of which were carried out in Colorado Springs during 1899. The device is described in his U.S. Patent as an "Apparatus for Transmitting Electrical Energy" [4] and claims some unusual characteristics among which were the propagation of waves faster than the speed of light, the transmission of signals, not around the earth, but through the earth, and doing this by eliminating as much as possible electromagnetic waves - the only electrically related waves known today capable of transmitting signals.

Tesla did this using a coil with 10,000 – 11,000 feet of cable [5], with what he claimed to be little or no resistance. This last fact, giving rise to the belief that in addition to tits other unusual characteristics, the device had the property of maintaining its current for a measurable period of time after disconnection from an outside power source.

Taking these ideas together – that the ether is equivalent to quantum wave energy, that Tesla held a belief in a physical ether, and that Tesla build a device capable of maintaining an electrical current without an external prime mover, a conclusion that can be reached, is that the quantum wave theory can be used to understand the dynamics of Tesla's magnifying transformer. This follows from the work of Dr. Andrija Puharich who, in a 1976 paper, put forth the idea that the magnifying transformer could not be explained by the laws of classical electrodynamics, but, rather in terms of high energy particle transformations [6].

The wave theory of matter gained its present popularity in 1923 through the efforts of deBroglie. When experiments showed that light could be considered both a particle and a wave he reasoned that an electron, clearly a particle, could behave like a wave. He deduced the wavelength of the electron from the equation $E=hf$ which equates the energy of a particle to the product of Planck's constant times the frequency. (Lambda works out to be 2.4×10^{-12} meters, which is the classical wavelength for the electron.)

In analyzing the Tesla magnifying transformer, this mathematical relationship can be used to determine the quantum energy of a wave in the transformer's operating frequency (here we use the pulse repetition rate of 7.5 Hz, following Corum [10] instead of the author's originally suggested kilohertz oscillation frequency – Ed. note) and putting that value into the equation gives:

$$E = hf = (6.63 \times 10^{-34} \text{ Js}) (7.5 \text{ Hz}) \qquad (1)$$

$$E = 4.97 \times 10^{-33} \text{ J/e}$$

which would be the radiated energy <u>per accelerated charge carrier</u> (electron) in the conductor.

If the magnifying transmitter were operating at a current $I = 100$ amperes, the total charge can be found. Current is charge per time $(I = q / t)$ and by definition, 1 Ampere = 1 Coulomb / second. This relationship can be used in turn to determine the number of charge carriers per second in the conductor for a 100 A current:

$$I = \frac{100 \quad \text{C/s}}{1.6 \times 10^{-19} \text{ C/e}} \qquad (2)$$

$$I = 6.25 \times 10^{20} \text{ electrons per second} \qquad (3)$$

The total number of charge carriers times the emitted energy per charge carrier would equal the quantum energy of the wave at a given frequency (7.5 Hz in this case):

$$E_Q = E I = (6.25 \times 10^{20} \text{ e/s}) (4.97 \times 10^{-33} \text{ J/e}) \qquad (4)$$

$$= 3.1 \times 10^{-12} \text{ J/s} = 3.1 \text{ picowatts}$$

If the highest reported current that Tesla used, 1000 amperes, is put into the calculation, the energy range would be 3.1×10^{-12} J/s to 31×10^{-12} J/s.

Converting to a more commonly used system of measures, the energy of a quantum wave at 7.5 Hz would be:

$$e = (6.2 \times 10^{12} \text{ Mev} / \text{J})(3.1 \times 10^{-12} \text{ J/s}) \qquad (5)$$

$$= 19 \text{ Mev}$$

If the highest current of 1000 amperes is put into the calculation, the energy of a quantum wave would be 190 Mev.[1]

In order to generate a wave of this energy, an electron would have to undergo a potential difference in the range of 19 to 190 million volts.

Tesla's magnifying transformer was reported to operate in the range of tens of millions of volts. At 20 million volts there would be more than sufficient electrical force to create a

---

[1] Compare to Corum [10] who calculate about 225 coulombs in a volume of 10,000 cubic meters of glow discharge. Using 2.5 eV per molecule of air, the amount of power Tesla used for a pulse repetition rate of 7.5 Hz is found to be only 6.5 hp, consistent with what Tesla reported. For reasons explained in the article, the Corums find that Tesla generated <u>10 MeV</u> electrons at 1000 amperes.

vacuum wave for the amount of charge in motion at 7.5 Hz. At 200 million volts there would be enough force to produce such a wave for a current of 1000 amperes at that frequency.

The generation of a quantum wave by the magnifying transformer goes a long way in explaining some of the properties Tesla claimed for the device. For one, he said that electromagnetic waves were reduced to a minimum and, indeed, it would seem hard to propagate any e.m. radiation with the blunt topped tower used in his transmission experiments. If, however, the waves that were being emitted were quantum waves, or waves of the ether, his claims for radiating energy from one point to another without the use of electromagnetism becomes clear.

Also, Tesla's statement that electromagnetic radiations were similar to the waves transmitted by an ordinary whistle through the air [7] makes sense. According to his view, e.m. waves would be nothing but undulations in the atmospheric gases, while his transmissions were taking place in a wholly different medium, that of the ether.

Tesla's claim for instantaneous transmission of energy has a basis in modern theory too, for a quantum wave is non-local in nature. That is, its effect is not limited to one particular point, but, through a physical process still not completely agreed upon, the effect can be measured at great distances from the point of origin at the moment of origin.

The Superconducting State

As to maintaining a current in the transformer without an external power source, the only condition known today for achieving this, is the state of superconduction, which seemed to be ruled out in the case of Tesla's device which operated far above the almost zero temperatures needed for superconduction. However, what is understood as the superconducting state in today's science is in fact a description of the conductor. If a material has a certain type of atomic configuration and is cooled to a certain temperature, a superconducting condition exists in which a perpetual current can be maintained. The superconducting state, though, can exist without there being a current in the conductor. The state is a characteristic of the conductor.

Tesla may have discovered that superconductivity can be a property not of the conductor but of the current itself.

To examine how this might be the case, a specific model of electrical activity will be used. Instead of picturing an electric current composed of billiard ball particles of of little satellites of nuclear suns, or as an electron gas, or as electron plasma, it can be imagined as an electron liquid. At this point the make up of the liquid is not as important as is its fluid nature and that the fluid is electrical.

The model of a liquid is useful because it provides an easy example of how a substance can remain the same and yet become radically different under certain conditions. With water, when heat is removed from it, a phase change takes place which transforms it into solid ice. When thermal energy is added to water, it undergoes a different phase change and becomes a gas. The substance remains the same, but it exists in three difference states.

One of the extreme states that a fluid can achieve is superfluidity during which a liquid will move up the walls of its container. This, of course, is a property of the liquid, not of the container.

Perhaps the same phase change phenomenon takes place in the electron liquid. Under certain conditions, high voltage and or high current, the electron liquid will remain the same substance but will take on radically different properties, similar to the state of superfluidity. This condition would be a state change in the current, not in whatever material is serving as the conductor.

A state of superfluidity in an electron liquid would explain how Tesla was able to send a current through the earth. When in its commonly known state a current does not travel far through the earth's resistance, but if the current has undergone the proper phase change, it could easily travel with no resistance.

Likewise, a phase changed current would travel through a generator coil with no resistance. Having undergone the change it would become a super-current in a non-superconducting conductor. Such a condition would allow a generator to maintain a current without an external power source.

This particular solution, which of course has to be tested, of Tesla's self-sustaining generators, is not an explanation of all the other similar devices such as the Figuera, Hubbard, and Herdershot devices [8]. There are probably as many engineering solutions to such generators as there are inventors of them.

One characteristic all the other devices have in common in contrast to Tesla's magnifying transformer, is that they did not require the high voltage and currents Tesla used. They do not, though, represent an engineering advancement over Tesla's engineering methods.

Tesla put his main efforts into high energy devices as a matter of mere practicality in marketing a product. A year after his Colorado Springs experiments, he wrote in his Century magazine article, 1900, that he had spent a great deal of time on a smaller generator but realized that negative market pressures would not allow such a machine to see the commercial light of day [9]. And he was right; it is not possible yet to by a Hubbard or a Hendershot generator to light our homes.

Tesla believed he had a greater chance for introducing a new electrical technology if it made use of the generators then being sold, but which used their output in novel ways – which is why he concentrated on the wireless power transmission project, though even that idea proved too much for his time.

A careful study of his later writings shows that many of his more advanced concepts were based on earlier work with lower voltage versions of generators capable of maintaining a super-current. These designs appear to be based on intricate configurations of coil geometries. The peak of this line of research might have been just before the fire of his New York City laboratory in which, many of his prototypes and papers were lost. The task of uncovering the precise nature of these designs becomes very complex, because after the fire, Tesla spoke of his more advanced work only obliquely and never in detail.

Recovering these earlier designs would bring about the second stage of electrical technology – one that Nikola Tesla started, here, a century ago.

References

1. There have been several such paradigm changes in western ideas about nature. Theories alternate between a one substance universe out of which everything is made and a many substance universe in which the constituent particles are separated by a vacuum.

2. Robert Fludd, *Mosaical Philosophy*. London, Humphrey Moseley, 1659, p. 221.

3. Fludd, p. 221

4. U.S. Patent #1,119,732 of December 1, 1914; application filed January 18,1902.

5. Nikola Tesla, *Colorado Springs Notes*, 1899-1900. Beograd: Nolit. 1978, p.43.

6. Andrija Puharich, "The Physics of the Tesla Magnifying Transmitter and the Transmission of Electrical Power without Wires". Planetary Association of Clean Energy, (Ottawa, Ontario, 1976).

7. Nikola Tesla, quoted in the *NY Herald Tribune*, Sept 22, 1929, pg. 21.

8. C. Bird and O. Nichelson, "Nikola Tesla, Great Scientist, Forgotten Genius", *New Age*, Feb. 1977. p. 41.

9. Nikola Tesla, "On the Problems of Increasing Human Energy", in Nikola Tesla, Lectures, Patents, Articles, Biograd, Nikola Tesla Museum, 1956, p. A-143. (Also in *Century*, June, 1900 – Ed. note)

10. Corum, James & Kenneth, "Critical Speculations Concerning Tesla's Invention and Applications of Single Electrode X-Ray Directed Discharges for Power Processing, Terrestrial Resonances and Particle Beam Weapons," *Proc. Inter. Tesla Symposium*, 1986, p.7-21

Oliver Nichelson can be reached at 670 W 980 N, Provo, UT 84604

11 Self-Sustained Longitudinal Waves

Robert W. Bass, Ph.D.

Reprinted from *Proceedings of the Tesla Centennial Symposium*, 1984

SELF SUSTAINED NON-HERTZIAN LONGITUDINAL WAVE OSCILLATIONS AS RIGOROUS SOLUTIONS OF MAXWELL'S EQUATIONS FOR ELECTROMAGNETIC RADIATION

In contradiction to common belief, Maxwell's classical theory of electromagnetic radiation does predict the existence of longitudinal electromagnetic waves *in vacuo*. (Longitudinal non-Hertzian compression waves are very likely to be the type that Tesla produced with the TMT. – Ed. note) This preliminary note contains what appears to be the first rigorous proof of the theoretical existence of such non-Hertzian self-sustained longitudinal wave oscillations. The E-fields and B-fields in such waves are everywhere parallel (so that the Poynting vector vanishes identically and no energy is transported along the waves, though it does appear possible in theory to transport force at a distance without attenuation via such waves *in vacuo*). The waves have the following geometry. Choose a static vector potential (A) parallel to the static magnetic field configuration of an arbitrary <u>force-free</u> magnetic field (B) of the type discovered in 1952 by Schluter and Lust.[1] Then pick an arbitrary frequency $\omega = \lambda$ c > 0 corresponding to an arbitrary wave number λ > 0 where c is light speed, and multiply the vector potential A by λ cos ωt to obtain the longitudinal wave B-field whose oscillations at frequency ω generate an E-field obtainable by multiplying the vector potential by ω sin ωt. The vector potential and both fields are parallel and typically in infinite cylindrical or toroidal configurations as depicted in Figures 1 & 2.[2]

To derive, let R denote an unbounded or bounded open connected region of real Euclidian 3-space E³. Let the boundary ∂R of R consist of the union of piece-wise smooth 2-surfaces. Let λ > 0 denote an arbitrary

---

[1] R. Lust and A. Schluter, Axial symmetrische magnetohydrodynamische Gleichgewichtskonfigurationen, *Z. Naturforsch.* 12a (1957), 850-854.

[2] <u>Ed. note</u>: Compare with 'Curl-Free Vector Potentials,' R. Gelinas, *Proc. ITS 1986*, p.4 –43, & Gelinas patents #4,447,779, #4,429,288, #4,429,280 on modulating/demodulating scalar waves, as well as J. Corum patents #4,622,558, #4,751,515 using toroidal fields.

positive constant wave number. Then according to potential theory, there exists on R, corresponding to every set of arbitrarily assigned Dirichlet or Neumann boundary condition on ∂R, a unique real analytic solution $\varphi = \varphi(r)$ of Helmholtz's scalar wave equation.

(1) $\nabla^2 + \lambda^2 - 0$, (r in R), on R

Next, let u be an arbitrary constant unity vector in E^3, and define frequency $\omega = \lambda c$ where c denotes light-speed.

Now define a static vector field $A = A(r)$ on R by

(2) $A = \nabla \times (\varphi u) + (1/\lambda) \nabla \times (\nabla \times (\varphi u))$.

Finally define

(3a) $B = \lambda A \cos \omega t$,

(3b) $E = \omega A \sin \omega t$.

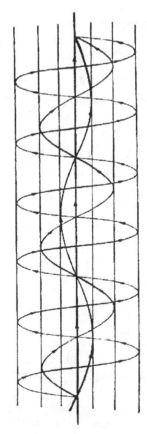

Figure 1. Illustrating a force-free magnetic field having axial symmetry. The axis is itself a line of force of the field. The other lines of force shown here are helices around this line as axis (with pitch angles decreasing away from the axis)

Then, using elementary vector calculus, it is easy to verify the following relationships :

(4) $\nabla \times A = \lambda A$

(5) $\nabla \cdot A = 0$

(6) $\nabla \times E = - \partial B / \partial t$

(7) $\nabla \times B = (1 / c^2) \partial E / \partial t$

(8) $\nabla \cdot E = 0$

(9) $\nabla \cdot B = 0$

The thus-constructed (E, B) fields on R are genuinely Maxwellian self-sustained, <u>non-Hertzian,</u> longitudinal electromagnetic oscillation *in vacuo.*

At each point R_b on the boundary ∂R of the radiation region R, the vector potential A has a multipole expansion (analogous to a Laurent series) of the form,

(10) $A = A (r) = C (r - r_b),$ (r in R near r_b) ,

defining a singular field $A = A (r)$ near r_b.

This field defines a new field,

(11) $J = J (r) = \nabla^2 A + \lambda^2 A,$ (r in R near r_b) ,

such that

(12) $J = J(r) = 0$ (r on R)

but such that a calculable limiting boundary distribution

(13) $J = J (r_b) \neq 0$

exists. Now on the boundary, define the current field

(14) $I = I(r_b, t) = J(r_b) \cos \omega t$.

This current, I on ∂R will generate the previously defined longitudinal radiation waves on R.

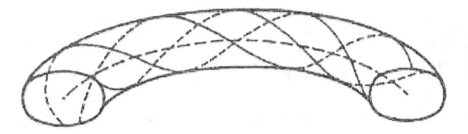

Fig 2. The lines of force of a force-free field must therefore lie on nexts of torus-like surfaces, one inside the other, with a limiting curve which is itself a line of force.

Ed. note: The Bass-cited paper by *Schluter and Lust* is required reading in magnetohydrodynamic courses, such as the one at the University of Montana (PHYS 515: Plasma Physics & MHD). Sample homework assignment is below:
(http://solar.physics.montana.edu/martens/plasma/calendar.html)

1. Working with the MHD equations. Consider an inviscid plasma of uniform resistivity with azimuthal symmetry, that is in a state of steady (but possibly non-uniform) rotation about the z-axis and permeated with a magnetic field that has no azimuthal component.
   a) Show that

   curl(j) = [curl(v X B)]/(eta*c)

   where j, v, and B are vectors with the usual meaning, eta is the resistivity, and c as usual the velocity of light.
   b) Prove that the current is wholly azimuthal.
   c) Use div.B=0 to show that the plasma has constant angular velocity.
      (omega=v_phi/r)

190

2. Force-free fields. Consider the Lust and Schlueter expression for force-free fields with cylindrical symmetry (i.e. axial plus azimuthal symmetry).

a) For the constant twist field, that we covered in class, derive an expression for alpha, the ratio between current and magnetic field vector.

b) The magnetic field vector is tangent to the magnetic field line at any point. Hence the tangent vector satisfies

$$\frac{rdphi}{B\_phi} = \frac{dz}{B\_z} = \frac{dr}{B\_r},$$

which is the field line equation. From the solution for a constant twist field derive the number of turns in the fieldlines on an axial segment of length L.

c) Now assume alpha is constant and use that to simplify the *Lust and Schluter* differential equation to derive Bessel's equation. Find and sketch the solutions for B_z and B_phi.

Robert W. Bass is a physicist working for an aerospace firm and can be reached at 45960 Indian Way #612, Lexington Park, MD 20653, donquixote@radix.net

12

TESLA & THE MAGNIFYING TRANSMITTER
A Popular Study for Engineers

K.L. Corum   *and*   Dr. James Corum

Give me a long enough lever and a place to stand and I will move the Earth.
Archimedes, 287-212 BC

The 'Magnifying Transmitter' is a peculiar transformer specially adapted to excite the Earth.
Nikola Tesla, 1917

The power networks of the future may have little resemblance to those of today.
Richard P. Feynman, 1964

Introduction - Machines That Magnify

Archimedes was one of civilization's most gifted and creative intellects. His striking contributions spanned mathematics, science, invention, and optical and mechanical engineering. He is commonly recognized to have been on a level with Isaac Newton; and even Galileo claimed him as a mentor. Over a century after his death, the Roman statesman Cicero sought out Archimedes' grave stone and was touched and saddened to find that it had been overgrown with thorns and brambles.[1]

Mechanics is commonly understood to be that branch of physical science which treats the effect of forces on ponderable bodies. As a science, it appears to have begun with Archimedes, who is the first one known to have worked out the theoretical principle of the lever. Other civilizations had used levers, of course, but Archimedes of Syracuse (on the Island of Sicily) was the first to state the principle.

Archimedes, the son of the astronomer Pheidias, had studied in Alexandria, but returned to his home at Syracuse for a lifetime of creative technical achievement. He solved many difficult problems in geometry, established the conceptual foundations of calculus, calculated the perimeter of the earth (yes, the world was known to be round at that time), determined that *pi* is approximately 22/7, authored a number of books on mathematics and theoretical mechanics, and spent his life in vigorous intellectual activity. He created and developed great instruments for agriculture and for military applications. Plutarch, Polybus, and also Livy provide colorful accounts of Archimedes, his directed energy weapons, his thundering mechanical engines of war, and the utter astonishment of both his own king (who was delighted) and the attacking Romans (who were soundly frustrated), during the siege of Syracuse.[2,3] Apparently, Archimedes was slain while in the midst of solving an analytical problem of great significance: he didn't respond fast enough to satisfy the whims of a Roman soldier, but begged not to be interrupted for a few more minutes in order to complete a formal solution.

Machines are devices which are used to transmit and modify force and motion and to do work. Machines don't create energy, they receive energy from a source (the prime mover) and bring about an advantage for the source in doing work on the load. There are six fundamental "machines" in classical mechanics. Do you remember them from grade school science?[4]

In their venerable introductory text on Physics, Sears and Zemansky have expressed that "a machine is a force-*multiplying* device."[5] They state the "force multiplication" factor of a machine as the ratio of the weight lifted, W, to the applied force, F, and call it the *actual mechanical advantage, R*:

$$(1) \quad R = \frac{W}{F} = \frac{\text{Force exerted by machine on load}}{\text{Force used to operate machine}} = \frac{\text{Output Force}}{\text{Input Force}}$$

This simple *magnifying ratio* reflects the incredible relief which machines have brought to the brows of multitudes of toiling laborers in the progress of civilization. Many orders of magnitude may be gotten, even in practical situations, so wonderful are these remarkable devices. No energy is created, but tremendous advantage is devised for that which is available. In this paper we will show the obvious reason why Tesla called his finest invention a "Magnifying Transmitter".

The Lever

As recounted above, Archimedes was the first to state the principle of the lever. Although his treatise on "The Lever" has been lost to antiquity, the principle is discussed in his works which are extant.[6]

The **rigid**[7] bar and the lever are common mechanical coupling devices for connecting mechanical systems together. The bar provides a way of transmitting forces and realizing *mutual* mass in coupled mechanical systems.

The lever is a pivoted rigid bar used to *multiply* force or motion. Consider Figure 1a. If force F_1 pushes down on one end and it moves with velocity v_1, then the force and upward velocity at the other end, F_2 and v_2, can be found either from conservation of energy ($\Delta W = F_1 r_1 \Delta \theta = F_2 r_2 \Delta \theta$) or by summing the moments about the immovable fulcrum (or pivot):

(2) $\qquad F_1 r_1 - F_2 r_2 = 0$.

Rearranging algebraically gives the remarkably simple expression

(3) $\qquad F_2 = (r_1/r_2) F_1$.

The rigid bar lever is a simple force *multiplying* mechanism.[8] The force multiplication, or magnification, is given by the mechanical advantage

(4) $\qquad R \underset{\Delta}{=} F_2/F_1 = r_1/r_2$

As an astonishing result, which still delights us today, it is seen that a very great weight can be moved with a very small force. Concerning the simplicity of Archimedes' proofs, Plutarch observed, "No amount of investigation of yours would succeed in attaining the proof, and yet, once seen, you immediately believe you would have discovered it yourself."[9]

In Archimedes' time, many asserted that it was as though he was reaching over into a mist enshrouded realm and drawing out Nature's hidden secrets so that anyone of slower intellect could comprehend them. Plutarch indicates that some of his contemporaries believed that incredible effort, toil, and intense contemplation on the part of the inventor had produced these wonderful results. Such conflicts to discover the hidden possibilities of Nature bring to mind the tragic struggles of Dr. Faust.[10] [The tragedy of Dr. Faust is that, in spite of his great natural talent and his years of academic study, he gave up a sacred and holy quest, and "sold out for guns, girls and a good time" (in the words of C.S. Lewis). Archimedes didn't. Nor did Tesla.]

The mechanical input power ($F_1 v_1$) and output power ($F_2 v_2$) must be equal if there is no frictional loss. Consequently, the velocities are related as

(5) $\qquad v_2 = (1/R) v_1$.

How remarkable the contributions of Archimedes are for mankind. His geometrical proofs, mechanical devices, and famous hydrostatic principle, so simple and obvious to us today, were, apparently, obtained at great price.

The Transformer - A Lever For 20th Century Civilization

The modern power transformer was invented and developed by William Stanley (1858-1916), and used at the first AC power plant in North America, at Great Barrington, Massachusetts, in 1886.[11] The physical principles of the electrical transformer go back to Michael Faraday and Joseph Henry. Stanley's plant demonstrated that electrical power could be generated at a low voltage, transformed to a higher voltage for efficient transmission, and retransformed back to a lower voltage for end-user applications. Tesla, of course, did not "invent" AC. But he did create and patent the entire system which has made AC commercially viable as a means for powering the 20th century.[12] He contributed the

AC motor, and also the polyphase power generation and distribution system. His first AC patents were issued in January of 1886. It was his invention of an AC motor, along with AC's more efficient generation and simpler distribution, which made alternating current of such great commercial value, brought about its triumph over Edison's DC system in the 1890's, and has powered the wealth and progress of the 20th century.

In assessing the importance of electric power generation and distribution, and the great accomplishment of the Niagara Falls power plant, Dr. Charles F. Scott, Professor Emeritus of Electrical Engineering at Yale University, and past President of the AIEE (now the IEEE) has said, *"The evolution of electric power from the discovery of Faraday in 1831 to the initial great installation of the Tesla polyphase system in 1896 is undoubtedly the most tremendous event in all engineering history."*[13] Certainly, not since the days of Archimedes had civilization experienced such a giant step forward in (to use the ECPD[14] definition for the profession of Engineering) "utilizing the resources of the earth for the benefit of mankind".

The ideal transformer is the electrical analog to the frictionless mechanical lever and also to the ideal gear train.[15,16,17,18,19] (See Figure 1b.) The operation of the ideal transformer follows from the definition of magnetic flux, Φ,

$$(6) \qquad \Phi = \iint B \cdot n \, da$$

where the integration is carried out over the transformer leg cross-sectional area, and from Faraday's law of induction,

$$(7) \qquad \xi = -N \frac{d\Phi}{dt}$$

where ξ is the emf induced in an N turn coil through which the flux passes. Since the flux is the same in both legs, we have

$$(8) \qquad \xi_1 = -N_1 \frac{d\Phi}{dt}$$

$$(9) \qquad \xi_2 = -N_2 \frac{d\Phi}{dt} = \frac{N_2}{N_1} \xi_1 = N \xi_1$$

where one commonly defines the turns ratio as:

$$(10) \qquad N = N_2/N_1 .$$

The similarity between equations (3) and (9) completes the analogy if force and AC voltage are taken as analogs and if

$$(11) \qquad N_2 \sim r_1 \qquad \text{and} \qquad N_1 \sim r_2 .$$

Depending upon the mechanical advantage, R, the lever magnifies the applied force: $F_2 = R F_1$. Depending upon the turns ratio, N, a transformer steps the primary AC voltage up to a higher (or down to a lower) secondary voltage: $\xi_2 = N\xi_1$.

No power is dissipated in an ideal transformer, so that

$$(12) \qquad P = \xi_1 I_1 = \xi_2 I_2 .$$

Consequently, the current must be transformed in a manner analogous to equation (5)

$$(13) \qquad I_2 = \frac{N_1}{N_2} I_1 = (1/N) I_1 .$$

The analogy illustrated above is called the f-v (force-voltage) analog. An alternative analogy[20] has been advanced by Firestone.[21]

As with the lever, the AC voltage has been *multiplied* by the turns ratio, and the current has been multiplied by the reciprocal of the turns ratio. **What magnificent leverage has been afforded 20th century civilization by this wonderful appliance.**

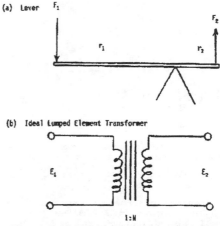

(a) Lever

(b) Ideal Lumped Element Transformer

Figure 1. Classical devices which provide advantage.

Conventional power transformers have been splendid electrical devices in the application of electrotechnology. Cascaded transformers, working at power line frequencies have even been used to produce voltages in excess of 1 megavolt.[22,23] Such transformers behave as lumped elements and depend entirely upon the turns ratio to achieve voltage rise. They are, however, incapable of attaining the voltage rises produced through the phenomenon of resonance. The ultimate limitation on lumped-element high voltage transformers is due to the conflicting requirements that many tightly wound turns are necessary to produce flux leakage and obtain large step-up, yet great turn-to-turn spacing becomes necessary in order to avoid high voltage breakdown in the appliance.

The Mechanical Oscillator

Mechanical oscillators come in a variety of configurations: mass on a stretched spring, the simple pendulum, torsional oscillators (twisted bars, watch flywheels), floating objects, U-shaped liquid columns, compressed air columns (pistons, shock absorbers, organ pipes, wind instruments). Consider the simple linear mechanical oscillator. The dynamical behavior is to be determined when it is driven by a variable frequency forcing function.

Let the system be assumed to have a linear restoring force (Hooke's law) and only viscous damping, i.e.- the damping force is proportional to the velocity of the body.[24] McLean and Nelson point out two other types of mechanical damping:

"1. Coulomb damping is independent of the velocity and arises because of sliding of the body on dry surfaces (its force is thus proportional to the normal force between the body and the surface on which it slides.

2. Solid damping occurs as internal friction within the material of the body itself (its force is independent of the frequency and proportional to the maximum stress induced in the body itself)."[25]

In view of the fact that every basic physics course discusses dry friction, it is remarkable how little space is devoted to Coulomb damping in advanced texts on classical mechanics. Feynman has observed that the Coulomb friction law "... is another of those semi-empirical laws that are not thoroughly understood."[26] We have discussed these issues in greater detail in a previous paper,[27] and so we will not consider them further at this time.

Consider the simple mechanical system shown in Figure 2. Suppose that viscous damping is the only loss mechanism present, and the above two types of mechanical damping are negligible. Further, suppose the system is driven by a sinusoidal forcing function of amplitude F_0. Equating the sum of the forces to ma leads to the well-known second order linear differential equation for the displacement[28]

$$(14) \qquad \frac{d^2x}{dt^2} + 2\alpha \frac{dx}{dt} + \omega_n^2 x = \frac{F_0}{m}\cos\omega t$$

where $2\alpha = g/m$ and $\omega_0^2 = k/m$

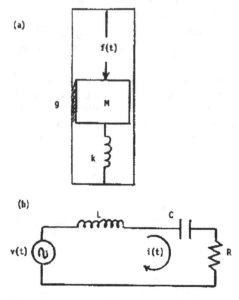

(a)

$f(t)$

g

M

k

(b)

$v(t)$ L C $i(t)$ R

Figure 2. (a) A one-coordinate translational mechanical system, consisting of a spring supported mass constrained to vertical motion by fixed guides. f(t) is a driving force between the frame of reference and the mass. (See Gardner and Barnes, Reference 8, pg. 57.)

(b) The electric analog of the mechanical system shown in (a). This circuit is based on the f-v (force-voltage) analog. (See Gardner and Barnes, Reference 8, pg. 62.)

and   g = the viscous damping constant
      m = the mass of the oscillating body
      k = system spring constant.

The steady state solution of Equation (14) is

(15) $x(t) = \text{Re}\{\mathbf{X}(\omega)\ e^{j\omega t}\}$
          $= \text{Re}\{X(\omega)\ e^{-j\phi(\omega)}\ e^{j\omega t}\}$
          $= X(\omega)\ \cos[\omega t - \phi(\omega)]$

where the response function $\mathbf{X}(\omega)$ is a complex frequency dependent quantity whose magnitude is given by

(16) $X(\omega) = \dfrac{F_o/m}{\sqrt{((\omega_o^2 - \omega^2)^2 + 2\alpha\omega)^2}} = \dfrac{F_o}{k}\left[\dfrac{\omega_o/\omega}{\sqrt{(\omega_o/\omega - \omega/\omega_o)^2 + 1/Q^2}}\right]$

and whose phase is given by

(17) $\phi(\omega) = \tan^{-1}\left[\dfrac{2\alpha\omega}{\omega_o^2 - \omega^2}\right] = \tan^{-1}\left[\dfrac{1/Q}{(\omega_o/\omega - \omega/\omega_o)}\right]$

where it is customary to introduce the selectivity Q, defined at ω_o as

(18) $Q = \dfrac{\omega_o}{2\alpha} = 2\pi\ \dfrac{\text{Energy stored}}{\text{Energy dissipated per cycle}}$.

Q is a measure of the spectral spread of, Δf, of the system response about the resonant frequency. The phase is the angle by which the displacement *lags behind the driving force. Referring to his classic text on mechanical vibrations,* Den Hartog has called Equations (16) and (17) **"the most important equations in the book".**[29]

It is of interest to plot these quantities as functions of frequency. (See Figure 3a and 3b.) Note that the effect of viscous damping is to spectrally broaden the response, and to shift the frequency of its maximum from the undamped resonance frequency ω_o, where the phase passes through $\pi/2$, down to ω_m where the displacement is maximum. (ω_m is analogous to the frequency of maximum voltage across the capacitor in series RLC circuit.)

Quite often older authors in mechanical engineering[30] would introduce the term **"Magnification Factor"**, which was defined as the ratio of the amplitude of the steady state solution $X(\omega)$ to the static deflection $X(0) = X_o$:

(19) $M(\omega) = \dfrac{X(\omega)}{X_o} = \dfrac{(\omega_o/\omega)}{\sqrt{(\omega_o/\omega - \omega/\omega_o)^2 + 1/Q^2}} = \dfrac{kX(\omega)}{F_o}$

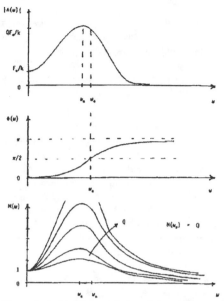

Figure 3. A plot of the magnitude and phase response of a linear mechanical oscillator driven by a time-harmonic forcing function. The *magnification factor* vs normalized frequency is also shown.

This is also the ratio of the magnitude of the force developed across the spring to the amplitude of the applied vibrating force, as shown in the last term. Once again, we have a force multiplication, or magnification, in the form of Equation (4) above. However, as with the transformer, the force is "AC".

A plot of the magnification factor verses normalized frequency for various amounts of damping, specified through the parameter Q, is shown in Figure 3(c). The maximum value of $X(\omega)$ is given by

(20) $X_m = \dfrac{QX_o}{\sqrt{1 - 1/(4Q^2)}} \approx QX_o = QF_o/k$

and force magnification at the maximum is simply

(21) $M(\omega_m) = \dfrac{Q}{\sqrt{1 - 1/(4Q^2)}} \approx Q$.

Further, the frequency for which this occurs is

(22) $\omega_m = \omega_o\sqrt{1 - 1/(2Q^2)}$

and the maximum stored potential energy occurs at this frequency.

How can such a system be built and the AC force magnification be used to mechanical advantage? Tesla described just such a system, in 1919, which can be driven by either a centrifugal pump or by a reciprocating pump. See Figure 4, which is taken from Tesla's series "My Inventions". The mechanism, an air powered jack-hammer, functions by means of the introduction of a nonlinear element — the "escape ports".

In the mechanical apparatus illustrated, an attempt is made to convey an idea of the electrical operations as closely as practicable. The reciprocating and centrifugal pumps, respectively, represent an alternating and a direct current generator. The water takes the place of the electric fluid. The cylinder with its elastically restrained piston represents the condenser. The inertia of the moving parts corresponds to the self-induction of the electric circuit and the wide ports around the cylinder, through which the fluid can escape, perform the function of the air-gap. The operation of this apparatus will now be readily understood. Suppose first that the water is admitted to the cylinder from the centrifugal pump, this corresponding to the action of a continuous current generator. As the fluid is forced into the cylinder, the piston moves upward until the ports are uncovered, when a great quantity of the fluid rushes out, suddenly reducing the pressure so that the force of the compressed spring asserts itself and sends the piston down, closing the ports, whereupon these operations are repeated in as rapid succession as it may be desired. Each time the system, comprising the piston, rod, weights and adjustable spring, receives a blow,

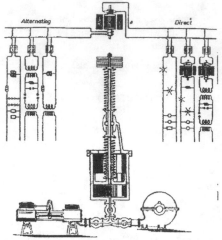

Figure 4. Mechanical analog for a lumped circuit Tesla coil.

it quivers at its own rate which is determined by the inertia of the moving parts and the pliability of the spring exactly as in the electrical system the period of the circuit is determined by the self-induction and capacity. If, instead of the centrifugal, the reciprocating pump is employed, the operation is the same in principle except that the periodic impulses of the pump impose certain limitations. The greatest energy of movement will be obtained when synchronism is maintained between the pump impulses and the natural oscillations of the system. [31]

It is not difficult to see the path in Tesla's thoughts leading to a mechanical conception of human energy: "Wherever there is life, there is a mass moved by a force."[32]

The Voltage Magnifier and Electrical Oscillator

Turning to the electrical case, we consider the analog of the mechanical oscillator. A series resonant RLC circuit driven by a sinewave generator has the same analysis as above, with $X \rightarrow q(\omega)$ and $F \rightarrow V(\omega)$. At resonance, the magnification, given by Equation (21), again ascends to $M = Q$ and the voltage across the capacitance increases to

(23) $V_c(\omega_m) = -jQV_o = QV_o|_{-\pi/2}$.

The $-j$ indicates that the capacitor voltage is in lagging phase quadrature with the source voltage. The voltage across the inductance rises to

(24) $V_L(\omega_m) = jQV_o = QV_o|_{+\pi/2}$

and is in leading phase quadrature with the source voltage. The current is limited only by the circuit resistance. At resonance, the current through the circuit rises until the voltage across the resistive loss is equal to the source voltage.

This simple circuit was a source of deep frustration to Edison because voltmeter readings taken

198

around the loop did not obey Kirchoff's law! As a result, he issued the famous warning,

> *Take Warning! Alternating currents are dangerous! They are fit only for powering the electric chair. The only similarity between an AC and a DC lighting system is that they both started from the same coal pile.*[33]

Ronald Scott, Dean of Engineering at Northeastern University has observed,

> *One of the reasons that Edison mistrusted alternating currents was that he didn't really understand them ... it wasn't until the work of Steinmetz at the General Electric Company that a good method for solving AC circuits became available. In 1897, complex algebra and phasors were applied to electric circuits in a classical paper by Steinmetz, and since that time no one has had any excuse for not understanding AC circuits.*[34]

The series resonant RLC circuit has been called "**a voltage amplifier**" or "**a voltage magnifier**" by many of the old electrical engineering texts. (See Ryder,[35] for example.) The dual network (the parallel or "anti-resonant" circuit) has been called a "current magnifier". In both cases, the magnification, M = Q.

The electrical analog of Tesla's jack hammer oscillator is found by introducing an electrical nonlinear element in place of the escape ports - the spark gap. When a spark gap is set across the circuit, the capacitor charges up to the gap breakdown and RF oscillations occur as long as the plasma arc across the gap conducts. (Contrary to many erroneous discussions of spark gap transmitters, the spark does not create the RF oscillations!)

Consider the push button "oscillator" shown in Figure 5. If the button is held down, current will flow and magnetic energy will be stored in the magnetic field of the coil. When the button is released, none of this energy can return to the battery, instead, as the magnetic field collapses, the current induced in the circuit will charge up the electrical potential energy stored in the capacitor. Energy, of course, will be dissipated in the resistive losses.

This interplay of kinetic energy and potential energy will occur at a fixed rate known as the natural resonant frequency of the system. The situation is not unlike a child on a swing, the pendulum in a grandfather's clock, or the torsional oscillator (or flywheel) in an old-fashioned pocket watch. In these mechanical oscillators the frequency may be maintained with considerable precision. However, a supply of energy must be periodically imparted to the system, at the right time in the cycle, to overcome the resistive damping. Otherwise the motion will simply ring down at the resonant frequency. In a watch, the energy was stored in a mainspring and periodically released to the flywheel through the use of a nonlinear mechanism called a ratchet and pawl.

In Figure 5, the energy exchange back and forth in the LC tank circuit provides an electrical flywheel for oscillations, but we need a creature the likes of Maxwell's demon to stand at the push button and press the button synchronously with the oscillations, in order to introduce enough energy each cycle to overcome the resistive losses.

Ryder notes that what is needed is a nonlinear device such as a vacuum tube:

> *The tube will operate as a synchronous switch, supplying dc power in pulses (of length $2\theta_1$) to the resonant LC circuit, in synchronism with the voltage across the load. After supplying a pulse*

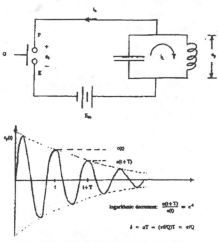

Figure 5. A "push-button" electrical oscillator.

of energy, the switch disconnects the energy source from the load, and the energy in the load continues to oscillate at the resonant frequency. The action is comparable to that of a pendu-lum. driven by short energy pulses, and swinging freely at its own rate for most of the cycle.[36] This type of operation occurs for class C amplifiers.

The Early Tesla Transformer - Coupled Lumped Oscillations

Coupled mechanical oscillators have been treated in considerable detail in many books. They are the counterpart to the tuned **lumped** coupled coils which describe Tesla coils during the duration of the primary spark. Since we have written many articles and an entire book on this (complete with software), we will refer the reader to the literature. A well written and lucid treatment of **lossless** lumped coupled tuned coils is presented by Finkelstein. Smythe includes losses. But, neither author recognizes the importance of spark duration, and both miss the slow-wave distributed resonator mode of the helical secondary.

TCTUTOR has a useful discussion of the whole topic. The bottom line on all this is that lossless tuned **lumped** coupled coils, assuming infinite spark duration, have a magnification given by

$$(25) \qquad M = \sqrt{C_1 / C_2}$$

This can be greatly improved upon by following the disclosures patented by Tesla in 1897 - operate the secondary as a quarter wave helical resonator, not a lumped coupled coil. The voltage rise will then be by VSWR, not lumped circuit Q. A true Tesla Coil is a distributed resonator.

Transmission Line RF Resonance Transformer

Transmission line resonance transformers utilize wave interference phenomena for the build up of extremely high voltage standing wave modes. It has long been recognized that RF resonance transformers possess the following capabilities:

1) high voltage generation (tens of megavolts, limited only by the electrical breakdown of the electrode geometry and surrounding dielectric medium)

2) high efficiency (limited only by the ac copper losses of the structure and RF dielectric losses of the primary capacitance)

3) high peak power performance (hundreds of megawatts per discharge)

4) high repetition rate, heavy duty cycle operation (hundreds of kilowatts of continuous average power processed)

5) broadband spectra (generated via high voltage RF discharge).

It is noteworthy that, in the days of Breit, Tuve, Cockroft, Walton and Van De Graaff, a complaint against spark gap transmitters and high voltage RF generation by distributed resonance transformers was that their output was too broadband! (These early pioneers of nuclear science merely required high energy — not high power.)

The relevant physical processes by which high voltage is produced on microwave resonators is common knowledge.[37] Consider the generic transmission line shown in Figure 6. The coordinate origin is taken at the load and a time harmonic generator drives the input end at $x = -\ell$. The voltage at any point along the line is then given by the expression

$$(26) \qquad V(x) = V_+ e^{-\gamma x} + V_- e^{\gamma x}$$

where x=0 at the load and $x = -\ell$ at the generator end. Physically, Equation (26) expresses the fact that the voltage at any point along the transmission line is the superposition of a forward travelling wave and a backward travelling wave.

The resultant analytical expression describes a spatially distributed interference pattern called a standing wave. As usual, γ is the complex propagation constant $\gamma = \alpha + j\beta$. The complex constants V_+ and V_- follow from the second order partial dif-

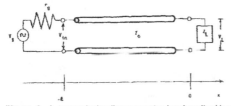

Figure 6. A transmission line resonator (as described by Tesla in US Patent # 593,138 - November 2, 1897).

ferential equation of which equation (26) is a solution (the "transmission line equation"), and depend upon the boundary conditions (the generator and the load).

Also, at the load end one may define the complex load reflection coefficient Γ:

(27) $\Gamma = V_+/V_- = |\Gamma|\,|\underline{\theta}$

For an open circuited line $\Gamma = 1|\underline{0}$.

From equation (26) we have, at the input end of the line,

(28) $V_{input} = V(-\ell) = \left[e^{\ell\gamma} + \Gamma e^{-\ell\gamma}\right]$

where, again, Γ is a complex quantity. Also from equation (26), we may write the voltage at the load end as:

(29) $V_{Load} = V(0) = V_+ + V_- = V_+[\,1+\Gamma\,]$.

Equations (28) and (29) may be combined in the following extremely useful expression which relates the load voltage to the input (generator end) voltage:

(30) $V_{load} = \dfrac{V_{in}[1+\Gamma]}{\left[e^{\ell\gamma} + \Gamma e^{-\ell\gamma}\right]}$

Now consider what happens on an open-circuited low-loss line one quarter wavelength long. Simple complex algebra gives the following well-known result:

(31) $V_{Load} = V_{in}\,/(j\alpha\ell)$ (for $\ell = n\lambda/4$ with n odd)

where, again, α is the attenuation per unit length of the transmission line, and the j implies that the voltages at the two ends are in phase quadrature. The structure is a lossy, tuned reactive resonator. Since the numerator of equation (31) is finite and the denominator is vanishingly small, the voltage standing wave will build up to very large values!

The transient build-up to the steady state expressions given above is easy to understand. Initially a forward wave of voltage is launched from the input end of the line. It propagates toward the open circuit (high impedance) end of the line where it is reflected with zero phase shift. This reflected wave travels back down the line to the low impedance input end, where, not only is it reflected with a 180° phase shift but, because it took a half cycle to travel

up and back along the quarter wave long line, this twice reflected wave is now in phase with the source of original energy being launched into the line.

The voltage wave now being launched into the line will add directly to the twice reflected wave as it travels back toward the open circuit load end. These coherent additions will continue to proceed, sloshing the load voltage higher and higher as a standing wave forms along the line with a voltage minimum at the input end and a voltage maximum at the open circuited load end. This growth process will continue until either a discharge occurs at the load end (i.e.,- a nonlinearity of the system), or the line's I^2R losses are equal to the power being supplied by the source (i.e.,- the generator can't push the system any higher).

What limits the maximum attainable voltage? The power driving the line, the line losses $\alpha\ell$, and the breakdown potential of the load geometry (which usually arises from the onset of cold field emission from the electrode).

In common parlance, this is called a Tesla coil. (The secondary of a Tesla coil is a helically distributed quarter-wave resonator, not a lumped tuned circuit. The voltage rise is by standing wave: $V_{max} = S\,V_{min}$, where S is the VSWR on the transmission line resonator. The actual measured voltage distribution on a Tesla coil or a transmission line resonator follows the first ninety degrees of a spatial sinusoid, much as it would on a quarter-wave vertical monopole antenna: V_{min} at the base and V_{max} at the top.)

It should be obvious that resonance transformers (or tuned transmission line resonators) do not have to be shock excited (link coupled) by spark gap oscillators. They perform equally well when driven by any high power master oscillator - spark gap, vacuum tube or solid state. Junction breakdown and device efficiency become major concerns with the latter two. However, the design philosophy is the same and, historically, the approach has been used with devices operating at frequencies as low as 60 Hz (HV power supplies for X-ray tubes) and as high as several GHz (RF plasma torches). The authors have published fairly extensive engineer-

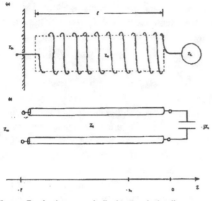

Figure 7. A slow-wave helical transmission line resonator (a Tesla coil).

ing analyses of such circuits in the past.[38,39,40,41]

There have been many practical applications of the above technique wherever high voltage impulse sources are needed. In particular, the Soviets have advanced the technology to a mature state for use with high current nanosecond pulsed beams.[42,43]

[It should parenthetically be remarked that, first, what some authors have called *Tesla Coils* in the past are simply lumped tuned coupled coils. Tesla was using these prior to 1892. When comparing the voltage rise produced by lumped coupled coils with that obtainable from distributed resonators, Tesla would write, "No such pressures - even in the remotest degree, can be obtained with resonat-·ing circuits otherwise constituted with two terminals forming a closed path."[44] The best engineering analysis which we have seen of the coupled coils configuration is given in Smythe.

Secondly, the "Tesla Coil" so commonly seen today is in fact a link-coupled distributed tuned resonance transformer. It is easily documented that Tesla was using the latter prior to 1898. As Sloan observes, the lumped analysis of this configuration totally fails.

Thirdly, Tesla's most famous high voltage RF experiments, the photographs of which the public at large is so familiar with, employed what he called his "Extra Coil." From his recently published Colorado Springs diary of 1899, it is clear that this structure is actually the slow wave helical transmission line resonator of Figure 7. The structure was excited at its base by a relatively narrow band RF signal generator.]

Cavity Resonators - Potential Magnification

As with both lumped and distributed circuits, it is possible to show that cavity resonators also are *magnifiers*. Both Smythe[45] and Condon[46] derive expressions for voltage magnification in cavity resonators excited by inductive loops and capacitive probes. The former is the ratio of the voltage across the cavity to the emf in the loop,

(32) $M = |V / \xi|$

and the latter is the ratio of the maximum potential across the cavity to that across the probe. Smythe's examples give voltage magnifications on the order of a hundred. The procedure could be formally executed for the Schumann cavity, of course. In all of the above, we have tacitly assumed that the elements are linear and passive. No external pumping is being done as with some possible ionospheric-TWT amplifier effect. There is yet one more surprise that distributed linear systems have in store - they can "multiply" power (without violating conservation of energy, of course).

Traveling Wave Ring Power Multiplier

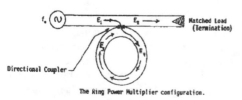

The Ring Power Multiplier configuration.

Figure 8. A ring power multiplier.

In addition to quarter wave transmission line resonators and cavity resonators which step up voltage, as just described, there is also a novel technique for stepping up real RF power (within practical limits). This unusual circuit geometry, invented by F.J. Tischer in 1952,[47,48] actually makes it possible to obtain practical power level *multiplication* of 10 to 500 times the transmitter output. (The

202

Poynting's vector power flow, $\frac{1}{2}Re\{\mathbf{E}\times\mathbf{H}^*\}$, is actually pumped up within this linear, passive distributed storage network. No physical laws are violated, and energy conservation is preserved in the process, of course.)

The closed ring structure shown in Figure 8 is connected by means of a directional coupler to a second transmission line which is excited by a source and terminated in a matched load. The distributed network is characterized by a real Poynting vector circulating within the ring, where unidirectional progressive traveling waves build up in time. The physical explanation advanced by MIT-Lincoln Lab's Stanley Miller is as follows:

A wave proceeding from the source is partially coupled into the ring (by the directional coupler), and propagates around the ring in one direction as shown. When this wave passes the coupling region, a small fraction is coupled back into the main transmission line with the remainder proceeding around the ring again. At the same time, more energy is being coupled from the main line into the ring. If the wave proceeding around the ring and the wave coupled into the ring are in phase, then it is evident that the wave in the ring can be reinforced and can become quite large in magnitude.[49]

The build up of the power in the ring will continue with each cycle until the losses around the loop plus the loss in the termination is equal to the power supplied by the source. Hayes and Surette report actual ring power levels of 200 kW with a 10 kW generator at VHF, with a 6-inch coaxial line resonant ring.[50] Miller reported 100 MW levels from a 200 kW source driving a WR 2100 waveguide at 425 MHz. (He also reported spectacular pyrotechnical and acoustical results when a slight mismatch was inserted into the guide.) The primary use of the traveling wave ring power multiplier has been to test high power RF components which fail due to power dissipation as opposed to voltage breakdown. (The latter could be tested with the transmission line resonance transformers described above.)

It is evident that a great variety of multiply connected geometrical enclosures[51] with the appropriate electrical constitutive parameters can serve as "*ring*" power multipliers. The necessary condition is that the successively added field components, from the ensuing circulations of field energy, possess the correct phases for constructive interference (coherent buildup). Such multiply connected structures can be operated either in the sinusoidal steady state (cw) or in a pulsed mode. Coherence and synchronization are particularly critical for the operation of the latter. We will now present a technique by which the fundamental ring resonator modes of an electrically large resonator may be determined by probing with a DSB (double sideband) modulated carrier of frequency much greater than the self resonant frequency of the resonator. [The theory may be applied to the probing of terrestrial (Schumann type) resonances with a VLF transmitter in place of an ELF source.]

Multiple Beam Interference

Optical multiple-beam interference is perhaps best known in the classical case of the monochromatically illuminated parallel plane geometry, such as dielectric plates and the parallel mirror geometry of Fabry and Perot. As a result of the multiple reflections, the wave is split into many partial waves reflecting back and forth from the parallel boundaries, which interfere with one another as they linearly superpose together. Knowing the optical path difference between successive partial waves and the phase of the reflections, the plane wave summation is formed and the intensity (the squared modulus of the field strength) is determined. The result is the famous expression obtained by G.B. Airy in 1831.

Let us repeat the analysis for the case of around the circuit propagation for TEM waves. We will consider two cases: the pure carrier wave and the DSB modulated wave. For the sake of simplicity we will employ an unpretentious ray optic model. The model is appropriate for optical and TEM transmission line ring resonators. It is clear and lends itself to simple prediction and experimentation.

Case I: Monochromatic Carrier

Consider the geometry shown in Figure 9 with $f_m = 0$. The transmitter at wavelength $\lambda = c/f$ is located at the top of the ring and multiple peripheral propagation paths are shown. In the region near the source, the nth time around propagating wave can be expressed as

(33) $E_n(\lambda) = E_o A(L)^n e^{-j2\alpha nL/\lambda} = E_o e^{-n\alpha L} \cdot e^{-j2\alpha nL/\lambda}$

where $A(L)$ is the propagation attenuation, α is the attenuation constant at the wavelength λ, and L is the circumference of the ring $L = C_e = 2\pi R_e$. For plane wave propagation, $A(nL) = e^{-n\alpha L}$. The total field observed is

(34) $E_T(\lambda) = E_o + E_1 + E_2 + \ldots$

That is, near the source

(35) $E_T(\lambda) = E_o [1 + Ae^{-j2\alpha L/\lambda} + Ae^{-j4\alpha L/\lambda} + \ldots]$

The terms on the right bear resemblance to the geometric series

(36) $1/(1-x) = 1 + x + x^2 + x^3 + \ldots$

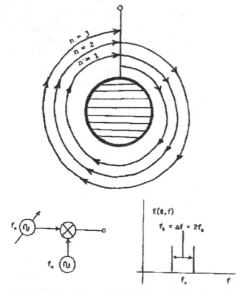

E(θ,f)

$f_b = \Delta f = 2f_o$

f_o f

Figure 9. Ray optic model of VLF transglobal propagating TEM waves and a DSB transmitter. The procedure is choose an f_o appropriate for propagation around the ring, and then to tune the modulating frequency to produce beats between the USB and the LSB such that envelope resonances are observed.

Consequently, Equation (35 may be expressed as

(37) $E_T(\lambda) = E_o [1 + Ae^{-j2\alpha L/\lambda}]^{-1} = |E_T| |\Phi$

where

$|E_T| = E_o [1 + A^2 - 2A \cos\beta L]^{-1/2}$

$|\Phi = \tan^{-1} [(-A \sin\beta L)/(1-A \cos\beta L)]$

This is a complex phasor associated with the time-harmonic field. The time domain expression for the field is found from the expression:

$E_T(t) = Re\{E_T(\lambda)e^{j\omega t}\} = Re\{|E_T(\lambda)| e^{j\Phi(\lambda)} e^{j\omega t}\}$.

The ring is resonant when $\Phi = 2n\pi$. The wave power density is proportional to the squared modulus, so that

(38) $P \sim EE^* = |E(\lambda)|^2 = E_o^2 \dfrac{1}{[1 - A(e^{-j2\pi L/\lambda} - e^{j2\pi L/\lambda}) + A^2]}$.

Remembering that $\cos\theta = \frac{1}{2}[e^{j\theta} + e^{-j\theta}]$, we can write

(39) $|E(\lambda)|^2 = E_o^2 \dfrac{1}{[1 + A^2 - 2A \cos(2\pi L/\lambda)]}$

It is customary to employ the trigonometric identity $\cos\theta = 1 - 2\sin^2(\theta/2)$ and write

(40) $|E(\lambda)|^2 = E_o^2 \dfrac{1}{[(1-A)^2 + 4A \sin^2(\pi L/\lambda)]}$

which can be manipulated into the form

(40') $|E(\lambda)|^2 = E_o^2 \dfrac{1/(1-A)^2}{[1 + (4A/(1-A)^2)\sin^2(\pi L/\lambda)]}$

which is analogous to Airy's formula for multiple beam interference in a parallel plate geometry. Fabry called the Airy denominator factor $F = 4A/(1-A)^2$ the *finesse*. Born and Wolf[52] define the finesse as $\Im = 1/2\pi\sqrt{F}$. When a polychromatic beam is present, as in a spectrum analyzer, the latter expression has the advantage of being equal to the ratio of the line separation (Δf) to the half power width of the line (δf).[53] Finesse is a measure of the resolution of an interferometer. The term $[1 + F \sin^2(\xi)]^{-1} \Delta f(\xi)$ is called the Airy function and it represents the transmitted power density. Figure 10 is a plot of $|E|^2/|E_o|^2$ verses ξ, the phase shift arising from the path-length difference between successive rays ($\xi = 2\pi\Delta\ell/\lambda$).

204

In the case where we have plane wave propagation, Equation (40) becomes

$$(41) \quad |E(\lambda)|^2 = E_o^2 \frac{1}{\left[\left(1 - e^{-\alpha L}\right)^2 + 4e^{-\alpha L}\sin^2(\pi L/\lambda)\right]}$$

A plot of the square root of Equation (41) against frequency shows the magnitude of the electric field strength as we tune the carrier. Resonances occur at the frequencies where $\pi L/\lambda$ is an integral multiple of π, i.e. at frequencies equal to $f = nc/L$. The ratio of the separation between peaks to the half power width of a resonance line is called the resonator's "finesse."

A direct application of Equation (41) to the terrestrial resonator gives resonances whenever $f = 7.5n$ for $n = 0,1,2,3,...$ Presumably this formula would be acceptable for carriers at VLF, however down at ELF (near 7.5 Hz and up to 1 kHz) the waves don't possess planar phase fronts and a full solution requires the Schumann analysis. At ELF the world behaves as a reactive resonator (E and H are in phase quadrature) instead of a ring power multiplier (for which E and H are in-phase). The ray optics and plane wave assumption is a fairly crude physical model for the VLF waveguide. (From a practical standpoint, propagation losses and phase inhomogeneities might hide terrestrial ring resonances). Basically, the form of ray optic assumption used above is that the earth-ionosphere waveguide can be approximated as a great circle TEM transmission line. A more formal treatment could be given,[54,55,56] of course, and microwave engineers

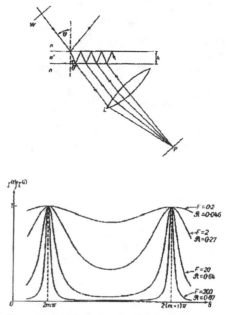

Figure 10. (a) Formation of multiple beam fringes in a parallel plate of index of refraction n', focused to give the transmitted intensity (power density) I'. (b) The Airy function - a plot of the ratio of the transmitted and incident intensities as a function of the propagation path-produced phase difference $\delta = 4\pi hn'\cos\theta/\lambda_0$. [Ref. Born and Wolf, pp. 326-7. Script R is the power reflection coefficient rr* and F = 4R/(1-R)².]

will recognize that the same results could easily be gotten from an S-matrix calculation instead of the Airy formalism, but we will explore this as a working hypothetical model.

Case II: DSB Modulation (Beats)

When the carrier is DSB/SC modulated one can employ carrier frequencies high enough for practical directional couplers, and lower the *modulation* frequency to the point where the wavelength associated with the modulation envelope is exactly the circumference of the ring. As shown in Appendix I, the interference pattern fringe spacing (the distance between the intensity maxima) is exactly equal to the circumference of the ring (L) when the beat frequency between the USB and the LSB is an integral multiple of c/L. We desire to obtain the variation of the wave intensity near the transmitter as the modulation frequency is varied. Consider the DSB waveform

$$(42) \quad E(t) = B\cos(\omega_m t)\cos(\omega_0 t)$$
$$= \tfrac{1}{2}B\cos(\omega_0 + \omega_m)t + \tfrac{1}{2}B\cos(\omega_0 - \omega_m)t$$

where ω_0 is the carrier frequency and ω_m is the modulation frequency. The spectral separation between the USB and the LSB is the "beat" frequency, which is equal to twice the modulation frequency: $\omega_b = 2\omega_m$.

Analogous to Equation (33) above, the nth global traversing DSB wave, measured at a site near the transmitter, can be represented as

(43) $E_n(f_o, f_m) = E_o A(L)^n [e^{-j2\pi(fo-fm)nL/c} + e^{-j2\pi(fo+fm)nL/c}]$.

where c is the speed of light. Again, $A(L)$ is the amplitude attenuation per global traverse. Algebraic manipulation leads to the expression

(44) $|E|^2 = \dfrac{E_o^2}{\left[1 - A \cos(2\pi f_m L / c)\right]^2 + 4A \sin(\pi f_o L / c)}$

(45) $|E|^2 = \dfrac{E_o^2}{\left[1 + e^{-2\alpha L} \cos(2\pi f_m L/c) - 2e^{-\alpha L} \cos(2\pi f_m L/c)\cos(2\pi f_o L/c)\right]}$

or, equivalently, as

(46) $|E|^2 = \dfrac{E_o^2}{\left[1 - e^{-\alpha L} \cos(2\pi f_m L / c)\right]^2 + \left[4e^{-\alpha L} \sin^2(\pi f_o L / c)\right]}$

where α is measured at the carrier frequency. This expression reduces to Equation (41) when the modulation frequency is zero. Equation (46) is a rather remarkable result. Admittedly, it is rigorous only in the ring optics and TEM transmission line case, but it does describe the ring resonator physics. It implies that ring resonances (and power multiplication) can occur in a "more-or-less" carrier independent manner. That is, holding the carrier frequency constant and **varying the modulation frequency** will lead to field strength maxima whenever $f_m = \frac{1}{2}nc/L$. The greatest finesse of the resonator will occur when the carrier frequency is tuned such that $\pi f_o L/c = n\pi$. That is the case whenever $f_o = nc/L$.

[Terrestrial resonances might be observed (depending on propagation losses) whenever the **beat frequency** is $f_b = 2 f_m = 7.5n$ for n = 0,1,2,3, ... Equation (12) of Appendix I gives the spatial distance between intensity maxima, Λ, for $f_b = 7.5$ Hz as $\Lambda = 40,000$ km = the circumference of the earth. If the beat frequency is tuned to 7.5 Hz, the travelling modulation envelope peak of the DSB wave will propagate all the way around the globe and return to the position of the source exactly in synchronism with the next peak of the modulation envelope. Although the result is "more-or-less" independent of the carrier frequency, critical to detecting the phenomenon is the use of a carrier frequency low enough so that great circle global propagation can occur (VLF), and it is necessary to tune the beat frequency for ring resonance.]

Example

Suppose $f_o = 5$ kHz (i.e., $\lambda = 60$ km). Further, suppose that we tune the modulation oscillator to $f_m = 3.75$ Hz. This will produce two signals:

$f_1 = 5.00375$ kHz (i.e., $\lambda_1 = 59.955$ km)
$f_2 = 4.99625$ kHz (i.e., $\lambda_2 = 60.045$ km).

And the beat frequency is $\Delta f = f_1 - f_2 = 7.5$ Hz.

The result is that the observed fringe spacing, given by Equation (I.12) is exactly the circumference of the earth: $\Lambda = 40,000$ km $= C_c = 2\pi R_e = 2\pi(6.37 \times 10^6)$ meters. If the beat frequency is tuned to 7.5 Hz, the travelling modulation envelope peak of the 5.0 kHz DSB wave in Figure 9 **propagates all the way around the globe and returns to the position of the source exactly in synchronism with the next peak of the modulation envelope**.

Calculation of Field Strength at VLF

While there exists extensive theoretical literature on the computation of electric field strengths at VLF, we shall employ an **empirically** determined engineering VLF formula. The most famous VLF empirical formula was obtained by Dr. Louis W. Austin and Cohen for the U.S. Bureau of Standards.[57,58] Although now considered more-or-less as a "museum piece", the empirical formula gives daylight VLF signal strength for ranges of 2,000 to 10,000 km with fair accuracy. According to Austin, when the transmitted power is known, the daytime field strength may be calculated as[59]

(47) $E_{RMS} = \dfrac{300\sqrt{P_{kw} G}}{d} \sqrt{\dfrac{\theta}{\sin\theta}} \exp(-0.0014 d_{km} / \lambda^{u(in)})$ mV / m.

where d is the range in km, λ is the wavelength in km, and θ is the angular separation between transmitter and the receiver. The formula can be used for wavelengths from 100 down to 10 km, corresponding to frequencies from 3 to 30 kHz. At ranges beyond 10,000 km, the Austin-Cohen formula gives values well below what is actually observed. It is particularly poor in the region of the antipode, and consequently it cannot be used to examine transglobal propagation.

During World War II, from 1939 to 1945, the German Navy collected extensive experimental data for global VLF transmission. Zinke deduced an empirical formula which represents correctly the daytime and nighttime observations of the German Navy.[60, 61] The result was an "Austin-Cohen like" formula which is far more accurate for propagation in the antipodal region (20,000 km).

$$(48) \quad E_{RMS} = \frac{15\sqrt{P_{kw}}\,G}{\sqrt{d_{km}}} \sqrt{\frac{\theta}{\sin\theta + 0.008}} \exp(-\alpha d_{km}/\lambda^x)\,mV/m$$

where

$$\alpha = \begin{cases} 0.003 & (day) \\ 0.0005 & (night) \end{cases}$$

$$x = \begin{cases} 1.0 & (day) \\ 0.5 & (night) \end{cases}$$

and both d and λ are in kilometers. The geocentric angle between transmitter and receiver, $\theta = d/R_e$, is in radians. The parameters α and x are multivalued because the propagation reflections are affected by the lower D region on the daylight side of the globe and by the E region on the night side. Physically, the exponential factors in Equations (47) and (48) are traceable to energy leakage out of the earth-ionosphere waveguide (attenuation is primarily due to the upper boundary).

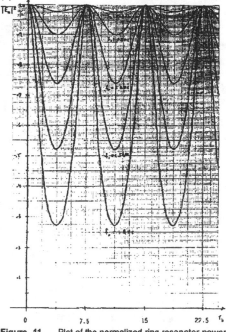

Figure 11. Plot of the normalized ring resonator power density (for various VLF carrier frequencies) as a function of the DSB beat frequency. As the modulation is tuned, the detected fields go through peaks and valleys.

Examples at VLF

The empirically based VLF model which we will employ assumes that the A(L) in Equation (43) above can be expressed by Zinke's exponential damping factor

$$(49) \quad A(L) = \exp(-\alpha d_{km}/\lambda_{kmx}).$$

where, for nighttime propagation $\alpha = 0.0005$ and $x = \frac{1}{2}$. For the sake of argument, we will assume that the carrier frequency (wherever it may be set) is tuned such that $f_0 = nc/L$ for maximum finesse. A plot of $|E|^2/|E_0|^2$, using Zinke's formula *for nighttime experiments*, verses DSB beat frequency is given in Figure 11 for several different carrier frequencies across the VLF band. *The result clearly demonstrates the ability of a VLF station to detect the terrestrial ring resonances*.

Historical Observations

In a previous publication we showed that if Tesla transmitted a series of impulses at 7.5 Hz (i.e. launched baseband ELF - a truly heroic accomplishment!) then due to the range ambiguity inherent in pulse range measuring devices, a pulse train of echoes would be received 0.08484 seconds after each transmitted pulse.[62,63] It should now be clear that it would **not** be necessary for Tesla's 250 kW Colorado Springs transmitter of 1899 to "radiate" at ELF to obtain the same result. All that is necessary is that his transmitter power and carrier frequency be capable of round-the-world propagation. In a subsequent newspaper interview, Tesla said,

With my transmitter I actually sent electrical vibrations around the world and received them again, and I then went on to develop my machinery.[64]

A high power pulsed transmitter operating in the 1 to 20 kHz VLF band would readily lead an ob-

TABLE I.   SUMMARY OF CLASSICAL LINEAR PASSIVE SYSTEMS THAT MAGNIFY

SYSTEM	SCHEMATIC	MAGNIFICATION
1. Lever	r_1 ... r_2, F_1 —— Δ —— F_2	$M = r_1/r_2$
2. Transformer	E_1 ⊢⊣ E_2 $\quad$ 1 : N	$M = E_2/E_1 = N$
3. The rest of the "simple machines"		$R = F_2/F_1$
4. Lens	0 ... I, $\longleftarrow r_1 \longrightarrow\!\!<\!\!-r_2\!\!-\!\!>$	$M = -r_1/r_2$
5. Telescope	$\longleftarrow\!\! f_1 \!\!\longrightarrow\!\!|\!\!<\!\!-f_2\!\!-\!\!>$	$M = -f_1/f_2$
6. Resonant Circuit	—WW R —mm L— C—	$M = \dfrac{|E_L|}{|E_g|} = \dfrac{|E_c|}{|E_g|} \sim Q$
7. Lumped Tuned Coils (Tesla Coils: t ~ 1891)	V_o —C_1— L_1 $\}$ $\{$ L_2 C_2	$M = \dfrac{V_{C2}}{V_o} = \sqrt{\dfrac{C_1}{C_2}} = \sqrt{\dfrac{L_2}{L_1}}$
8. Transmission Line Resonators (Tesla Coils: t > 1894) "Resonance Xfmrs"	$V_{min} \longleftarrow\!\!-\!\!\cdots\!\!- \sim \frac{1}{4}\lambda \cdots\!\!-\!\!-\!\!\longrightarrow V_{max}$	$V_{max} = S\, V_{min}$ $\quad S \approx 1/\alpha\ell$
9. Cavity Resonators	$V_c = \int E\!\cdot\!d\ell$	$M = \dfrac{V_{cavity}}{\mathcal{E}_{probe/loop}}$
10. Ring Power Multipliers	V_1^+ —ww→ V_4^-	$M = \dfrac{|V_4^-|^2}{|V_1^+|^2} \sim 1/q^2$ $\quad M = \dfrac{P_{ring}}{P_{gen}}$
11. Light Recycling		

server to the following conclusions:

- The terrestrial globe resonates at multiples of 7.5 Hz. (See Figure 11 above.)
- The received pulse train has echoes which occur 0.08484 seconds after the transmitted pulses. (See References 62 and 63.)
- These conclusions are more-or-less independent of carrier frequency for $f_0 \leq 25$ kHz. **Above 25 kHz the global response "washes out"**. (See Figure 11 above.)

Tesla's exact words from May of 1900 were,

I would say that the frequency should be smaller than twenty thousand per second, though

shorter waves might be practicable. The lowest frequency would appear to be six per second...[65]

On July 14, 1905, Tesla wrote that the appropriate wavelengths to transmit were in the range of 25 - 70 km. These correspond to frequencies of 12 kHz -4.29 kHz. On July 6, 1912, Tesla indicated that his experiments indicated that the transmitted wavelength should be no shorter than 12 km (25 kHz), and he went on to state, "... on this fortunate fact rests the future of wireless transmission of energy." Certainly, these would be consistent with the excitation of the whole world as a ring power multiplier-resonator.

Tesla Speaks

Tesla has spoken of his global wireless communication system on many occasions. When reviewing his "World system of Wireless Transmission of Energy" Tesla once wrote,

The chief discovery which satisfied me thoroughly as to the practicability of my plan, was made in 1899 at Colorado Springs, where I carried on tests with a generator of 1500 KW

capacity and ascertained that under certain conditions the current was capable of passing across the entire globe and returning from the antipodes to its origin with undiminished strength.

Have we overlooked the obvious? Will the Magnifying Transmitter be a lever for 21st Century civilization?

Conclusion

In one of the most eloquent sections of his enchanting "Lectures on Physics," Richard Feynman describes the wonderful art of electrical technology and how electrical power generation and transmission rests solidly on Faraday's remarkable discovery that an emf is produced by a time varying magnetic field. Listen as he describes the generators at Boulder Dam and the cities to which it is connected:

What is Boulder Dam? ... an exquisitely intricate mess of copper and iron all twisted and interwoven ... A revolving monster thing. A generator ... And everything must be enormously efficient ... The power for a metropolis is going through ... All done with specially arranged pieces of copper and iron ... [In the cities] thousands of little wheels, turning in response to the turning of the big wheel at Boulder Dam. Stop the big wheel and all the little wheels stop; the lights go out.[66]

Recognizing that today's electro-technology is not the ultimate, Feynman was moved to prophesy,

Power networks of the future may have little resemblance to those of today.

If there be *any* truth in Tesla's disclosures, then we all must certainly agree. In September of 1915, Tesla wrote,

*The present limitations in the transmission of power to distance will be overcome ... through the introduction of the wireless art. .. we have now the means for the economic transmission of energy in any desired amount to the distances only limited by the dimensions of the planet. In view of the assertions of some misinformed experts to the effect that in the wireless system I have perfected **the power of the transmitter is dissipated in all directions, I wish to be emphatic in my statement that such is not the case. The energy goes only to the place where it is needed and no other.** ... With the full development [of water and solar power] and a perfect system of wireless transmission of the energy to any distance, man will solve all the problems of material existence.*[67]

Finally, in 1927, Tesla said,

... the experts must come to the same conclusions I have reached long ago. Sooner or later my power system will have to be adopted in its entirety and so far as I am concerned it is as good as done.[68]

As we approach the end of the century, we can't help but wonder if something has been overlooked in electrical science. We seem to hear the words of Faust again,

What to your spirit she's unwilling to reveal, You cannot wrestle from her with levers, screws and wheel.

In Aeschylus'[69] famous tragedy,[70] Prometheus,[71] the Titan, is fettered with unbreakable adamantine chains to a lofty ragged rock "at the end of the world" amid roaring thunder, flashing lightning, blasting storms, fiery whirlwinds, and an earthquake. His sin:

I gave to mortals the gift of fire ... and they will learn from it all kinds of arts ... I set mortals on the path of a science hard to judge.

Greek translator Rex Warner has identified Prometheus as the patron of the mathematical sciences and engineering:

In recognizing that intelligence, not brute force, was the governing power of the universe, he initiated men into all the arts and sciences which make civilization possible.[72]

Artist Robert Kendall has captured kindred thoughts on the famous Battelle Mural in Columbus, Ohio.

The scientist is truly a Prometheus; he still continues to steal fire from heaven in seeking to comprehend the secrets of nature ... he is mankind's greatest benefactor ... he holds out the greatest single hope that they ultimately may also live together in peace.[73]

Somehow, in spite of the promise of electrical fire, our engineering science still seems to be chained.

THE TELESCOPE

"The 'Magnifying Transmitter'... is in the transmission of electrical energy what the telescope is in astronomical observation."   Nikola Tesla, June 1919

A refracting telescope consists of an objective converging lens of long focal length, f_1, and an ocular (or eye-piece) of short focal length, f_2.

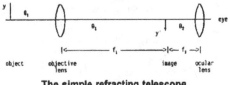

The simple refracting telescope.

The purpose of the objective lens is to increase the amount of light collected from the object in order to make a bright image at the focal point. The objective lens can collect a great deal more light than the eye's pupil. Let θ_1 be the angle subtended by the object at the objective lens, and let θ_2 be the angle subtended by the image at the ocular.

The eye-piece is essentially a microscope to magnify the angular dimension of the image. The *magnification* of the telescope is defined as the ratio of these angles:

$$M = -\frac{\theta_2}{\theta_1} \approx -\frac{f_1}{f_2}$$

The latter follows from the small angle approximation. (The sign is a consequence of the inverted image.) Thus, for great *magnification* it is desirable to have a large f_1 and a small f_2. The similarity to equation (4) had not escaped Tesla.

APPENDIX I: BEATS

Beats are fluctuations in amplitude produced by two oscillations of **slightly different** frequency. Consider a single, omnidirectional source emitting two frequencies f_1 and f_2. (The case of spatially separated sources is treated in the references.) The wave equation gives the resultant field at some distance, r, from the source as the **superposition** of two monochromatic travelling waves:

(1) $E(r,t) = A(r)\cos(\beta_1 r - \omega_1 t + \theta_1) + B(r)\cos(\beta_2 r - \omega_2 t + \theta_2)$

where $\beta = 2\pi/\lambda$ and $\omega = 2\pi f$. (In a dispersive medium $\beta(\omega)$, and the propagation will be different for f_1 than for f_2.) We will let the amplitudes be equal and constant, and assume zero epoch angles. In order to manipulate the expression for the field into a simple formula, we note the trigonometric identity

(2) $\cos a + \cos b = 2 \cos \frac{1}{2}(a+b) \cos \frac{1}{2}(a-b)$.

Equation (1) can then be written as

(3) $E(r,t) = 2A\cos\frac{1}{2}[(\beta_1+\beta_2)r - (\omega_1+\omega_2)t]\cos\frac{1}{2}[(\beta_1-\beta_2)r - (\omega_1-\omega_2)t]$.

Let us define the following quantities
(4a) $\beta_0 = \frac{1}{2}(\beta_1 + \beta_2)$ = average carrier wave number
(4b) $\beta_m = \frac{1}{2}(\beta_1 - \beta_2) = \frac{1}{2}\Delta\beta$ = modulation propagation wave number.
(4c) $\omega_0 = \frac{1}{2}(\omega_1 + \omega_2)$ = average carrier frequency
(4d) $\omega_b = (\omega_1 - \omega_2) = \Delta\omega$ = "beat frequency"
(4e) $\omega_m = \frac{1}{2}(\omega_1 - \omega_2) = \frac{1}{2}\Delta\omega$ = modulation frequency

Then, equation (3) may be written as the simple expression

(5) $E(r,t) = 2A \cos(\beta_m r - \omega_m t) \cos(\beta_0 r - \omega_0 t)$.

We recognize this as a propagating modulated wave,[74] and identify the modulation envelope as

(6) $m(r,t) = 2A \cos[\beta_m r - \frac{1}{2}(\omega_b t)]$.

The intensity, or power in the wave, is proportional to the square of the modulation envelope:

(7) $m^2(r,t) = 2A^2 [1 + \cos(2\beta_m r - \omega_b t)]$

which will oscillate about $2A^2$ with a frequency of ω_b (the beat frequency), i.e.- at twice the modulation frequency. There are two "beats" per cycle of the modulation wave.[75,76] (See Figure I2.)

Each of the separate waves in Equation (1) has a phase velocity found by writing

(a) For Equation (1)

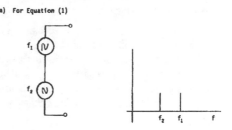

(b) For equation (5')

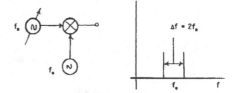

Figure I1. The combination of two wave trains of slightly different frequencies, as shown in (a), is equivalent to a wave of the average frequency DSB modulated by the difference frequency, as shown in (b). In reality, each spectral component has a finite line-width δf.

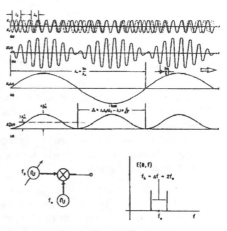

Figure 1.2.   Propagating DSB wave.
(a) Two separate wave components verses spatial position, x.
(b) Spatial distribution of the propagating DSB wave.
(c) Spatial distribution of the propagating modulation waveform.
(d) Distribution of the square of the modulation.
(e) The DSB transmitter.
(f) The transmitter's DSB spectrum.

(8) $\Psi = \beta r - \omega t = \text{const}$

and differentiating with respect to time

(9) $v_\phi = dr/dt = \omega/\beta = f\lambda$.

In a nondispersive medium, both spectral components propagate with the same phase velocity, $v_{\phi 1} = v_{\phi 2}$, and in the electromagnetic case in free space both phase velocities are equal to the speed of light, c. The interference of the two waves in Equation (1) is seen most simply by examining Equation (5). The carrier propagates with velocity $v_\phi = \omega_0/\beta_0 = f_0\lambda_0$.

The modulation envelope, $m(r,t)$ also propagates. To find its velocity, we again we set the argument equal to a constant:

(10) $\Psi_m = \beta_m r - \tfrac{1}{2}\omega_b t = \text{const}$.

Differentiating gives the modulation envelope velocity as

(11) $v_g = \dfrac{\omega_m}{\beta_m} = \dfrac{\Delta f}{1/\lambda_1 - 1/\lambda_2} = \Delta f \Lambda$

The interference pattern fringe spacing is simply the spatial distance between intensity maxima, (see Figure I3)

(12) $\Lambda = \dfrac{c}{\Delta f} = \dfrac{\Delta f}{1/\lambda_1 - 1/\lambda_2} = \dfrac{\lambda_1 \lambda_2}{\lambda_2 - \lambda_1} \approx \dfrac{\lambda_0^2}{\Delta \lambda}$

That is

(13) $\dfrac{1}{\Lambda} = \dfrac{1}{\lambda_1} - \dfrac{1}{\lambda_2}$

Figure I2 shows the relations between the various wave forms. In particular, notice that the intensity waveform oscillates with two "beats" per modulation waveform. Further, the spatial separation between maxima of the propagating modulation (envelope) wave is given by Λ.

The human ear is a square law detector responding to sound intensity, E^2. The modulation frequency, corresponding to $m(r,t)$, is ½ of the beat frequency $\omega_m = \tfrac{1}{2}\omega_b$. There are two beats per cycle of the modulation wave.

For the case of acoustic waves, Crawford points out that if f_1 and f_2 differ by more than about 6% from f_0, then one "hears" the sound as two separate tones, or oscillators, with significantly different notes [Equation (1)]. In music, this is called a

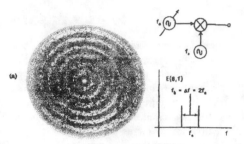

Figure I3. Free space DSB transmitter radiating the outward propagating beat pattern of Equation 1.5. (a) corresponds to Figure I1(b). The white bands are the propagating envelope maxima and the white rings in each band correspond to the peaks of the oscillating carrier. The spacing between the outward propagating beats is $\Lambda = c/\Delta f$.

"chord". However, if f_1 and f_2 are closer, then one "hears" the sound as a single oscillator of frequency f_0, modulated with a slowly varying amplitude $m(t)$ [Equation (5)].[77]

Finally, we mention that in a dispersive medium, e.g. a waveguide near cutoff, or a cavity resonator near its resonant frequency, $v_g \neq v_\phi$ and the beats don't propagate at the speed of the carrier wave. The modulation appears as a disturbance propagating along the carrier wave with velocity v_g through space.

Beats as Moving Interference Patterns

Several discussions of "beats", produced by two sources of slightly differing frequencies, have been published.[78,79,80,81,82] Such phenomena may be described in terms of *a traveling pattern of interference fringes.* In the discussion above, we have examined only the case where the two point sources coincide. The above referenced articles treat the more general case, which reduces to our discussion when the spectrally different sources are at the same point, and reduces to Young's two slit diffraction when the two sources are at the same frequency but different locations (**as with a phased array antenna**). When both spatially separated sources and different spectra are employed, the resulting interference pattern can result in slowly propagating envelopes and in stationary (standing) waves, i.e. waves for which the envelope velocity is zero, just as Tesla said.

REFERENCES

[1] Tusculan Disputations, V, xiii, 64-66.

[2] Plutarch records that the Roman general in charge of the invasion even issued a statement wanting to know why his scientists and craftsmen weren't equal to those of Syracuse.

[3] Plutarch's Lives, translated by John Dryden (1686), The Modern Library, Random House. See the chapter entitled "Marcellus".

[4] The six simple machines are: the lever, the inclined plane, the wedge (a double inclined plane), the screw (a helically wound inclined plane), the wheel and axle (a wheel rigidly attached to an axle, i.e.— a crank and axle), and the pulley (a wheel that turns on an axle).

[5] University Physics, by F.W. Sears and M. W. Zemansky, Addison-Wesley Press, 2nd edition, 1957, pg. 134.

[6] Treatise on Plane Equilibria or Centers of Gravity of Planes: "Proposition 6 - Two magnitudes balance at distances reciprocally proportional to the magnitudes."

[7] A *rigid* bar is one which will propagate disturbances instantaneously throughout its length. It is the mechanical equivalent of a lumped-constant electrical system. By considering lumped-constant mechanical systems, the wave propagation phenomenon associated with distributed-constant systems is excluded.

[8] Transients in Linear Systems- Volume I: Lumped Circuits, by M.F. Gardner and J.L. Barns, Wiley, 1942, pg. 80. (Apparently, Volume II, on distributed circuits, was never published.)

[9] Plutarch loc cit.

[10] In his burning frustration and desire to pierce the secrets of the physical universe and discover what holds the worlds together Goethe's Faust, surrounded in his laboratory by the classical simple machines, utters:

> *Mysterious, even in the light of day,*
> *Nature's enshrouded secrets, denied, stay;*
> *What to your spirit she's unwilling to reveal,*
> *You cannot wrestle from her with levers,*
> *screws and wheel.*

These lines were quoted by Pauli in the early days of quantum mechanics.

[11] "Some Contributions To The Electrical Industry",
by C.C. Chesney, Electrical Engineering, Vol. 53, May, 1934, pp. 726-729. (Reprinted in Turning Points in American Electrical History, J.E. Brittain, editor, IEEE Press, 1977, pp. 102-106.)

[12] If the reader visits the "electrical exhibit" at the Smithsonian Institute he will find photographs of the Niagara Falls power plant and a large brass "boiler plate", from one of its original huge generators, listing Tesla's patents covering the electrical power system. In the midst of the Niagara Falls section stands a lifesize replica of Edison with the comment, "While the Niagara AC plant was being built by Westinghouse, Edison was busy with other important things." One is not told what these things, considered by the curator of the Smithsonian to be so important, were, nor why it was thought that this was relevant to the Niagara AC power plant.

[13] "Tesla's Contribution to Electric Power", C.F. Scott, Electrical Engineering, August, 1943, pp. 351-355.

[14] Engineering Council for Professional Development: the organization which accredits the engineering curricula at colleges and universities in the USA.

[15] Transients in Linear Systems- Volume I: Lumped Circuits, by M.F. Gardner and J.L. Barns, Wiley, 1942, pp. 76-81.

[16] Acoustics, by L.L. Beranek, McGraw-Hill, 1954, pp. 56-58.

[17] Analysis Of Linear Systems, by D.K. Cheng, Addison-Wesley, 1959, pp. 109-112.

[18] Linear Circuits, by R.E. Scott, Addison-Wesley, 1960, pp. 555-556, 565-566.

[19] Electrical Engineering, by A.B. Carlson and D.G. Gisser, Addison-Wesley, 1981, pg. 87.

[20] The analogy advanced by Firestone is called the f-i (force-current) analog. With the latter, $F \sim I$, $v \sim x$, mass $\sim$ capacitance, spring $\sim$ inductance. In the f-i analog, not only are the equations formally similar, but mechanical elements in series are represented by electrical elements in series (electrical nodes are analogous to mechanical junctions). In the f-i case $N_1 \sim r_1$ and $N_2 \sim r_2$. It is recognized that the f-i and f-v analogies are duals to one another.

[21] "The Mobility Method of Computing the Vibration of Linear Mechanical and Acoustical Systems: Mechanical-Electrical Analogies", by F.A. Firestone, *Journal of Applied Physics*, Vol. 9, June, 1938, pp. 373-387.

22 R.W. Sorenson, Jour. A.I.E.E., Vol 44, 1925, pp. 373.

23 "A New High Potential X-Ray Tube", by C.C. Lauritsen and R.D. Bennett, Physical Review, Vol. 32, December, 1928, pp. 850-857.

24 Vibrations and Waves, by A.P. French, W.W. Norton & Co., N.Y., 1971, pp. 83-101.

25 Engineering Mechanics, by W.G. McLean and E.W. Nelson, Schaum Publishing Co., 2nd edition, 1962, pg. 313.

26 The Feynman Lectures on Physics, by R.P. Feynman, R.B. Leighton, and M. Sands, Addison-Wesley Publishing Co., 1963, Vol. I, section 12-2.

27 "Mechanical Oscillations", by K.L. Corum and J.F. Corum, Appendix X (1986) to our Tesla Handbook. The appendix is 59 pages in length.

28 The astute reader will recognize that the electrical analog corresponds to a zero impedance source.

29 Mechanical Vibrations, by J.P. Den Hartog, Dover, 1985, pg. 53.

30 Analytic Mechanics, by V. Fairs and S. Chambers, MacMillan Co., 3rd edition, 1958, pg. 493.

31 Tesla, Nikola, "My Inventions, Part IV - The Discovery of the Tesla Coil and Transformer", Electrical Experimenter, May, 1919, pp. 16, 17, 64, 65, 89.

32 "The Problem of Increasing Human Energy", by Nikola Tesla, Century Illustrated Magazine, June, 1900, pp. 175-211. (Unabridged republication by J.A. Hayes and S.R. Elswick, High Energy Enterprises, Inc., 1990, ISBN 1-882137-00-0).

33 Thomas Edison, Pamphlet, of 1887.

34 Linear Circuits, by R.E. Scott, Addison-Wesley Press, 1960, pg. 637.

35 Networks, Lines and Fields, by J.D. Ryder, Prentice-Hall, Inc., 2nd edition, 1955, pg. 60.

36 Electronic Fundamentals and Applications, by J.D. Ryder, Prentice-Hall, 4th edition, 1970, pp. 388-393.

37 "Microwave Designer's Quiz: How Much Do *You* Know About Connectors?", by J.F. Corum, K.L. Corum, V.G. Puglielli and J.F.X. Daum, Microwave Systems News, June, 1990.

38 "The Application of Transmission Line Resonators to High Voltage RF Power Processing: History, Analysis, and Experiment", **Proceedings of the 19th Southeastern Symposium on System Theory**, Clemson University, Clemson, South Carolina, 1987, pp. 45-49.

39 Vacuum Tube Tesla Coils, by J.F. Corum and K.L. Corum, Published by Corum & Associates, Inc., 1988, [100 page text], ISBN 0-924758-00-7.

40 TCTUTOR - A Personal Computer Analysis of Spark Gap Tesla Coils, by J. F. Corum, D.J. Edwards and K.L. Corum, Published by Corum & Associates, Inc., 1988, [110 page text + Disk], ISBN 0-924758-01-5.

41 "High Voltage RF Ball Lightning Experiments and Electro-Chemical Fractal Clusters," by K.L. Corum and J.F. Corum, Soviet Physics - Uspekhi, Vol. 160, No.4, April, 1990, pp. 47-58, (in Russian).

42 "Analysis of Transient Processes and Spectral Characteristics of Electron Beams in High-Current nanosecond Electron Accelerators with Forming Lines", by G.A. Mesyats and B.Z. Movshevich, Soviet Physics -Technical Physics, Vol. 34, No. 5, May, 1989, pp. 523-528.

43 Industrial Electron Accelerators and Applications, by E.A. Abramyan, Hemisphere Publishing Corp., 1988.

44 Colorado Springs Notes, 1899-1900, by Nikola Tesla, Nolit, Beograd, Yugoslavia, 1978, PP. 79, 103, 180, 345.

45 Smythe, W.R., Static and Dynamic Electricity, McGraw-Hill, 2nd edition, 1950, pp. 545-547.

46 Condon, E.U., "Forced Oscillations in Cavity Resonators", Jour. Appl. Phys., Vol. 12, Feb., 1941, pp. 129-132. (See pg. 131 for voltage "multiplication".)

47 "Resonance Properties of Ring Circuits", by F.J. Tischer, I.E.E.E. Trans MTT, Jan., 1957, pp.51-56.

48 Microwave filters, Impedance Matching Networks, and Coupling Structures, by G. Matthai, L. Young, and E.M.T. Jones, Artech House, pp. 846-847.

49 "The Traveling Wave Resonator and High Power Microwave Testing", by Stanley Miller, Microwave Journal, September, 1960, pp. 50-58.

50 "Methods of Producing High Levels of RF Power for Test Purposes", by P. Hayes and R. Surette, National Association of Broadcasters Engineering Conference Proceedings, 1988, pp. 380-386.

51 The term *"multiply connected"* implies that there exists a *closed path* in the resonator which cannot be shrunk to a point without part of the path passing through regions outside the resonator. A toroidal shaped ring resonator and the earth-ionosphere cavity are examples. The latter has the added feature that it focuses converging energy at the antipode.

22 R.W. Sorenson, Jour. A.I.E.E., Vol 44, 1925, pp. 373.

23 "A New High Potential X-Ray Tube", by C.C. Lauritsen and R.D. Bennett, Physical Review, Vol. 32, December, 1928, pp. 850-857.

24 Vibrations and Waves, by A.P. French, W.W. Norton & Co., N.Y., 1971, pp. 83-101.

25 Engineering Mechanics, by W.G. McLean and E.W. Nelson, Schaum Publishing Co., 2nd edition, 1962, pg. 313.

26 The Feynman Lectures on Physics, by R.P. Feynman, R.B. Leighton, and M. Sands, Addison-Wesley Publishing Co., 1963, Vol. I, section 12-2.

27 "Mechanical Oscillations", by K.L. Corum and J.F. Corum, Appendix X (1986) to our Tesla Handbook. The appendix is 59 pages in length.

28 The astute reader will recognize that the electrical analog corresponds to a zero impedance source.

29 Mechanical Vibrations, by J.P. Den Hartog, Dover, 1985, pg. 53.

30 Analytic Mechanics, by V. Fairs and S. Chambers, MacMillan Co., 3rd edition, 1958, pg. 493.

31 Tesla, Nikola, "My Inventions, Part IV - The Discovery of the Tesla Coil and Transformer", Electrical Experimenter, May, 1919, pp. 16, 17, 64, 65, 89.

32 "The Problem of Increasing Human Energy", by Nikola Tesla, Century Illustrated Magazine, June, 1900, pp. 175-211. (Unabridged republication by J.A. Hayes and S.R. Elswick, High Energy Enterprises, Inc., 1990, ISBN 1-882137-00-0).

33 Thomas Edison, Pamphlet, of 1887.

34 Linear Circuits, by R.E. Scott, Addison-Wesley Press, 1960, pg. 637.

35 Networks, Lines and Fields, by J.D. Ryder, Prentice-Hall, Inc., 2nd edition, 1955, pg. 60.

36 Electronic Fundamentals and Applications, by J.D. Ryder, Prentice-Hall, 4th edition, 1970, pp. 388-393.

37 "Microwave Designer's Quiz: How Much Do *You* Know About Connectors?", by J.F. Corum, K.L. Corum, V.G. Puglielli and J.F.X. Daum, Microwave Systems News, June, 1990.

38 "The Application of Transmission Line Resonators to High Voltage RF Power Processing: History, Analysis, and Experiment", **Proceedings of the 19th Southeastern Symposium on System Theory**, Clemson University, Clemson, South Carolina, 1987, pp. 45-49.

39 Vacuum Tube Tesla Coils, by J.F. Corum and K.L. Corum, Published by Corum & Associates, Inc., 1988, [100 page text], ISBN 0-924758-00-7.

40 TCTUTOR - A Personal Computer Analysis of Spark Gap Tesla Coils, by J. F. Corum, D.J. Edwards and K.L. Corum, Published by Corum & Associates, Inc., 1988, [110 page text + Disk], ISBN 0-924758-01-5.

41 "High Voltage RF Ball Lightning Experiments and Electro-Chemical Fractal Clusters," by K.L. Corum and J.F. Corum, Soviet Physics - Uspekhi, Vol. 160, No.4, April, 1990, pp. 47-58, (in Russian).

42 "Analysis of Transient Processes and Spectral Characteristics of Electron Beams in High-Current nanosecond Electron Accelerators with Forming Lines", by G.A. Mesyats and B.Z. Movshevich, Soviet Physics - Technical Physics, Vol. 34, No. 5, May, 1989, pp. 523-528.

43 Industrial Electron Accelerators and Applications, by E.A. Abramyan, Hemisphere Publishing Corp., 1988.

44 Colorado Springs Notes, 1899-1900, by Nikola Tesla, Nolit, Beograd, Yugoslavia, 1978, PP. 79, 103, 180, 345.

45 Smythe, W.R., Static and Dynamic Electricity, McGraw-Hill, 2nd edition, 1950, pp. 545-547.

46 Condon, E.U., "Forced Oscillations in Cavity Resonators", Jour. Appl. Phys., Vol. 12, Feb., 1941, pp. 129-132. (See pg. 131 for voltage "multiplication".)

47 "Resonance Properties of Ring Circuits", by F.J. Tischer, I.E.E.E. Trans MTT, Jan., 1957, pp.51-56.

48 Microwave filters, Impedance Matching Networks, and Coupling Structures, by G. Matthai, L. Young, and E.M.T. Jones, Artech House, pp. 846-847.

49 "The Traveling Wave Resonator and High Power Microwave Testing", by Stanley Miller, Microwave Journal, September, 1960, pp. 50-58.

50 "Methods of Producing High Levels of RF Power for Test Purposes", by P. Hayes and R. Surette, National Association of Broadcasters Engineering Conference Proceedings, 1988, pp. 380-386.

51 The term "*multiply connected*" implies that there exists a *closed path* in the resonator which cannot be shrunk to a point without part of the path passing through regions outside the resonator. A toroidal shaped ring resonator and the earth-ionosphere cavity are examples. The latter has the added feature that it focuses converging energy at the antipode.

[52] Principles of Optics, by M. Born and E. Wolf, Pergamon Press, 5th edition, 1975, pp. 327-328.

[53] Introduction to Optical Electronics, by A. Yariv, Holt, Rinehart and Winston, 2nd edition, 1976, pp. 60-66.

[54] Those familiar with Van der Pol and Bremmer's "rainbow expansion" will recognize that Equation (41) can be gotten from the inclusion of higher terms of the asymptotic expansion in the residue series representation for the fields on a conducting sphere.

[55] "The Diffraction of Electromagnetic Waves from an Electrical Point Source round a Finitely Conducting Sphere, with Applications to Radiotelegraphy and the Theory of the Rainbow - Part II", by B. Van der Pol and H. Bremmer, Philosophical Magazine, S. 7, Vol. 24, No. 164, Suppl. Nov., 1937, pp. 825-864.

[56] "Propagation of Electromagnetic Waves", by H. Bremmer, in Handbuch der Physik, S. Flugge, editor, 1958, Vol. 16, pp. 423-639. (See pg. 604.)

[57] "Some Quantitative Experiments In Long Distance Radio Telegraphy", by L.W. Austin, Bureau of Standards Bulletin (J. Res. NBS), Vol. 7, 1911, No. 3, October, pp. 315-366.

[58] "Preliminary Note on Proposed Changes in the Constants of the Austin-Cohen Transmission Formula", L.W. Austin, Proceedings of the IRE, Vol. 14, June, 1926, p. 327.

[59] Propagation of Radio Waves, by M. Dolukhanov, Mir Publishers, Moscow, 1971, pp. 275-284.

[60] "Ausbreitung langer Wellen um die Erdkugel", by O. Zinke, Frequenz, October, 1947, pp. 16-23.

[61] Propagation of Waves, by P. David and J. Voge, Pergamon Press, 1969, pp. 229-235.

[62] "The Transient Propagation of ELF Pulses in the Earth-Ionosphere Cavity", by J.F. Corum and A. Hamid Aidinejad, Proceedings of the 2nd International Tesla Symposium, Colorado Springs, Colorado, 1986, pp. .

[63] "The Global Response of the Earth to Schumann Pulse Repetition Frequencies", by J.F. Corum, K.L. Corum and A.H. Aidinejad, Proceedings of the International Scientific Conference on Energy and Development, Zagreb, Yugoslavia, 1986, pp. 1-18.

[64] "A Noted Inventor: Nikola Tesla", interview by Frank G. Carpenter, The Illustrated Weekly Magazine, (a section of The Los Angeles Times), December, 1904, pp. 3-4.

[65] "Art of Transmitting Electrical Energy Through the Natural Mediums", Nikola Tesla, U.S. Patent #787,412. Applied for May 16, 1900; Issued April 18, 1905.

[66] The Feynman Lectures on Physics, F.P. Feynman, R.B. Leighton and M. Sands, Addison-Wesley, 1964, Section 16-4.

[67] "The Wonder World To Be Created By Electricity, by Nikola Tesla, Manufacturer's Record, September 9, 1915, pp. 37-39.

[68] World System of Wireless Transmission of Energy", by Nikola Tesla, Telegraph and Telephone Age, October 16, 1927, pp. 457-460.

[69] 525-456 B.C.

[70] Prometheus Bound (460 B.C.).

[71] The name means "forethought".

[72] Ten Greek Plays, edited by, L.R. Lind, Houghten Mifflin Company, 1957, pg. 3.

[73] "The Battelle Mural", a brochure, Battelle Memorial Institute, Columbus, Ohio.

[74] At r = 0, the disturbance is expressed as

$$(5') \quad E(0,t) = 2A \cos(\omega_m t)\cos(\omega_o t)$$
$$= A \cos(\omega_o + \omega_m)t + A \cos(\omega_o - \omega_m)t$$

$$\Updownarrow \qquad\qquad\qquad \Updownarrow$$

$$\text{USB} \qquad\qquad\qquad \text{LSB}$$

which we identify as a double sideband suppressed carrier wave (DSB/SC), with modulation frequency $\omega_m = \frac{1}{2}(\omega_1 - \omega_2) = \frac{1}{2}\omega_b$. The DSB spectrum is simply $E(0,f) = \frac{1}{2} M(f-f_o) + \frac{1}{2} M(f+f_o) = \frac{1}{2}\delta[f-(f_o-f_m)] + \frac{1}{2}\delta[f-(f_o+f_m)]$, where $M(f)$ is the spectrum of the baseband message, $m(t)$.

[75] Optics, by E. Hecht and A. Zajac, Addison-Wesley, 1974, pp. 202-205.

[76] Vibrations and Waves, by A.P. French, Norton & Co., 1971, pp. 22-27, 213-215.

[77] Waves, (Vol. 3 of the Berkeley Physics Course), by F.S. Crawford, McGraw-Hill, 1968, pg.30.

[78] Physical Optics, R.W. Wood, Dover, 1967 (1934 edition), pp. 180-181.

[79] "Interference Pattern of Beats", L.G. Hoxton, American Journal of Physics, Vol. 31, No. 10, October, 1963, pp. 794-801.

[80] "Demonstration of Beats as Moving Interference Patterns", by T.S. Stein and L.G. Dishman, American Journal of Physics, Vol. 50, No. 2, February, 1982, pp. 136-145.

[81] "A Ripple tank Demonstration of the Conditions for Interference of Waves", by Y. Hao, X. Qi-cheng, and L. Zhen-di, American Journal of Physics, Vol. 56, No. 8, August, 1988, pp. 745-747.

[82] "Beats as Moving Interference Patterns", by T. Stein and L. Dishman, American Journal of Physics, Vol. 57, No. 7, July, 1989, pg. 584.

About the Authors

The guiding principle behind the work of the Corum brothers has been to reduce the publications and RF experiments of Dr. Tesla to conventional modern engineering terms. Together, they have published 3 books and several dozen technical articles about Tesla!s work, and contributed to several other books on Tesla, ball lightning, and the history of electroscience. Their engineering analyses and replications of Tesla's slow-wave resonator and fire ball research has gained an international interest in the scientific community.

Kenneth L. Corum holds a B.A. in Physics from Gordon College (1977) and has done graduate work at the University of Massachusetts. He has taught computer electronics and digital techniques in England, France, Germany, Netherlands, and the United States. He is listed in **American Men and Women of Science** and **Outstanding Young Men of America**. Mr. Corum was Director of the Commercial TV Satellite Division of Pinzone Communications, and he is now a consultant with Hewlett-Packard. He has spoken at Zagreb and at The Tesla Museum at Belgrade as a guest of the Serbian Academy of Sciences and Arts.

Ken Corum can be reached at 104 River Rd., Plymouth, NH 03264

Dr. James R. Corum has a BSEE (University of Lowell), MSEE and PhD in Electrical Engineering from Ohio State (1974). Dr. Corum was active in academia for 18 years, serving as an Associate Professor in the Department of Electrical Engineering at West Virginia University, and as a Professor in the Department of Electronic Engineering Technology at the Ohio Institute of Technology. While a faculty member in West Virginia, he received 11 outstanding faculty recognition awards, prizes and citations from the University, the Faculty, the Students and the Alumni.

He was a Senior Research Scientist with the Electromagnetics Department and a Research Leader in the Applied Physics Group at the Battelle Memorial Institute in Columbus, Ohio. Dr. Corum is a Senior Member of the IEEE, and a member of the American Geophysical Union, the American Association of Physics Teachers, the American Society for Engineering Education, the Research Society of North America, and the Society of Motion Picture and Television Engineers.

He is listed in **Who's Who In Engineering, American Men And Women of Science, Leading Consultants In High Technology**, and 14 other biographical dictionaries. He currently holds five patents and has in excess of 90 technical publications.

86 Weirton Mine Rd., Morgantown, WV 26508

Ed note: For those interested in the specific equations in this article that may not be legible, visit www.IntegrityResearchInstitute.org or email: iri@erols.com

13   Tesla's ELF Oscillator for Wireless Transmission

James Corum, Ph. D.
Kenneth L. Corum
Reprinted from *Tesla: A Journal of Modern Science*, 1997

Introduction

For a scientist, Tesla was a prolific but abstruse and poetic writer. He left behind a considerable volume of technical and enigmatic descriptive writings, lectures, patents, articles and newspaper interviews. Additionally, his Colorado Springs diary and his Long Island notes provide a showcase through with the wealth of his creative mind may be viewed from our present technological perspective. Such remarkable situations rarely present themselves with such prolific documentation to future generations. In reading the Colorado Springs and Long Island notes, one feels as though he has just blown away the dust of the years and opened diaries of Columbus or DaVinci. Before him sit the thoughts and experimentations of the powerful intellect which invented for modern civilization the electrical equivalent of the wheel—the rotating magnetic field.

It would be an understatement to say that an electrical engineer can certainly empathize with the excitement described by Carter as he peered through the chink in the wall of an Egyptian tom and saw before him the treasures of Tutankamen. What we must ascertain, however, is whether Tesla's words, unlike Tutankamen's trinkets, are of any practical value to the scientific community today. Certainly they do reflect how he thought and interpreted his experiments and how his physical concepts led him to make the remarkable statements published subsequent to his Colorado Springs experiments.

The fact that in addition to his Colorado Springs Diary, we also have the associated patent wrappers, special articles and later recollections, is of great importance. Any interpretation or speculation made today about his experiments must not only be internally consistent with these documents, but must also cement them together. This we take as a first requirement of speculation.

Additionally, as a second requirement, such conjecture must be made within the bounds of accepted and verifiable physical principles. Tesla was, apparently, experimenting with potentials in excess of 12 MV. Not unlike Columbus, he was sailing in uncharted seas. The possibility of "peculiar physics" notwithstanding, our efforts have been to attempt to discuss Tesla within the framework of modern electrical theory. Whether the experimental results of Tesla were, in fact near-field induction coupling, or perhaps Schumannn resonance excitation (as we believe) or some sort of magnetospheric stimulation, or even some peculiar presently yet unknown physical phenomenon—the fact remains, a significant portion of his apparatus was constructed of wire, capacitors, spark gaps, and tuned circuits. It ought to be comprehensible in its intent and physical operation. That he got high voltage RF from his circuits is clearly understandable. It follows from a straightforward network analysis. How he obtained the incredible ELF results which he subsequently claimed, however, is the chasm to be bridged.[1]

---

[1] J. Corum also holds patents #4,622,558, #4,751,515 (ELF toroidal antennae) – Ed. note.

Issues to be Resolved

Because of his repeated insistence on terrestrial extra low frequency resonances, it does not seem unreasonable to hypothesize that Tesla's experiments were actually carried out at this end of the electromagnetic spectrum. However, his indoor Colorado Springs apparatus clearly operated in the range of 30 KHz to 100 KHz, and the tower described in his diary was only 145 feet high. [1] With such an electrically short tower threw could be no radiation at ELF. But, on the other hand, if it was merely a waveguide or cavity resonator probe, one would expect its radiation resistance to be zero. The radiation resistance of a cavity resonator probe is zero because the resonator fields are purely relative.

It has been observed the the photographs of the Colorado Springs Laboratory offer not clue as to how one might generate ELF energy. There is an even more fundamental issue to be resolved, however. The excitation of Schumann resonances by means of a vertical probe requires a considerable current moment. The question is, "Even if Tesla successfully generated ELF voltages how did he ever get significant ELF currents to flow on the vertical structure?" [2] One could not just connect an ELF source onto the base of a 45 meter tower – the feedpoint capacitive reactance would be so large that not current would flow. Yet, Tesla maintained on several occasions that his "antenna" current was well in excess of 1000 amperes! [By the way, there is a similar issure to be resolved for the reception of power. The Thevenin equivalent of a vertical receiving antenna may have a substantial ELF. However, the series capacitive reactance of vertical tower at ELF will preclude substantive power transfer in the sinusoidal steady state].

Lastly, we mention the curious inscription in Tesla's own handwriting along the side of a now famous photograph of the Colorado Springs Laboratory:

"Experimental Station fully developed. Activity [power delivery] one hundred thousand horsepower" [3]

How could he possibly deliver 74.6 megawatts? The 60 Hz power mains to the laboratory were operated at 1 kV, but his Westinghouse transformer was only rated at around 40 kVA.

In the remainder of this article, we wish to speculate on how these issues could be resolved within the bounds of consistency with the historical documents mentioned above. The reader will have to judge whether we have successfully met the second requirement – that of conjecture based on acceptable physics

Tesla's Descriptions of His Oscillator

Tesla described his electrical oscillator on many occasions and in many different places. It is clear that, as early as 1893, he was considering terrestrial responses:

"If ever we can ascertain at what period the earth's change, when disturbed, oscillates...we shall know a fact possibly of the greatest importance to the welfare of the human race...I propose to search for the period by means of an electrical oscillator." [4]

In the years between 1893 and 1900, he developed the coupled tuned transformer [or Tesla Coil], published the results of his extensive experiments with x-rays, contributed to the conceptual development of cosmic ray, patented a variety of circuit controllers [rotary spark gaps] and was sought out by members of the scientific and social communities, both of which he continued to dazzle with his latest electrical discoveries. These were the golden years of his professional career, and they found a focal point in his experiments at Colorado Springs. It is here, where he finally was able to assemble the apparatus which, he maintained to his dying day, permitted him to ascertain terrestrial natural resonant frequencies.

In 1900, he disclosed that this apparatus could be operated in a variety of configurations to perform many different types of desired functions:

"Thus a transformer or induction coil on new principles was evolved, which I have called 'the electric oscillator'... the essential parts of which are shown in Fig. 6. For certain purposes a strong inductive effect is required; for others the greatest possible suddenness; for others again, an exceptionally high rated of vibration or extreme pressure; while for certain other objects immense electrical movements are necessary... I have produced electrical movements occurring at the rate of one hundred thousand horsepower..." [5]

By the way, the "essential parts" shown in the photograph referred to appear to be his Westinghouse transformer, a rotary break, a capacitor bank and a circular fence upon which the secondary was wound.

Perhaps the most curious of all his descriptions of the terrestrial resonance oscillator was published in 1919:

"It is a resonant transformer with a secondary in which the parts, charged to a high potential, are of considerable area and arranged in space along ideal enveloping surfaces of very large radii of curvature, and at proper distances from one another thereby insuring a small electric surface density everywhere so that not leak can occur even if the conductor is bar. It is suitable for any frequency from a few to many thousand of cycles per second, and can be used in the production of currents of tremendous volume and moderate pressure or of a smaller amperage and immense electromotive force. The maximum electric tension is merely dependent upon the curvature of the surfaces on which the charge elements are situated and the area of the latter.

"Judging from my past experience, as much as 100,000,000 volts are perfectly practicable. On the other hand, currents of many thousand of amperes may be obtained in the antenna... the Hertz-wave radiation is an entirely negligible quantity as compared with the whole energy... an enormous charge is stored in the elevated capacity. Such a circuit may then be excited with impulses of any kind, even of low frequency and it will yield sinusoidal and continuous oscillations like those of an alternator... it is a resonant transformer...accurately proportioned to fit the globe and its electrical constants and properties, by virtue of which design it becomes highly efficient and effective in wireless transmission of energy". [6]

What we are to make of this? Based upon the available electrical output of his extra coil and the reports of spark lengths measured 100 meters away from the Colorado Springs Laboratory, we estimate that the charge stored in the elevated capacity was probably on the order of 20 millicoulombs. But how was this to be used to excite terrestrial resonances? The apparatus so furtively described in 1919 is probably a near relative of that for which he sought protection by a patent application in 1902, and which subsequently issued at the close of 1914. [7]

Directed Energy Devices

We believe that the evolution of these ideas continued to be a central activity of Tesla's later years. It is merely our opinion , but we find it difficult to accept the senility hypothesis concerning his motivation to reach for the goals of these final years. It seems probable that the apparatus which so concerned his thoughts at this time, not only was a successive conceptual development of his prior oscillators, but any credible knowledge gained about these later structures would probably throw light on the operation of the earlier experiment – no matter how improbable his final research might have been. We believe that Tesla's surprisingly detailed 1934 analysis of the Van Der Graaft's [then] new machine, published in Scientific American, lends support to the hypothesis that he was still technically alert and deeply engaged in high voltage research. [8]. Perhaps we are not being specific enough for the reader. The apparatus which Tesla's final disclosures concerned has come to be known as the "Death Ray". Whether there be any actual merit to such contraptions, we leave for others to speculate upon. Our interest in Tesla's thoughts on "directed energy devices" rest upon the proposition that they might shed light upon his terrestrial resonance oscillator. In 1927, Tesla said:

"More than twenty five years ago my efforts to transmit large amounts of power through the atmosphere resulted in the invention of a great promise, which has since been called 'Death Ray'... The underlying idea was to render the air conducting by suitable ionizing radiations and to convey high tension currents along the path of the rays. Experiments conducted on a large scale showed that with pressures of many millions of volts virtually unlimited quantities of energy can be projected to a small distance, as a few hundred feet..."
[9]

From a variety of published references, spanning the years from 1934-1940, we gather that Tesla envisioned a machine, which required the cooperative action of four separate entities. Again, our interest is not in the feasibility of such an apparatus, but rather in how he hgouth such a device was to work and what, if any, light it might shed on his terrestrial resonance oscillator experiments. The four elements specified by Tesla are as follows:

1. "A Method and an Apparatus for producing rays and manifestations of energy in free air (eliminating the necessity of the usually required vacuum tubes)."
2. "A Process and an Apparatus for producing very great electrical force (50 MV). This is necessary to power the first mechanism."
3. "A method of intensifying and amplifying the force developed the by second mechanism"
4. "A new method for producing a tremendous repelling force."

It is perhaps not unremarkable that these components are quite similar to the description provided by John G. Trump, when he examined Tesla's estate. He described the "Death Ray" as:

"An electrostatic method of producing high voltage, capable of very great power... As a component of this apparatus there is an open end vacuum tube... A beam of high energy electrons is the... means by which energy is transmitted through natural media" [10]

It should be parenthetically remarked that Tesla explicitly denied that his apparatus was a "Ray" as indicated below:

"This invention of mine does not contemplate the use of any so called 'Death Ray.' Rays are not applicable because they cannot produce in requisite quantities and diminish rapidly in intensity with distance... My apparatus projects particles..." [11]

We started out discussing electrical oscillators and now find ourselves confronted with "direct energy devices". Perhaps this is not surprising when we observe that Tesla's early x-ray researches involved the use of his "single electrode X-ray tube" attached to the top of a resonant Tesla coil.

The Single Electrode X-Ray Tube

Between March 11, 1896 and August 11, 1897, Tesla wrote at least 10 articles about his x-ray experiments. There is an explanation for the development of a single electrode tube in the March 11, 1896 issue of *Electrical Review*. Tesla saw that in order to attain the most intensive effects, one should use the greatest voltages available.

"Clearly, if we put two electrodes in a bulb, or use on inside and another outside electrode, we limit the potential... Thus, to secure the result aimed at, one is driven to the acceptance of a single electrode bulb, the other terminal being as far remote as possible". [12]

Tesla later hinted at the manner in which the tube was excited: "...in 1896, I brought out a new form of vacuum tube capable of being charged to any desired potential and operated it with effective pressures of 4,000,000 volts [13]

And in a 1913 newspaper interview, Tesla said:
"As far back as 1897, I disclosed before the New York Academy of Sciences the discovery that Roentgen, or x-rays projected from certain bulbs have the property of strongly charging an electrical condenser at a distance. The energy so accumulated readily can be discharged." [14]

We know today, of course, that x-rays are high energy photons and have neither rest mass nor charge. The question before us, however, is "How might these single electrode tubes produce x-rays?"
In a standard x-ray tube of the Coolidge type, a hearted filament provides electrons which are then accelerated and strike an anode target. If an AC supply is employed, x-ray emission occurs only during the positive half cycle. However, Tesla's tube had only one electrode. We hypothesize that the tube's operation probably depended upon the quantum mechanical phenomenon of High Field Emission. One might suppose that during half of the RF high voltage cycle field emission could possibly occur into the region of high vacuum elongated bulb, and on the positive half cycle the cloud of electrons might be swept back into the "plain polished surface on the front side of a hemispherical aluminum electrode" with an ensuing emission of hard x-rays. This is only a hypothesis and certainly its acceptability needs to be closely examined.
There is evidence that, during the course of his New York city experiments, Tesla took to surrounding these tubes with an insulated shield in order to reduce corona losses. He called this "static screening". The configuration is quite similar to that discussed in the Colorado Springs notes on June 6, 1899:
"Arrangements with single terminal tube for production of powerful rays. There being practically no limit to the power of an oscillator, it is now the problem to work out a tube so that it can stand any desired pressure... The best results will probably be obtained in the end by static screening of the vulnerable parts of the tube. This idea was experimented on in a number of ways... In each case there would be an insulated body of capacity so arranged that

streamers cannot manifest themselves. The capacity would be such as to bring about maximum rise of e.m.f. on the free terminal" [15]

The associated figure shown in the diary entrance indicates a [square!] container with the comments:

"Metallic Enclosure but insulated so that observers can step inside" [15]

Little more can be inferred from the diary about x-ray experiments until November 23, 1899. Finally, on January 2, 1900, Tesla states,

"…my conviction is growing stronger every day that, with apparatus such as the present, wonderful results must be secured provided only that a tube is constructed capable of taking up any amount of energy…Many tubes have been worked here from the secondary." [16]

In a later interview, Tesla said concerning the Colorado Springs experiments:

" At the time of those test, I succeeded in producing the most powerful x-rays ever seen. *I could stand at a distance of 100 feet from the x-ray apparatus and see the bones of the hand clearly with the aid of a fluoroscope screen*… I now have apparatus designed whereby this tremendous energy of hundred of kilowatts can be successfully transformed into x-rays."[17]

What he was doing with these x-rays? In light of the comments by Tesla regarding charge transfer by x-rays, it does not seem unreasonable to hypothesize that a small aperture in the conducting enclosure would permit the emission of x-rays to the exterior region, causing x-ray photoionization of the atmosphere near the enclosure, these ions providing a short conducting discharge path for the charged "insulated body of capacity". It is clear from Reference 13, that Tesla was observing "coronal discharges" exterior to his single electrode tubes. This process could clearly be employed to instantaneously lower the disruptive potential of an isolated spherical capacitor, and to initiate a discharge into the air of to a nearby isolated electrode at a lower potential than the given sphere. Perhaps it is not surprising that this process has been of recent interest in x-ray and UV laser triggered switching of high voltages. The latter being particularly interesting because capacitances charged into the megovolt range can be triggered with nano second switching delays and with subnanosecond jitter.

On January 4, 1900, Tesla experimented with a ball on the top of the extra coil, "…very brilliant and thick sparks passing from the ball to the hood above". Tesla continued that the discharge was "highly sensitive" to, among other things, "Roentgen rays" [another name for x-rays]. Is it possible that Tesla was employing x-rays as a switching mechanism to statically charge his tower? The tower appears to be an "elevated insulated body of capacity". This, we hypothesize, is the x-ray charging mechanism which Tesla sought to protect in his two U.S. patents disclosures No. 685,957 and No. 685,958.

The X-Ray Patents

In March of 1901, Tesla filed two patent applications concerning x-ray devices. One was for a "Method" and one was for an "Apparatus" for the "Utilization of Radiant Energy". They describe in considerable detail a remarkable technique for switching high voltages and for charging and discharging and "elevated insulated body of capacitance".

Consider, for example, an isolated sphere. Such a body may be charged to a certain electrical potential with respect to a zero potential reference, taken as infinitely distant. In such a system the spherical conductor may be charged to a certain potential before the electric field intensity gives rise to a force great enough for the surrounding air to break down and "disruptively' discharge the sphere. Tesla found by experiment that the disruptive potential, in volts, for a sphere at sea level could be approximately calculated as 7,540,000 R, where R is radius of the sphere in meters. Tesla, in fact, reported that he kept a variety of spheres around to use both as capacitors and in order to estimate the voltages used in his experiments.

A practical form of high voltage capacitor may be constructed by elevating an insulated spherical conductor above the surface of the earth. It is an elementary problem in electromagnetics to calculate the field and capacitance of such a charged system. In this configuration the "capacitor" effectively has "true ground" as one terminal and the conducting spherical ball itself forms the second "armature" or terminal of the distributed capacitor. This form of capacitor may be charged up by bodily conveying charge of one sign to the elevated sphere. Alternatively, it may be discharged simply by bringing the grounded conductor close enough to the sphere for arcing to occur.

Bearing this in mind, Tesla's x-ray patents take on a meaningful interpretation:

"It is well known that certain radiations such as... Roentgen rays... possess the property of... discharging conductors... They ionize or render conducting the atmosphere through which they are propagated... they may at any rate discharge an electrified conductor... by carrying off bodily its charge... When rays of the above kind are permitted to fall upon an insulated conducted body connected to one of the terminals of a condenser... a current flows into the condenser... an indefinite accumulation of electrical energy in the condenser takes place. This energy after a suitable time interval, during which the rays are allowed to act, may manifest itself in a powerful discharge ... taking every possible precaution in insulating the armatures, so that the instrument may withstand great electrical pressure without leaking and may leave no perceptible electrification when discharging instantaneously... the above precautions should b more rigously observed the slower the rate of charging and the smaller the time interval during which the energy is allowed to accumulate in the condenser... A simple way of supplying... electricity is to connect... to an insulated conductor supported at some height in the atmosphere... I usually connect the second terminal of the condenser to ground.... in order to utilize... the energy accumulated in the condenser, I furthermore connect to the terminals of the same... another instrument or device for alternately closing and opening the circuit... if the device ... be of such character that it will operate to close the circuit... when the potential in the condenser has reached a certain magnitude, the accumulated charge will pass through the circuit... The controller may consist of two fixed electrodes separated by a minute air gap... which breaks down more or less suddenly when a definite difference of potential is reached at the terminals of the condenser and returns to its original state upon the passage of discharge". [18]

Tesla then describes the manner of excitation of his single electrode x-ray tube:

"....the source of radiant energy is a special form of Roentgen tube divised by me, having but one terminal K, generally of aluminum, in the form of a half sphere, with a plain polished surface on the front side from which the streams are thrown off. It may be excited by attaching it to one of the terminals of any generator of sufficiently high electromotive force." [18]

Tesla continues, describing the operation of the apparatus:

"The... discharge circuit connected to the terminals... of the condenser includes in this case... a circuit controller comprising a fixed terminal or break... and a movable terminal.... in the shape of a wheel, with a conducting and insulating segments, which may be rotated at an arbitrary speed by any suitable means.... When the tube ... is excited... streams of matter ... convey a positive charge to... the condenser terminal... This results as before explained in an accumulation of electrical energy in the condenser, which goes on as long as the circuit is opened. Whenever the circuit is closed owing to the rotation of the wheel the stored energy is discharged... The source may be any form of Roentgen or Lenard tube; but it is obvious from the theory of action that in order to be very effective the electrical impulses exciting it should be wholly or at least preponderantly of one sign. If ordinary symmetrical alternating currents are employed, provision should be made for allowing the easy to fall upon the condenser plate only during those periods when they are productive of the desired result." [18]

What we make of this is that Tesla is describing a technique to take the high voltage RF output of the secondary and use it to chare up an "elevated insulated body of capacitance" – in essence, an open air switch or diode rectifier. After charging the capacitor, at RF rates, he subsequently discharges the capacitor, at relatively low pulse repetition frequencies (PRF's) for example at perhaps 6 or 8 discharges per second – or any other that he might desire. The companion patent is also interesting. [19]

It is also somewhat revealing that Tesla said in his Van de Graaf article in 1934 that:

"My wireless tower on Long Island erected in 1902, carried a sphere which had a diameter of 67.5 feet... It was to be charged to 30,000,000 volts by a simple device supplying static electricity and power" [8]

After analyzing the Van de Graaf machine he concedes that it produces large static voltages but concludes that its power performance is trifling – the rate of charge delivery to spherical electrode being on the order of a few tens of milliamperes.

"As far back as 1899, I made experiments with 18,000,000 volts and in some tests I was able to pass a current of 1100 amperes through the air. With my transformers a potential difference of 30,000,00 volts or more, could easily be obtained and in the present state of the technical arts a tube or other device capable of taking up very great energy might be manufactured". [8]

By the way, diary schematic diagrams notwithstanding, it is evident that neither the elevated tower at Colorado Springs nor the 67.5 food diameter sphere on the tower at Wardenclyffe were electrically connected to the extra coil when in operation. This is also borne our by the Long Island notes of May 29, 1901 where Tesla shows an elevated insulated body of capacitance being charged through space from a ball on the top of an extra coil. The spacing between these elements is shown to be a controllable distance- perhaps this was the purpose for the stream elevated shaft at the center of the Wardenclyffe tower. The geometry is not unlike that shown in Figure 5 of Reference 20. It is well known that drawn out electrical discharges will affect rectification much as point to plane discharge. However, the process is usually considered too inefficient.

An ELF Generator

Whatever the rectification mechanism might have been one might hypothesize an ELF generator which employs the charging mechanism just discussed. Such a charging technique could have been used to electrostatically charge the tower with "small" pulses of charge occurring on the positive half cycles of the RF coil oscillations. This would build up the static charge of the tower at some large Q at a very high DC voltage. When discussing the upper hood configuration on Nov. 28, 1899, Tesla said:

"This arrangement permitted the charging of the pole easily up to a million volts." [21]

If Tesla were to discharge the condenser at a much slower frequency, the discharge current could be extremely large, being limited only by his ground bed resistance. This hypothesis is consistent with a public statement made by Tesla in 1934 in *Scientific American*:

"...Under proper conditions, it is possible to discharge spheres in a time interval incomparably shorter than consumed in charging them, and so amplify enormously the intensity of action." [8]

This is to say, the rate of flow of energy during the charging cycle might be at 75 kilojoules per second over 1 sec... but the rate of flow of energy during the discharge cycle could be at a rate of 75 Megajoules per second over a time interval of 1 millisecond. In both cases, the average power is 75 kW but the peak power during the discharge activity would be about 100,000 HP.

As early as 1893, in the Franklin Institute lecture, Tesla described an electrostatic pulse generator which was repetitively charged with a small amount of energy per charge, at a high pulse repetition rate, and then rapidly discharged but at a low pulse repetition rate.

This would make possible extremely large peak powers on the discharge cycle. In his speech, Tesla is describing the situation where a large condenser has been charged up to its disruptive potential by a small machine supplying static charge:

"When the condensers are charged to a certain potential, air gives way and a disruptive discharge occurs. There is then a sudden rush of current and generally a large portion of accumulated electrical energy spends itself. The condensers are thereupon quickly charged and the same process is repeated in rapid succession... It is evident that if the rate at which the energy is dissipated by the discharge, is very much greater than the rate of supply to the condensers, the sudden rushes will be comparatively few, with long time intervals between. This always occurs when a condenser of considerable capacity is charged by means of a comparatively small machine." [22]

Several paragraphs later, Tesla continues the description with a hydromechanical oscillator analogy:

"...Imagine a tank with a wide opening at the bottom, which is kept closed by spring pressure, but so that it snaps off suddenly when the liquid in the tank has reached a certain height. Let the fluid be supplied to the tank by means of a pipe feeding at a certain rate. When the critical height of the liquid is reached, the spring gives way and the bottoms of the tank drops out. Instantly the liquid falls through the wide opening and the spring, reasserting itself, closes the bottom again. The tank is now filled and after a certain time interval, the same process is repeated." [23]

Thus it appears that Tesla had conceived of a technique for obtaining large discharge currents with controlled pulse repetition frequencies. The vertical discharge current would produce a vertical current of moment I ·dl. This signal, we hypothesize could be controlled at an appropriate pulse repetition frequency for Schumann Cavity excitation. The controller, as described in Tesla's patents quoted above could either be "operated by a given rise of potential in the condenser." (Effectively an ELF relaxation oscillator) or "by rotation of the wheel" (break device).

In spite of the fact that our hypothetical ELF generator has some merit for satisfying the internal consistency hypothesis which we stated earlier as a ground rule, its acceptability must be measured against the second requirement of sound physics Tesla said that he got over 1000 amperes in his "antenna." Schumann's solution is in the sinusoidal steady state and even 1000 amperes in a 45-meter tower would seem to make possible relatively weak global field strengths. We have taken up this issue in another research document.

(Interestingly, however, if one looks at this hypothetical ELF generator as a fundamental form of the "switched capacitor" devices now of such great interest, the switched charged $\Delta Q = C \Delta V$. Over a period which is much larger than the switching period T_s the charge may be assumed to be quasi-continuous so that an equivalent current flow is equal to ΔQ divided by T_s. The equivalent resistor is T_s divided by C. [24] The application of the theory, however requires careful attention in Tesla's case if damped waves are assumed at the RF output of the extra coil.)

We observe that if our hypothesis is correct, then it is not remarkable that Tesla would have said:

"such a circuit may then be excited with impulses of any kind, even low frequency and it (the magnifying transmitter) will yield sinusoidal and continuous oscillations like those of any alternator." [6]

If our conjecture has any substance in fact, then *the tuned circuit of his magnifying transmitter was the whole earth-ionoshpere cavity resonator*! (This should help the reader appreciate why source dissipation will be experienced only when a load is engaged in a tuned receiver somewhere within the earth-ionosphere cavity. – Ed. note)

Corona Effects

There is one other observation to make about his "Magnifying Transmitter" and that is that its upper regions were engulfed in a coronal glow. In Colorado Springs and at Wardenclyffe he employed hoods to reduce corona. At Wardenclyffe, he had apparently planned to employ inverted hemispherical bowls to cover the spherical ball. In 1921, he said that "the underlying principle" and the "practical significance" of his 1914 patent #1,119,732 [7] was a technique "for confining the highest tensive flow to the conductors." He stated that the idea was to construct a conductor:

"…so that its outer surface has itself a large radius of curvature, or is composed of separate parts, which, irrespective of their own curvature, are arranged in proximity to one another and on an ideal enveloping symmetrical surface of large radius. These parts my be in the shape of shells, hoods, discs, cylinders or strands…[25]

We take it that the role of all the hemispherical shells in the 1914 patent was perhaps to physically bring about a more uniform distribution of charge over the sphere than could have been gotten with a lower portion missing because of the supports. If this be so, then they apparently would function in a distributed manner much like resistive dividers in a power supply capacitor chain, more or less causing a uniform charge distribution over the effective area of the sphere, and raising its disruptive potential to a maximum possible value. This would mean that a given size ball on a support could be charged to a greater maximum voltage.

Speaking of corona, we should also point out another curious feature of the Colorado Springs experiments.

From the patent wrappers associated with U.S. patent # 645,576, it is apparent that Tesla included a remarkable description of a rather extensive corona sphere surrounding his "elevated and insulated" antenna terminal, sometime before November 25, 1899. [26]

"... a conductor or terminal, to which impulses such as those here considered are supplied, but which is otherwise insulated in space and is remote from any conducting bodies, is surrounded by a luminous, flamelike brush or discharge, often covering many hundreds of even as much as several thousands of square feet of surface... This influence is not confined to that portion of the atmosphere which is discernable by the eye as luminous and which, as has been the case in some instances actually observed, may fill the space within a spherical of cylindrical envelope of a diameter of sixty feet or more but reaches out to far remote regions, the insulated qualities of the air being, as I have ascertained, still sensibly impaired at a distance of many hundred times that through which the luminous discharge projects from the terminal and in all probability, much further....I have noticed that his region of decidedly noticeable influence continuously enlarges as times goes on... in some instances the area covered by the flame-discharge mentioned, was enlarged more than six-fold by an augmentation of the electrical pressure amounting scarcely to more than 50%" [27]

Tesla apparently observed a corona sphere in excess of sixty feet in diameter. The space charge distribution apparently was due to the extremely high static or DC voltage on the elevated electrode. We conjecture that both its mode of production and its use were as outlined above.

The Tesla Tower

During the mid 1930's Tesla's work on a defense weapon apparently went so far as to be actually considered for construction. From file at the Tesla museum, it is apparent that Tesla had several "artist conceptions" made of a building with a tower in the form of a cylinder 16.5 feet in diameter, 115 feet tall. The structure was capped at the top by a 10-meter diameter sphere (covered with hemispherical shells as in the 1914 patent). The sketches were prepared by on Titus de Bobula of New York City. There is also correspondence with Alcoa Aluminum Company between July 29 to September 24, 1935, concerning fabrication, the last letter in essence saying that Alcoa was ready to start as soon as Tesla advanced the funds.

Whether the project would have been another disaster or not, we have no ideas. Since we have already gone this far out on a limb of speculation, permit us to conjecture what Tesla might have had in mind. We listed four components that Tesla maintained were essential. With the first, one might associate the Method (Patent #685,958) and Apparatus (Patent #685,957) for producing x-rays and providing rectification. With the second, one might associate the Process (#649,621) for producing high voltage RF – i.e. the Tesla Coil patents. Certainly Tesla powered his x-ray tubes from the top of Tesla Coils.

These four components are mentioned in at least four references during 1934 – at time when he was thinking and writing about the Van de Graaf machine. Perhaps it is not surprising to find the same language as appears above in component three also appearing in reference 8:

"...under proper conditions, it is possible to discharge spheres in a time interval incomparably shorter than consumed in charging them and so amplify enormously the intensity of the action. [8]

Certainly as pointed out above, this would be consistent with the second item.

To guess what the fourth component is would be shooting in the dark. However, let us go even further out on the limb and suggest that Tesla was perhaps employing a technique to rapidly lower the disruptive potential of a statically charged elected electrode. As is evidenced by the diary entrance of June 6, 1899, Tesla had already experimented along these lines. For example, suppose that one had a charged, insulated spherical shell in static equilibrium, and then rapidly punctured the shell with a very slender highly conductive track or path. (Or equivalently introduced, a charge of like sign immediately external to the sphere). The question to be answered is "Would a 32-foot diameter sphere charged to 50 MV produce sufficient repelling force for the contemplated weapon?." This question can probably be answered but, we have not yet performed the calculation. The answer might be no.

Final Comments

We have gone well beyond the bounds of propriety in our speculations. However, we believe that considerable light may have been thrown upon the intent and operation of Tesla's terrestrial resonance oscillator. If we have been able to provoke the reader to probe more deeply into Tesla's research, then we feel that we have attained some degree of success.

Lastly, no matter what the results or scientific merit of our research, whether every speculation be false or perfectly true, we all must never lose touch with the central fact that Tesla was a man whose creative intellect was set free to soar.

Truly, he touched the Holy Fire – and the world community is better off because of this good and decent and noble gentleman, whom we honor at this Tesla Centennial Symposium.

References

1. Colorado Springs Notes, by Nikola Tesla, Nolit, Beograd, 1978pp. 192;226-227

2. I am indebted to Professor Vojin Popovic of Belgrade University for asking this insightful question.

3. Reference 1, p. 364.

4. "On Light and other High Frequency Phenomena" by Nikola Tesla, delivered before the Franklin Institute, Philadelphia, PA February 1893, and the National Electric Light Association, St Louis March 1893. Repuplished in Inventions, Researches and Writings of Nikola Tesla. By T.C. Martin, Omni Publications, (1977) p. 347

5. 'The problem of increasing Human Energy" by Nikola Tesla, The Century Illustrated Monthly Magazine, June 1900, p. 208.

6. "My Inventions, Part V –" The Magnifying Transmitter" by Nikola Tesla , The Wireless Experimenter. June 1919, pp 112, 113, 148, 173, 176, 177, 178.

7. "Apparatus for Transmitting Electrical Energy", Nikola Tesla, U. S. Patent # 1119,732. Application filed January 18, 1902; Patented December 1, 1914.

8. "Possibilities of Electrostatic Generators" by Nikola Tesla, Scientific American, March 1934, pp.132-134; 163-165; and April 1934 p. 205.

9. "World System of Wireless Transmission and Energy" by Nikola Tesla, Telegraph and Telephone Age, October 16, 1927, pp.457-460. See page 459.

10. Abstract of notes by John G Trump, quoted in Tesla, Man out of Time by Margaret Chaney, Prentice Hall, 1981 p. 275.

11. " A Machine to End War" by Nikola Tesla, Liberty Magazine, February 1935, pp.5-7

12. "On Roentgen Rays" by Nikola Tesla, Electrical Review, March 11, 1896 pp 131, 134, 135.

13. "Nikola Tesla tells of New Radio Theories" NY Herald Tribune, Sunday, September 22, 1929. pp. 1, 21.

14. "Nikola Tesla Plan to keep Wireless Thumb on Ships at Sea" New York Press, November 9, 1913.

15. Reference 1, Page 29 (June 6, 1899).

16. Reference 1, Page 365 (January 2, 1900).

17. "Tesla's Views on Electricity and the War" by H.W. Secor Electrical Experimenter, August 1917, pg 270.

18. "Apparatus for the Utilization of Radiant Energy" Nikola Tesla, US Patent # 685,957. Applied for March 21, 1901 granted November 5, 1901.

19. " A Method of utilizing Radiant Energy" Nikola Tesla, US Patent # 685958. Applied for March 21, 1901, granted November 5, 1901.

20. "Lightning Protector" Nikola Tesla , US Patent # 1266,175 Applied for May 6, 1916, granted May 14, 1918.

21. Reference 1 p. 322 (November 26, 1899).

22. Reference 4 p. 304

23. Reference 4 p. 309

24. "Switched –Capacitor circuit design" by R Gregorian, K.W. Martin and G.C. Temer, Proceedings of the IEEE vol 71, No.8 August 1983 pp. 941-966

25. "Nikola Tesla on Electrical Transmission" Letter to the Editor, New York Evening Post, September 26, 1921. Republished in Tesla Said, edited by John T Ratzlaff, Tesla book Company, 1984, p. 224.

26. Dr. Nikola Tesla- Selected Patent Wrappers compiled by John T Ratzlaff, Tesla Book Company, 1980, Vol 1 pp. 166, 168-187

27. "System of Transmission of Electrical Energy" Nikola Tesla , U.S. Patent # 645,576. Applied for September 2, 1897, granted on March 20, 1900.

James Corum can be reached at 86 Weirton Mine Rd., Morgantown, WV 26508
Kenneth Corum can be reached at 104 River Rd., Plymouth, NH 03264

14 Harnessing Earth-Ionosphere Cavity Energy for Wireless Transmission

Elizabeth Rauscher and **William Van Bise**
Reprinted from *Tesla: A Journal of Modern Science*, 1997

Fundamental Excitatory Modes of the Earth and Earth-Ionosphere Resonant Cavity

Some of the principles of geologic precursor and meteorologic frequencies in the extremely low frequency (ELF) range of the electromagnetic spectrum and the possible relationship to the occurrence of earthquakes and volcanoes are explored. Monitoring of electromagnetic waves and magnetic fields has indicated the presence of characteristic natural and unique ELF frequencies which precede *seismic events.*

We have gathered extensive ELF magnetic field data from 1979 to the present time in many locations on the North American continent before the eruptions of Mt. St. Helens. The pre-eruptive and eruptive phases of Mt. St. Helens in the state of Washington were observed and analyzed in detail. Our system was on-line in the Portland, Oregon area, 40 miles south west of Mt. St. Helens, from 1979 through 1983 and on line in the San Francisco Bay Area from 1984 to the present.

Field measurements have augmented the permanent station data. The detection system utilizes a 150,000 foot antenna wound on a coil form adjacent to a very high permeability mu metal and the signal is passed into unique electronic processing elements which amplify and smooth the signal for flat response and permit readout and analysis in the time and frequency domains. The coil is electrically shielded so that pure magnetic field intensities are observed. The long-axis coil-core system allows directivity as well as high sensitivity. These are the main elements in the T-1050 detection system.

We have observed that characteristic ELF magnetic field oscillations with Earth rotational periods from 1.2 to 1.8 Hz, determined theoretically and subsequently measured at around 1.56 Hz with first harmonics of 2.9 to 3.8 Hz appearing in the Americas which grow greater in amplitude and then disappear from 24 to 72 hours preceding a geologic event. The amplitude of these oscillations is roughly proportional to the distance from measurement to event site and event magnitude. Multi-station detection could forecast locations, time and magnitude of impending events.

We also present some of our theoretical calculations related to the description of coherent collective modes of oscillation in the earth and earth-ionosphere resonance media. We will also examine some of our work in relation to Tesla's wireless energy transmission concepts of harnessing earth-ionospheric cavity energy.

Introduction

Extensive monitoring in areas of the Pacific Northwest during the period of time from early 1979 through late 1983 was conducted by Van Bise. The measured signals showed significant correlation between the volcanic activity of Mt. St. Helens and a range of frequencies between 0.1 and 30 Hz, with the frequency of approximately 3 Hz corresponding to, presumably, magmatic pulsations which preceeded eruptive events. Researchers at Portland State University examined the volcanic ash after the May 18, 1980 eruption and found the ash contained 30% of a material similar to magnetite. In the state of Washington on Sunday morning, 8:32 AM, Pacific Daylight time, May 18, 1980, Mt. St. Helens erupted in a cloud of fire, ash, steam and particulate matter that launched a half a cubic mile of this matter laterally and one quarter of a cubic mile of the volcano's mass was ejected vertically, to a height of 10 miles. When this event was complete, 1,000 feet of the mountain had disappeared and 60 people were dead. Future deployment of detection equipment such as is described here could prevent such a loss of life. [1,2,3]

Since then Rauscher and Van Bise have monitored ambient field impulses in California and many other areas of the United States and Canada. The data show a significant correlation with specific signatures which preceded earthquakes and volcanic eruptions. The pattern of signatures always ceased some 24 to 72 hours before such an event occurred.

Equipment used consisted of a calibrated T-1050-L-H magnetic field detector with a lower frequency range from 0.01 Hz to 300 Hz and a sensitivity factor of 10^{-10} gauss (Low pass system) and a higher frequency range from 1.0 Hz to 50 KHz at 10-6 gauss sensitivity (High pass System) was developed and employed at Tecnic Research Laboratories. The detector specifications are given in more detail later in this paper. Other equipment included a custom designed electrostatic voltmeter, a field intensity meter and two spectrum analyzers. [4,5,6]

The natural planetary impulses and vibrations preceding geologic events suggest that work with multi-station detection can lead to the successful development of an earthquake-volcanic eruption early warning system. We use our magnetic field detector to measure magnetic field changes, some of which reflect oscillatory modes of the earth. These modes of oscillation can be detected as seismic magnetic and electromagnetic pulsations of the earth and earth's surface which move in the earth's normal static magnetic field and the Earth's ionosphere resonance cavity. Movement of magmatic material with ferromagnetic (magnetite) inclusions and corresponding ionospheric changes in turn affect and produce flux fields which affect the entire earth ionospheric processes. [5,7]

In this report, the authors present experimental field data and their analysis and theoretical models demonstrating possible mechanisms of the dynamic earth processes. We also examine the relationship between the results of these data and Tesla's wireless energy transmission concepts.

The ground wave and the ionospheric wave are set up in such a manner as to produce the predicted 1.57 ratio to the velocity of light which was stated by Tesla in one of his 1905 patents. [8] In his model, Tesla treated the earth as a finite capacitive reactive component surrounded by an ion shell of variable altitude, beginning at about 50 km in height, which represents a system whereby a resonant ringing signal can be set up and transmitted. Although the system represents a leaky capacitor with a Q of about 4 to 5, it is possible to set up a resonant state so that it appears as though a signal is transmitted and received from any two points on the earth's surface. In actuality, according to the Rauscher-Van Bise model, the signal is not "transmitted and received," but represents a non-local global coherent state. Any event which can "wiggle" the static earth-ionospheric magnetic flux is transmitted as both a local and non-local influence.

In 1966, Rauscher determined the relationship expressed by Tesla in the Colorado Springs Notes in which he utilizes the dimensions in centimeters to represent the units of inductance, "L" in henries and capacitance, "C" in farads. This conversion factor system has been found to be crucial in understanding the principles involved in Tesla's Colorado Springs experiments. The purpose of the experiments and why, to this day, they have never been successfully completed is given. Also explained in detail is the interpretation of Tesla's work and the operation of his wireless energy device. Rauscher presents the mathematical principles germane to producing ball lightening from a fully ionized resonant stable plasma. [9] This research is summarized in this report and is detailed more completely in other papers. [9,10,11,12]

Tesla's Colorado Springs experiments are examples of a class of coherent state experiments and other experimental examples are discussed. Although much of Tesla's notes and data were lost, "confiscated" or presented briefly and in a cryptic manner, enough information exists to reconstruct some of Tesla's principles and his planned experiments so that we can describe the unfinished phase of Tesla's work. [13] The Tesla materials relevant to this presentation is from the time period of about 1897 to 1910.

Tesla's Vision

In 1905, Tesla described the earth as a finite small capacitance with regard to frequencies in the VLF region, and a resonant LRC system to ELF frequencies. He had hoped to utilize the VLF and ELF frequencies in concert simultaneously to provide a very large conduit through which nature's vast reservoir of electrical forces could be routed for the benefit of mankind. Tesla observed that nature's electrical system is activated by lightening storms or through other meteorologic and geologic activity. The type of system originally designed by Tesla could have acted as a "great energy siphon" by exciting the ionosphere and intervening media and then, by tapping into the flow of this immense reservoir of energy and tunneling it down to earth stations, mankind would today have all the "clean" energy necessary with which to put his machines to work. Tesla's visions, confirmed by his experiments at Colorado Springs in 1899 and by his life-long extraordinary ability in constructing electrical and mechanical devices, led him to develop ideas and concepts for his wireless energy transmission which he described in his patents and papers. In his words,

> *Now that I have discovered that, not withstanding its vast dimensions and contrary to all observations heretofore made, the terrestrial globe may be in a large part or as a whole behave(s) toward disturbances impressed upon it]n the same manner as a conductor of limited size; this fact being demonstrated by novel phenomena which I shall hereinafter describe. [8]*

With the formulation in this patent, Tesla treated the earth as a finite capacitor and as an element of a circuit. Through the legalities of patent law, Tesla had patented the earth! Use of his device to harness the energy of the earth was not to be, however, and we may be poorer for this undeployed natural resource. He had exclusive rights to the planet-ionospheric energy for 17 years and we, the people, re-own it by now as it is in the public domain. One wonders what our world would have been like had Tesla's vision come true and his "magnifying" transmitter had been deployed.

Tesla had developed the techniques and conducted experiments on the transmission of information through space before the turn of the century. Tesla, not Marconi, was the first to invent the radio and after his death in 1943 and after a review of the claims and dates given by Tesla relating to the invention of wireless communications, the Patent Office conceded that Tesla had indeed preceded Marconi and was actually the inventor of what we now call radio and television communications. Little serious research has been conducted on his effort to develop a wireless energy transmission grid or to examine the relationship between his work at Colorado Springs and Wardenclyffe, New York. Tesla's research from his Colorado Springs Notes and his work in the design and construction of the tower at Wardenclyffe are examined in relation to our current research. [14] We present our interpretation of these experiments and some of our data on measurements of earth resonant phenomena taken over the last eleven years.

Both authors have been interested in Tesla's research and related work since our early teens. It is interesting, in looking back over one's life, how various pieces of different puzzles began falling into place. [15] Our ideas and research on earth resonant phenomena and some possibilities for wireless energy transmission, both natural and man-made appear to complement those of Tesla.

We suggest that a system which involves a pulsed AC system in a high DC potential can create a "steady state." In Tesla's words in 1934

> *Most people, and not a few electricians, will think that very long and noisy sparks are indicative of great energy, which is far from being the case.*

In fact, at Colorado Springs, Tesla ran an AC system and raised and rotated a capacitor ball on a swivel utilizing the natural DC potential charge and discharge characteristics of the earth.

The Colorado air sustains a high potential before breakdown. The purpose of Tesla's experiments were to build up a voltage to achieve resonance; the necessary voltage was often not attainable from the local AC power generators since over voltage breakdown would occur before the necessary potential could be achieved. Needless to say, the problems created at the local power station by Tesla's experiments did not endear him to the power company or the people living in the area--even though he invented the power system!

Tesla carried on extensive correspondence with his laboratory workers in New York as his work progressed at Colorado Springs. His plan was to use the Colorado Springs laboratory as a resonance generating station and use the system to be built in New York as an amplifier and receiver. Work commenced at Wardenclyffe in 1901 at Shoreham, Long Island. Work on this project was never completed due to lack of funding. In Tesla's words,

> *My wireless tower on Long Island carried a sphere which had a diameter of 67 1/2 feet and was mounted in this manner. It was charged to 30,000,000 volts by a simple device for supplying static electricity and power.*

The key concepts are that it was a static, high voltage device. Later he compared it to a Van de Graaff generator. He also explained the purpose of Wardenclyffe to be that "one does not need to be an expert to understand that a device of this kind is not a producer of electricity like a dynamo, but merely a receiver or collector with amplifying qualities."

We have calculated the proper spacing to produce and receive a signal resonant with the earth. The location of Colorado Springs and the Wardenclyffe tower are in the proper relationship to produce earth-ionosphere resonant waves to achieve Tesla's desired results for worldwide communications and an enormous energy system.

Again in Tesla's words, this system would "not only (make possible) the instantaneous and precise wireless transmission of any kind of signals, messages or characters, to all parts of the world, but also allow the inter-connection of the system, telegraph, telephone, and other signal stations, without any change in their present equipment." Again, his stated purpose was to free the human race from forced labor and to create a time when "rich and poor no longer meant differences of materials conditions but of spiritual capacity and ambition--a time when inter-communication all over the earth should be immediate and universal and even when knowledge should be derived from sources now hardly imagined."

Rauscher and Van Bise formulate a simple model involving a resonant system which sets up a ground and air wave that would be simultaneously emitted and would add by resonant reinforcement. In the following and necessarily incomplete analysis we consider two interactive waves of similar but different frequencies. The analysis proceeds as follows.

Check Out Receipt

Davenport Fairmount Street Branch
563-326-7832

Thursday, November 3, 2022 12:54:09 PM
WINNE, MARY *A

Item: 30050046476080
Title: Harnessing the wheelwork of nature :
Tesla's science of energy
Material: Book
Due: 11/26/2022

Total items: 1

You just saved $16.95 by using your library.
You have saved $121.95 this past year and
1,770.80 since you began using the library!

Thank you for visiting the library!

Our phone renewal number is (563) 823-55
or (888) 534-6130

In calculating the velocity ratio of air and ground waves, one approach is to consider an air (earth ionosphere) wave travelling at v_2 and a through-the-earth wave traveling at v_1. Consider two waves emitted from the same location on the earth's surface, one in the air and the other through the earth and both traversing paths in the same time so as to come back to the emission location as reinforced. The path length for the air wave is πD and the through-earth wave is $2D$. For equal time of travel, the velocity becomes $v_2/v_1 = \pi/2 = 1.57$. In this analysis, the greater velocity wave, v_2, is the air wave. If v_1 is chosen to be the velocity, then the relative velocity (v_2) is $\pi/2$ = 1.57 times the speed of light. We could also consider the velocity v_2 as the velocity of light and then v_1 is $2/\pi = 0.64$ smaller than the velocity of light.

In Tesla's patents he makes it clear that the ground wave is the more rapidly travelling wave and the air wave is an electromagnetic wave travelling at the velocity of light. The above analysis is therefore not consistent with Tesla's model. In fact, there would be a mixing and reinforcing of a phonon/earth wave and an electromagnetic wave in the rarefied air and interaction. Therefore the above simple geometric problem does not apply. The problem, in fact, invokes phonon (longitudinal) and transverse electromagnetic wave interactions, as discussed in the next section.

Before proceeding further, any calculation involved in developing design parameters based on Tesla's work, needs examination in the light of two expressions in order to understand his calculations. In the Colorado Springs notes (1899-1900), Tesla obtained expressed quantities of capacitance (normally expressed in farads) and inductance (normally expressed in henrys) in terms of cm. For example, on June 28, 1899, he calculated the capacitance of the secondary with 26 turn windings as C = 1200 cm with the self inductance of $L_1 = 9 \times 10^6$ cm with a resonant frequency at 93,458 cycles/sec (Hertz). If we are to apply his calculations, we need to construct a system from E.A. Rauscher's research (from 1966) as capacitance C in cm is related to farads as $_{l(cm)} = \pm\sqrt{Q^2/mf^2C} \propto 1/\sqrt{C}$ so that $cm \propto 1/\sqrt{farads}$ where Q is the charge, m is the mass, f is the frequency of the system. Unit dimensions are given as $C[m^{-1}/l^{-2}t^4I^2]$ where I is the current.

For Inductance $_{l(cm)} = \pm\sqrt{Q^2/mf^2L} \propto 1/\sqrt{L}$ so that $cm \propto 1/\sqrt{henrys}$ for charge Q, mass m, frequency f, and inductance L, with unit dimensions $L[m^{-1}l^{-2}t^5I^2]$. In the square root relations we utilize the plus solution in both cases. For example, for calculation of capacitance of a hollow conducting sphere with a radius of approximately 250 cm, which can accomodate five million volts charge, the cm equivalent capacitance is 2.9 $\times 10^{-4}$ microfarads. The electricity created will be about 1.45×10^{-3} coulombs, depending on the material in the capacitor. [13,16,17]

Energy and Field Resonances in the Ionosphere and Tesla's Proposed Wireless System

As the earth rotates it carries with it all of the kinetic energy of the earth as well as its steady state magnetic field, particulate matter and the atmosphere in decreasing densities out to and through the most rarefied strata above the various layers of the ionosphere. At the same time, all of the natural and artificially generated mechanical, electric, magnetic, acoustic, thermal and gravitational energies at fixed or moving locations on or within the earth, are adding to or subtracting from each other for resonant and anti-resonant nodes as the rotation of the earth carries and drags these energies essentially past a fixed radiant zone illuminated by the sun and its solar wind, which we define as the magnetosphere and the ionosphere.

The ionosphere-magnetosphere-earth system can be treated as though it were in dynamic equilibrium over archeological time, but as subject to significant local and nonlocal effect perturbations waxing over intervening periods. Other treatments of the potential energy available from the earth up to the ionosphere have involved calculations based on theoretical models and measurements and which deal with the problem of the intervening short periods of time and address the local perturbations as observed and measured by relatively crude instruments within these perturbations. A satisfactory solution which resolves theory, observation and experiment in a self-consistent manner does not yet exist in the literature [17].

We therefore present the following simple calculations based on archeological time and a dynamically equilibrized earth-ionosphere system. The numbers given yield a rough approximation of the potential energy available but the figures are probably conservative since we have not taken into account the well known electrojet-Hall current contributions to the total energy.

We also have not taken into account the Peltier and Seebeck effects, the former occurring at the leading and the latter occurring at the trailing edges respectively, of the earth-ionosphere interface. Thermal energy from the SUN meeting the cold junction of the leading edge of the ionosphere would generate a potential difference and dynamic current as a Seebeck effect. Conversely, the trailing edge Seebeck voltage would be affected by the Peltier junction thermodynamic difference as they trail off into the night-side cold. The night-side hemisphere magnetospheric flux line excitations from the sunlit hemisphere ionosphere-magnetosphere-earth excitation would facilitate transfer of power at the night-side. These effects would give rise to significant local and non-local ionic current flows. An earth-ionospheric interface transverse Hall voltage would be a natural result of the Peltier-Seebeck effect generating earth-ionospheric circulating currents, and these factors are also left out of our potential energy calculations. Nevertheless, the amount of potential energy available within the earth-ionospheric system, if it could be harnessed, is surprisingly large.

The frequency differential between the North American power grid and the European/Asian power grid may also produce unique effects. The 60 and 50 Hertz differential produces a 10 Hz sum and difference frequency (2 x 60 = 120 - 50 = 70 Hz near the Navy project ELF Center frequency). [18]

A great deal of power is being transmitted or pumped into the atmosphere from power line losses as I^2R drop is emitted into the earth-ionosphere cavity, which acts as a leaky but extremely large capacitance. The earth-ionosphere represents two plates of a moving variable capacitor of roughly 24,000 by 24,000 miles area separated by an approximately average distance of some 108 miles.

A simple calculation based on the half sphere of the sunlit hemisphere of the earth shows that the capacitance of the hemisphere from the ground plate up to the D region "plate" of the ionosphere where the peak electron density exists to 65 km (about 108 miles) is approximately 7,568 microfarads.

From the formula C = (22.45 KA(N-1))/(10^8t) where C is in microfarads, K is the dielectric constant of free space with a value of 1, A is the area of one plate in square inches, N is the number of plates and t is the thickness of the dielectric in inches.

Using an average value of 100 volts per meter increases in the vertical field at the earth's surface up to 65 kilometers, we have 6.5 x 10^6 volts per unit meter. Twelve thousand miles is 14,400 meters, and for 14,400 meters squared the available potential is about 1.35 x 10^{15} volts.

Applying Ohm's law for power, we have P = E^2/R, where E^2 is the electric potential in volts for 14,400 meters squared and R is the free space impedance in ohms and since R is about 377Ω, we see that 3.575 x 10^{12} watts potential is available if it is possible to produce a dynamic resonance motion in the electrostatic potential. By definition, 746 watts is equivalent to 1 horsepower, and for a dynamic resonant earth-ionosphere, the potentially available horsepower on a sunlit hemisphere would be about 4.79 x 10^9 horsepower!

These factors and radio-television communication systems as well as satellite systems produce extremely complex energy production and re-radiation processes. Certain particular systems can become *locked*; that is, interaction of energy systems with each other and natural sources may become resonantly coupled or locked. Some of this energy resonates in the ionosphere and some is transmitted from this system which has a Q of 4 to 5. The Q is defined the "figure of merit" or the ratio of the energy stored over the energy dissipated. The transmitted power that does not escape forms frequency mixes such as 10Hz and the odd and even harmonics of the 50 and 60 Hz power systems. These frequencies form a complex based on physical areas of emission from the earth and day/night effects.

The observed ELF artificial impulses in the environment lead one to speculate that such pulses may be the result of a device [19,20] similar to the one envisioned by Tesla and which he described as a magnifying transmitter. By means of such a device, high potential stored electric charges should be able to be converted to propagating magnetic wave resonances between the earth's core, the ionosphere and the magnetosphere. Such conversion, if done with sufficient precision, would make it possible to realize a gain of acousto-electric energy by matching and utilizing the approximately 1.5 Hz rotational vibrations set up in and above the earth as it moves on its axis in its orbit around the sun.

The magnifying transmitter, according to Tesla, was to facilitate worldwide communications while at the same time it could be used to transmit electrical power without wires to ground stations on the globe which are suitably designed for resonance and are connected to a power generator. The local power generators would of course have step down transformers and meters and wires for distributing 97% efficient electric power to a convenient radius of customers. Power would still have to be metered and sold but at considerably lower rates. The I^2R losses would be minimal however, although maintenance to the ground generators would be necessary.

The earth's magnetic field lines describe minute motions due to micropulsations set up in them as a result of this rotational vibration. As is known, a moving magnetic field produces a current in a conductor. The earth's core is the likely conductor which would be expected to respond to these minute field variations.

Although the magnetic field of the earth is of small intensity, (about 0.5 gauss in the mid-latitudes) [21], the very large volume of the conductive core and the even larger radius of the surface magnetic field lines, provide a system with a great volume of electric current circulating in it.

Another potentially usable volume of electric current exists in the earth's magnetosphere. Some 10^{12} watts exist as the result of the magnetosphere. The combined electric power potential available from the ionosphere-core-magnetosphere is about 4.5 times the world's electric generation capacity! The major problem seems to be development of a method to gain access to these systems of electric currents.

We believe that Tesla had solved this problem in his experiments in 1899 at Colorado Springs. [13] By means of a specially constructed electrical detection system, he observed stationary waves showing that the earth behaved as a spherical conductor with finite dimensions and he also found that high potential, tuned circuits capacitively coupled to the earth developed two wavelengths when resonance occurred.

A spherical conductor mounted on an insulated pole served as the electrically elevated terminal which emanated radio waves that obeyed the ordinary formula for wave length, where the frequency divided into the speed of light yielded the length of the waves. However, the earth terminal, coupled to the secondary of a critically tuned inductor through a low value capacitance of special design, ostensibly propagated waves at the same time which were longer by a factor of about 1.57 times the velocity of light, or some 40% to 60% greater in length [19]. Tesla said that the Maxwellian electromagnetic component from the elevated terminal would become negligible as resonance of the earth's core and ionosphere developed. The magnetic flux from Earth's steady state can be "strummed" communicating such phenomena as "pearls" or vibrating magnetic flux density increases at nodes and anti-nodes and the vibrations propagate almost with no delays. [7]

The elevated terminal was to be specially constructed with a unique and very large radius of curvature in order to raise the electrical pressure extremely high and to store it there by virtue of its own electrical attraction until released into the air and ground terminals in a pulsed manner. The initial primary current would be of very large magnitude until the condition of resonance was struck on the earth's half sphere radius, after which the primary current could be expected to lower to a more practical value.

The earth-ionosphere was evidently envisioned by Tesla to be able to be treated, in certain electrical cases, as a lossless transmission line containing kinetic energy from its rotational motion which would be able to be utilized with a magnifying transmitter. The air above the elevated high potential terminal would offer the conductive path to the lower ionosphere by virtue of pulsed ionizations of the air molecules directly above this highly charged terminal. All vehicles could operate on electrical power and aircraft could fly on electrically driven motors and none would ever cross the ionization paths due to repulsion effects and thus all could be collision-free.

In order to gain access to the closed earth magnetic field core system, the period of the wave from the transmitter would have to be carefully controlled and would have to be somewhat below 20,000 Hz down to a low of 6 Hz or cycles per second, for practical utilization.

Furthermore, the time interval (on/off time) of the wave train excitation should be between one eighth and one twelfth per second. The electromagnetic component, free space, half wave length and longer magnetic half-wave length would thus be able to couple in a heterodyne manner--"mixing" in a constructive interference pattern at each half wavelength around the earth from the transmitting device producing larger magnitude effects. When these two waves couple, a lateral travelling wave plus a vertical standing wave should develop. By carefully adjusting the repetition rate and impulse duration, these transmitted dual waves may be "latched" onto or ride on the earth's magnetic field lines. The vertical wave might then begin to move in a path through the earth and out into space again, gaining kinetic energy (harmonic pendulum effect) from the earth's mechanical/rotational vibration system. These dual waves of the same period but of different lengths, interacting, may be sufficiently compressed to exhibit plasma-like wave circulation forms which could fit the criteria of a macrocosmic soliton-antisoliton. Solitons or solitary waves are dynamic entities that are localized in space and retain a fixed shape or form. Nonlinear recohering terms in the wave equation describing soliton-like behavior overcome dispersive losses so that the wave appears non-dissipative in space. The amplitude of these wave-like quantities is proportional to their velocity.

Tesla had stated as early as 1904 that the mode of excitation and the action of his magnifying transmitter may be said to be diametrically opposite to that of an electromagnetic transmitting circuit. He described the magnifying transmitter circuit as a device which acts like an immense pendulum, indefinitely storing the energy of the primary exciting impulses and impressing upon the earth and its conducting atmosphere, uniform harmonic oscillations of very great intensities. He also said that the electromagnetic radiations of a properly tuned magnifying transmitter would be reduced to an insignificant quantity. In addition, Tesla said that a number of distinctive elements put together in a manner analogous to the

human nervous system, would enable the magnifying transmitter to send, simultaneously, many thousands of encoded messages without serious mutual interference
[22].

Tesla had predicted 95% to 98% efficiency in the transmission of electrical energy
without wires via his magnifying transmitter. The primary currents, however, would
be very large and require a substantial amount of input energy. Once electrical
resonance was established with sufficient energy expenditure, Tesla felt that the
sustaining of large current-carrying standing wave ionized paths on the planet, could
provide almost lossless transmission channels.

The frequency of about 30 Hz is an interesting one in view of Tesla's writings
regarding earth-resonance and his magnifying transm]tter. He noted that the ground
terminal would produce waves with a length: $\lambda = (\pi/2)c\upsilon$, while the elevated
terminal would produce waves obeying the ordinary formula $\lambda = c/\upsilon$, where c can be
taken as the speed of light in free space {2.99 (10^8) meters/sec} and υ is the
frequency of the pulses. At 30 Hz then, the ground wave length would be 15,707.96
kilometers and the free space wave length 10,000 kilometers (see also Appendix I).
Monitoring has indicated that about 30 Hz waves exist over vast areas of the North
American continent and maximum intensities appear both from overhead and below
a vertical coil orientation as if the core and the ionosphere-magnetosphere were
being excited.

The distance from the surface of the earth to the inner core is about 6,370 kilometers
and, interestingly enough $\lambda/4$ at 30 Hz in free space is 2,500 kilometers while $(\pi/2)\lambda$
/4 is 3926.99 kilometers. The sum of these quarter length waves is 6,426.99
kilometers, almost exactly the distance from the earth's surface to the inner core for
$\lambda/2$. If one assumes that a phase velocity delay occurs both from the core up to the
surface of the earth and from the magnetosphere, at ten earth radii, down to the
earth's surface, a vertically oriented acousto-electrohydrodynamic cylindrical wave
may develop along a boundary layer near the earth's surface with its uppermost
boundary extended some $\lambda/2$ or 5,000 kilometers. The configuration would resem-
ble a "slow moving" standing wave. Heterodyne-like patterns may be expected to
occur which can be measured and the converging up waves and down waves would
be able to produce magnetosonic waves near the surface of the earth, which, at times,
might reach the audible range as "clicks" and "booms."

Analogies in physical optics might be applied to the concept of a single source
frequency with two different wavelengths existing simultaneously; they are the
phenomena of birefringance and double refraction [23,24]. It is also possible to
consider selective resonance absorption, also known as restrahlen and sometimes
referred to as "ghost rays," which we have observed in two of our midnight
(graveyard in Portland, Oregon and Skull Valley, Arizona) ELF measurements.
These residual rays can be produced by molecular rotation (which is related to
magnetic rotary power and magnetic rotary dispersion) [25]. Though resonance
absorption is usually associated with optically active absorption bands [26], it is
possible to extend the relationship of the optical equations into the ELF regions.

It is known that the index of refraction varies with the wavelength of refractive
waves [27]. The earth's core and ionosphere-magnetosphere may be able to be
treated as a single-system special case of a dense-rarified reflective medium which
may show an anomalous dielectric constant under the condition of resonance. The
usual description for absorption and selective reflection defining a complex angle of
refraction is $\alpha + i\beta$. Absorbing media can be described by the complex index of
refraction n(1-ik) and the wave traveling in it damps and introduces a phase shift
between the parallel and perpendicular polarized components of both the transmitted
and reflected wave producing elliptic polarization. If k<<1, the medium is weakly
absorbing, and if k~1 or k>>1 and sinh β>>1, wave penetration is only on the order
of a few wavelengths. The relationships are paradoxical, ie. a strongly absorbing

medium is one which rapidly attenuates the part of the electromagnetic wave that penetrates inside it, but since most of the electromagnetic wave is reflected from its surface, a strongly absorbing medium is really a poor absorber, if measured by the fraction of the total electromagnetic wave energy absorbed. At about 30 Hz, the earth elastic-acoustic wave and the ionosphere magnetic wave couple at the core. Thus about 30 oscillations are able to be sustained by the kinetic energy of the planetary rotation, which fits our observation.

During field measurements in 1979 and 1980, a curious anomaly was discovered about one mile east of the Bonneville Dam in Oregon, in a region where no 30 Hz waves were found. The complete area of absent waves was not plotted but the measurements indicated that a "hole" existed beginning at about the Eagle Creek fish hatchery and stretching approximately east south eastward. It is possible that a resonance of the earth-ionospheric cavity might cause resonance absorption bands at nodal locations on the earth's surface. In the case of ELF resonance absorption, the reflection-absorption-retransmission zone would be of large radius and could possibly result in a photon-phonon gyroscopic spin wave effect [28,29]. The spin processional frequency of waves may be sustained by a given absorption producing frequency as long as the driving pulse intensity is present at a given threshold level. The damping factor will be reduced, perhaps to the degree necessary for sustaining a spatial soliton wave [30,31,32]. Latent effect periods (delayed re-emission) would occur and depend upon pulse duration and repetition rate. When pulse transmissions change in frequency, the waves may damp momentarily and at times might precess to a rate that could match the vibrational rate of various piezoelectric geologic materials such as quartz, which might induce earth movements (volcanoes and seismic activity).

Since it is well known that a difference in phase in field intensity and polarization is always accompanied by energy absorption, it may be that ELF magnetic field resonance absorption effects cause far-field electric vector effects (earth to ionosphere E-field) and may produce a near-field, high-intensity magnetic component at even wavelength distances from an ELF source. At the same time, a high intensity electric field may produce as a far-field effect magnetic pulses at each quarter wavelength distance from the electromagnetic emitter source. If ELF absorption effects exist at areas on the earth's surface as a result of core-ionosphere-magnetosphere excitations, the energy may be retransmitted from those areas at substantially greater intensities than the intensity of the pulsed energy originally absorbed. This would be seen as abnormally high electric field intensities.

Examples of real time data covering a period of time from 1979 through 1986 are given in Reference 14, and a few illustrative examples are given in Appendices I and II. Experimental field measurements in the ELF-VLF frequency ranges were begun in 1972 and in that year, an approximately 10 Hz infrared frequency shift pulse was detected in the summer day-lit sky around noon and 4pm from a monitoring station in Portland, Oregon. The pulse was only on at intermittent intervals during the day for the next few months, then the pulse disappeared. In late 1975 and officially in July 1976, the 10 Hz pulses of the "Russian woodpecker" [33] came on the air and has remained on the air until the present time.

In the late summer of 1979 other artificial signals came on the air with repetition rates of 15 and 30 Hz. In November and December 1979, monitoring in the Pacific Northwest yielded magnetic signals of many different waveforms and frequencies mixed in with 10 Hz, 15 Hz and 30 Hz.

Many signals of natural and artificial origin coexist and synergistic effects between the natural resonances and man-made energies allow us to make a working hypothesis of the electromagnetic hydrodynamics of the earth-ionosphere, particularly with regard to geologic activity and weather.

Some of the artificial impulses have been interfering with lawful communications worldwide since 1976. Significant interfering electromagnetic signals were found on the 3 to 30 megahertz (MHz) bands and are usually pulsed at an on/off rate of 10 per second. The 10 Hz signals may have resulted from what the Soviets have admitted to be "radio wave experiments." They presumably originated somewhere east of the Baltic Sea. Some of these signals were seen to be phase and pulse width correlated with magnetic waves. Since July 1979, variable pulse width magnetic waves of approximately 10, 15, and 30 Hz resembling pulse time modulation (PTM), along with data-like impulses, were also observed with intensities exceeding an order of magnitude above the earlier observed natural signals in this frequency range, with an amplitude of about 100 to 150 microgauss (μG). As earlier stated, most of the artificial magnetic waves in the 10 to 30 Hz range were observed to propagate at maximum amplitudes in the vertical direction, suggesting the possibility of altered earth-ionospheric resonance excitations.

We have examined the characteristics of some of the natural waves and, as we have already noted, related these frequencies and wave forms to some of those hypothesized by Tesla. We have also examined some of the artificial, man-made signals as perhaps emanating from a modern variation of a device patented by Nikola Tesla in 1905 and which was termed a magnifying transmitter. (See data in Appendix I).

Measurements of Magnetic and Electromagnetic Pulsations in the ELF Frequency Range

The device we used in making our magnetic field measurements is described as follows. The T-1050 field detector operates on the principle that a coil of conducting wire, insulated and consisting of some 150,000 feet of #44 AWG wound on an insulating spool form with a high (μ) permeability mu-metal material adjacent to the inner windings, responds to a fluctuating magnetic field; or, if the coil is moved in a static magnetic field, it will respond to the field in a dynamic manner [34]. In either case, coil response to magnetic field fluctuations results in the generation of fluctuating voltages. The voltage is proportional to the number of turns of wire and the dimensions of the coil, the permeability of the mu-metal material and the magnitude of the magnetic field.

The sine of the angle of the "cutting" of magnetic field flux lines is another important factor in the sensitivity-frequency response characteristics of a coil-core type of magnetic field detection system. Slow moving magnetic flux changes in the coil, such as would be the case at frequencies below 1 Hertz, will induce a much lower voltage at the coil output than, for example, magnetic fluctuations moving at a 100 Hertz rate. There is an order of magnitude difference between 0.1 Hz and 1 Hz, but there are about three orders of magnitude of decreased sensitivity between these two frequencies and less than an order of magnitude (about three times less) between 1 Hz and 100 Hz.

In a coil system such as used in the T-1050, the resonant frequency of the coil-core is about 48 Hz; normally the resonant frequency is determined by $f_R = 1/(2\pi\sqrt{LC})$. Ordinarily the inductance L and the capacitance C in radio frequency circuits are directly calculable by the above formula. However, in a coil-core magnetic field detection system, other complicating factors are introduced, such as the permeability of the core, the distributed capacitance of the wire over its length, the magnitude of self induction relative to the induced back electromotive force and non-sinusoidal wave fronts acting on the coil-core, etc. These are some of the problems which need to be addressed when utilizing a coil-core type of magnetic detection device system. These are dealt with in the equalization, filtering and amplifying networks following the coil-core generated voltages. We take advantage of the fly wheel-like storage capabilities of a coil system and feed the coil generated voltages through carefully designed electrical networks to achieve extraordinary sensitivity and equivalent frequency response. The Faraday shielded coil allows us to "trap" the E-fields generated internally from the moving B-fields on the coil windings and utilize this energy to "MASE" (Magnetic Amplification of Stimulated Electrons) the sensitivity and frequency response. The T-1050 detection-equalization-amplification network following the detector coil-core sensor well accommodates real time wave shapes, forms and frequencies for ready analysis with spectrum analyzers, oscilloscopes and other instruments. In spite of the complexity of the above addressed problems which the T-1050 solves, its operation is straight-forward. We have examined the magnetic field flux with both shielded and unshielded coils. The shielded coil allows the pickup of varying magnetic lines of force only without electric field components. By using shielded and unshielded coils in simultaneous measurements, we can examine the E field contribution at ELF frequencies, even though externally generated E-fields contribute very little energy below about 300 Hz. [34]

We have been making magnetic field measurements for over twelve years. One of us (Van Bise) has been observing an approximately 31.5Hz ambient signal since 1979. We began monitoring of magnetic signatures on a regular basis, observing a signal of about this frequency since the end of 1983. We used a Spectral Dynamics model 335-B Spectrascope II real-time spectrum analyzer with a range of 0.06 Hz to 50 KHz to analyze the frequency components of detected ELF (extremely low frequency) and VLF (very low frequency) signals. With careful analysis, the 31.5 Hz signal was determined to be at 31.4 Hz. Note 2 x 1.57 = 3.14, which is π, which is what led us to consider the relationship of this frequency to the Tesla $\pi/2 = 1.57$ ratio. Even before making more accurate measurements with our spectrum analyzer, Rauscher predicted using Tesla's approach, that the 31.5 Hz signal was 31.4 Hz. The following analysis is intriguing but perhaps not definitive. Using the velocity ratio defined by Tesla as $v_1/v_2 = \pi/2 = 1.57$, for the two velocities for the same wave length, we can determine the two associated frequencies υ_1 and υ_2 for $v_2 = \lambda\upsilon_2$, and $v_1 = \lambda\upsilon_1$, and therefore $\upsilon_2/\upsilon_1 = \pi/2 = 1.57$. One assumption is that a signal associated with this frequency ratio activates oscillations in the earth and the earth-ionospheric resonant cavity and is associated with the so-termed 10 Hz "woodpecker." Then we may consider a ten times factor of the 1.57 frequency ratio or 15.7 Hz. Note that if this wave represents a wave length, then we have 2 x 15.7 = 31.4, which is the dominant frequency outside of the 60 Hz powerline, which we see in the western hemisphere.

We have measured the 31.4 Hz frequency in northern and central Oregon, central and southern California, central Arizona, southern Louisiana, in the New York area and in the Boston area. (See Appendix I.) Intensity of this frequency varies with location and time of day. From 1979 through March 1985 the frequency remained around 31.4 Hz and between March and April 1986, we observed a shift in this major intensity frequency, shifting from 31.4 to between 30.4 and 30.6. We sometimes observed simultaneous 31.4 and 30.6 signals or a cluster of signals in this range with side lobes up to four or five. Other clusters exist, around 48.5 Hz as detected by a "T" antenna measurement of electric field impulses. (See appendix I and II) [35]

We have considered a number of possible sources of this signal and the reason for the frequency shift in 1985. The wave forms are very regular displaying a surge in power from back EMF (electromagnetic force) as the signal goes away which indicates a man made source; perhaps a spurious side band of Project ELF which happens to be near one of the earth's natural resonances. It appears that the 31.4 Hz or 30.6 Hz signal does not involve a powerline subharmonic or mix of 30 Hz but occupies a more fundamental role. One hypothesis is that it involves an excitatory mode of the earth, activated by some specific external man-made source.

The 31.4 Hz signal can be analyzed as about a 30 Hz signal, heterodyning and mixing with the approximate 1.5 Hz earth rotational vector frequency. Other frequencies we have observed associated with natural phenomena, such as volcanoes and seismic activity, are complex, showing sine-waves with interspersed jagged waves unlike the 30 Hz signals and other artificial ELF frequencies. Some of the man-made signals display telemetry like characteristics.

Seismic activity has been occurring periodically in San Leandro in the San Francisco Bay Area of California. We have been observing irregular slow waves of 0.48 Hz and some clusters at 1.32, 1,56, 1.84 and 3.18 to 3.2 for the vertical coil configuration. For example, these frequencies were observed at 8pm on January 14, 1986. News reports later that evening indicated that earthquakes occurred near Salinas and San Jose, California measuring 4.3 and 5.2 respectively on the Richter scale.

We observed the onset of a 3.2 Hz wave maximum with an approximately north-south coil orientation from the 6th through the 13th of November 1986. Previously, this frequency and other specific frequencies had been associated with the volcanoes we observed with the Mt. St. Helens activity in 1980. On November 15th, 1986 the enormous volcanic eruption of volcano Nevada de Ruiz, near Bogata, Columbia was reported and this event occurred south of our observation station. This activity had been preceded by some steam eruptions before the major blast that killed over 23,000 people. A small quake of 3.2 on the Richter scale occurred that day in San Jose, which was consistent with our observation of the north-south coil orientation measurements of slow waves. Some other example predictions of seismic activity are our measurements in the period preceding August 1986, where we observed magnetic field oscillations around 1.5 Hz and 3.2 Hz which were most intense in the north-south direction.

During this period an earthquake occurred in Alaska measuring 5.0 on the Richter scale and two others occurred in China measuring 6.8 on the Richter scale. A "precursor" quake occurred near Mexico City and another occurred at Santa Barbara which measured 3.3 on the Richter scale. Earthquake signatures continued and we expected more and stronger activity. On the 19th of August, 1985, news reports stated that at 8:18am Pacific time, a quake occurred 250 miles west of Mexico City measuring 7.8 on the Richter scale which was felt as far away as Houston, Texas. Predictions of continued activity was made and the next day another quake occurred measuring 7.3 on the Richter scale which was called an "after shock." Some 20,000 lives were lost during that period. [14]

Theoretical Models of Collective, Coherent Resonant States in the Earth and Ionospheric Resonant Cavity

Our observations of frequency and time domain wave form similarity indicates the need to formulate a more complex dispersion relationship than the usual ones. That is, the relationship between wave number (or wavelength) and frequency is not of a simple form, ie. one cannot use the simple relation $c = \lambda \upsilon$. One can, however, proceed from a dispersion relation and then derive specific wave equations for specific applications. We proceed in an opposite manner and utilize our data, interpreting it as wave form solutions to a wave equation and then deduce nonlinear dispersion relations in which the leading order term is $c = \lambda \upsilon$.

The simultaneous occurrence of similar wave forms and frequencies over the globe would indicate a large portion of, or even the Earth as a whole, has been set into dynamic resonance by natural and man-made events. The existence of such electric and magnetic waves indicates the presence of a local and global resonance. The impulse waves observed and the on-off intermittency of the about 30 Hz and other ELF signals, would indicate that one or more local resonator-generators are activating very non-linear modes of oscillation in the Earth. Our data indicates that a mechanical-electrical system or systems can activate normal complex modes of oscillation in the earth and earth-ionospheric resonant cavity, some of which are naturally occurring and some are artificially induced.

We proceed in one or more of three ways to determine a wave equation which describes the observed generated wave forms. (1) One Way is to determine the complex and perturbations-expansion dispersion relationship from which a generalized wave is derived. (2) Another way is to introduce the formation of a new geometric space metric of more than four dimensions, which we term geomagnetic space. The prefix *Ge* is determined from the Greek term for earth and leads to such terms as geology or geography and geometry (earth measure). (3) A third procedure is to present a generalized, nonlinear wave equation and its solution which appears to fit the general form of the data.

Our experience with other electromagnetic and hydrodynamic systems leads us to proceed in the third manner and set down a general nonlinear equation and its solutions. The earth is a highly dynamic and enigmatic system whose origin and detailed structure remain a mystery. It appears in general to be a structure of a layered elastic sphere, as is evidenced by scattering of seismic waves by the Gutenberg discontinuity between the earth's mantle and core. Discreteness of the structure of the concentric zones appears to be due in part to different major compositional components of the various layers such as the crust, mantle (upper and lower) and core, resonantly locked together. The lithosphere, the stony outer portion, is elastic and flexible as determined from ice and geologic materials which show uplift and rebound. Temperature gradients produce convective processes within the earth and associated seismic waves travel at from 7 to 8 km/sec.

The earth forms certain normal modes of oscillatory states which are a function of its size, composition, inhomogeneties, elasticity, viscosity, capacitance etc. These normal modes can be activated by natural phenomena, as we mentioned before, such as volcanoes, meteorological activity, solar wind, etc. These states can also be activated by artificial or man-made systems. For example, nuclear testing, power line systems of 50 to 60 Hz, radio, television and other communications systems. A vast amount of power is pumped into the earth-ionosphere resonant cavity primarily during daytime and early evening hours on the sunlit hemisphere of the earth. Some of these frequencies interact with the natural electromagnetic fields of the earth and can either enhance or diminish these resonant modes. We strongly believe that the current power grids and other electromagnetic radiation will prohibit the design and use of the wireless energy system as Tesla perceived it. Certainly a system based on Tesla's design would have to be modified to accommodate current developments including aircraft travel and satellites.

There are some very striking and intriguing properties of soliton phenomena that lead us to formulate a model of earth resonance in terms of a dispersive-nonlinear wave equation having soliton-like solutions. The earth system is a media which has elastic rebound properties or acts fluidly as observed by the continental drift, and supports nonlinear coherent resonant wave modes that disperse, such as Love waves (or S-wave or stress wave-like) and Rayleigh waves (or P-waves or pressure wave-like) which can be activated from seismic adjustments. The earth acts as a dynamic nonlinear resonator with dispersion. (See Appendix I and II)

We have made extensive measurements of some of these resonant modes which are activated by man-made or other natural sources. Some of these modes may be self sustaining soliton-like waves similar to the process suggested by Tesla. A soliton is a pulse-like traveling wave solution of a linear dispersionless wave equation or a nonlinear equation with dispersion. The basic form of soliton wave equations have the classical wave equation as their leading order terms. If we have a linear equation with dispersion, ie. the usual classical wave equation, no soliton waves will occur as the Fourier components of any initial condition, and it will propagate at various different velocities and as the interface of Fourier as components in which energy will be lost. If nonlinearity is introduced without dispersion , again the possibility of soliton wave modes does not occur since a continuous source of pulse energy must be injected via harmonic generation into higher frequency modes. In the time domain, we often see such phenomena as a shock wave, ie. a wave of relatively short duration. Soliton waves can form with both dispersion and nonlinearity. The soliton wave can be quantitatively understood and interpreted as representing a balance between the effect of the nonlinearity and of the dispersion process. Phenomena amenable to this type of description involve nonlinear, coherent resonant states with dispersive losses.

Examination of natural phenomena, such as sun spot activity, ball lightening, hydrodynamic solitary waves and biological colonies including man, exhibit self-organizing approximately "non-dispersive" processes. These classes of phenomena that involve (1) non-linearity, (2) non-equilibrium, (3) coherent resonance and (4) collective particle states, can be described as self-organizing and non-dispersive. These phenomena can involve solid, liquid, gas and plasma states of matter-energy and can be mechanical and/or electrical (or electromagnetic, chemical or biological) in nature. The key element in such processes is that they do involve dispersion (or diffusion) but that this dispersion (diffusion) is overcome and recohered by the non-linear structure and or fields of the system under consideration. A prime example is that of the hydrodynamic soliton phenomenon, well described by the Korteweg-deVries equation developed in 1895 to describe the observations of John Scott Russel in 1834.

These equations describe phenomena which is dispersive in the third order derivative in space, $\partial^3 U/\partial x^3$, rather than the usual wave equation, which is dispersive in the second order in space, $\partial^2 U/\partial x^2$, which is "balanced" by the nonlinear term of the form $U(\partial U/\partial x)$, where U is a wave function amplitude dependent on space and time. There are also quantum analogies to this classical equation such as the sine-Gordon equation.

We have examined the structure and form of soliton equations applicable to a wide variety of physical, chemical and biological systems and demonstrate how these equations relate to the usual wave equation.

$$\frac{\partial^2 U}{\partial x^2} - \frac{1}{c_o^2}\frac{\partial^2 U}{\partial t^2} \tag{1}$$

where c_o is the velocity of the wave amplitude, U, and we have the usual dispersion relation for $k = \upsilon/c_o$ for wave number k and frequency υ. Each non-linear equation which exhibits a soliton wave-like solution has a different associated dispersion relation. We have presented ample examples of natural phenomena that exhibit these properties and demonstrate the application of these theoretical models to describing geologic phenomena [34,36].

We will now discuss the soliton model for the development of possible efficient energy devices. Although these devices will not violate the second law of thermodynamics, they are highly efficient and utilize some of the available ambient energy as efficient energy converters [37].

In deriving the form of a generalized nonlinear wave equation, one can usually proceed from a general dispersion relation where in the wave equations the dispersion term is independent of the nonlinear aspect. This method has its limitations when the above condition does not hold, ie. these terms are interrelated such as in the sine-Gordon equation, the method breaks down and becomes cumbersome to use.

Proceeding along the lines suggested by A.C. Scott's mechanical analogy [37], we examine the wave forms we have observed as describable by solitary wave-like phenomena and that these solitary waves are solutions to the sine-Gordon equation. We will also demonstrate the manner in which one can easily relate the sine-Gordon equation to the Korteweg-deVrles equation.

We consider the periodic variation of the amplitude of the earth's magnetic flux ϕ and governed by the nonlinear evolution equation. We proceed from the sine-Gordon equation

$$\frac{\partial^2 \phi}{\partial x^2} - \frac{\partial^2 \phi}{\partial t^2} = \sin \phi \tag{2}$$

which can be written in compact notation as $\phi_{xx} - \phi_{tt} = \sin\phi$. If diffusion as well as dispersion occurs, then additional terms in ϕ_{xx} and ϕ_t will exist on the right side of the above equation. The flux amplitude ϕ plays the role of the wave amplitude, U

$$L = \frac{1}{2}\left(\frac{\partial \phi}{\partial x}\right)^2 - \frac{1}{2}\left(\frac{\partial \phi}{\partial t}\right)^2 - \cos \phi \tag{3}$$

or in compact notation

$$L = \frac{1}{2}(\phi_{xx})^2 - \frac{1}{2}(\phi_{tt})^2 - \cos \phi \tag{4}$$

For x dependence only. For the more general case, x,y,z dependencies of the flux can be examined.

The wave equation and its Lagrangian can be modified by the introduction of certain pertubation terms that can account for fundamental and harmonic resonances [14]. These terms are taken as small and in the form of an exponential. Examples are given in the above reference.

For example, the linear stability of traveling wave solutions of the sine-Gordon equation have the form $\phi(x,t) = \phi(x - ut)$ which is expressed in terms of an elliptic integral with three arbitrary parameters, u, the traveling wave velocity, ϕ_o, the value of ϕ for $(x - ut) = 0$ and C an integration constant,

$$\int_{\phi_o}^{\phi} \frac{d\pi}{\sqrt{2(C - \cos\phi)}} = \pm \frac{x - ut}{\sqrt{1 - u^2}} \tag{5}$$

If C = 1 is chosen and $\phi_o = \pi$ is set and corresponds to ϕ rotating by 2π as $-\infty < x < \infty$, the integral yields a soliton solution,

$$\phi = 4 \tan^{-1}\left[\exp \pm \frac{x - ut}{\sqrt{1 - u^2}}\right] \tag{6}$$

If u = 0, the factor in the exponent reduces to x.

The plus sign in the exponent corresponds to a positive sense of rotation and the wave pulse can be conceded to be a soliton; the minus sign can be considered to be a negative sense of rotation yielding an "anti-soliton." Solitons and antisolitons are created or annihilated in pairs and the sine-Gordon equation is invariant to a Lorentz transformation which can be defined for luminal and sub-luminal velocities.

The soliton model yields a description of a very stable entity which exhibits both particle and wave like properties. The solution to the sine-Gordon equation are stable for example for u = 0 and C ≥ 1, and unstable for |C| < 1. For moving solutions where u ≠ 0, the dynamic symmetry of the nonlinear wave equation can be expressed by its invariance to the Lorentz transformation defined by $\phi(x,t) \longrightarrow \phi'(x';t')$) as

$$x \rightarrow x' = \frac{x - ut}{\sqrt{1 - u^2}} \quad \text{and} \quad t \rightarrow t = \frac{t - ux}{\sqrt{1 - u^2}} \tag{7a,b}$$

and the appropriate derivatives $\partial/\partial x$ and $\partial/\partial t$.

The point to be taken at this juncture is that although the sine-Gordon form of nonlinear equations yield solutions which appear to reflect some of the properties of elementary particle physics [38], these forms may be useful to our application as well. Scott's mechanical analogy description in reference [37] suggests to us that some of our observed wave forms might well described by a similar approach involving elliptic integrals with space, time, velocity and also frequency arguments; see Appendix II.

We can write a more familiar form of the soliton solution by writing our elliptical integral as

$$\int_{\phi_o}^{\phi} \frac{d\phi}{\sqrt{P(\phi)}} = x - ut \tag{8}$$

where the term $P(\phi)$ can be written as a polynomial expansion in terms of integration constants C and C_2, and velocity u up to third order in ϕ as

$$P(\phi) = 2C_2 + 2C_1\phi + u\phi^2 - (g/3)\phi^3 \tag{9}$$

The above integral can the be written (in the form which in general is not Lorentz variant) as

$$\phi(x - ut) = \frac{3u}{g} \text{sech}^2 \left[\frac{\sqrt{u}}{2}(x - ut) \right] \tag{10}$$

which now looks like the solution to the Korteweg deVries equation for u ≥ 0. The constant g then appears as the nonlinear term of the nonlinear equation of the form

$$\frac{\partial \phi}{\partial t} + g\phi \frac{\partial \phi}{\partial x} + \frac{\partial^3 \phi}{\partial x^3} = 0 \tag{11}$$

which is the Korteweg-deVries equation that occupies the role of a coupling constant of the nonlinear term which balances the highly dispersive term $\partial 3\phi/\partial x^3$.

For completeness we can write the Lagrangian density as

$$L = \frac{1}{2}\frac{\partial\theta}{\partial x}\frac{\partial\theta}{\partial t} + \frac{g}{6}\frac{\partial^2\theta}{\partial x^2} + \frac{\partial\theta}{\partial x}\frac{\partial\psi}{\partial x} + \frac{1}{2}\psi^2 \tag{12}$$

or as $L = 1/2\ \theta_x\theta_t + (g/6)\ \theta_{xx} + \theta_x\psi_x + (1/2)\ \psi^2$ where we define $\partial\theta_x = \theta_x = \phi$ and $\theta_{xx} = \partial^2\theta/\partial x^2$, etc.

We have observed similar wave forms in both the time and frequency domain. See Appendix I. From these data taken simultaneously in time and in frequency, we deduce that the dispersion relations governing these wave forms is a complex relationship between wave number (inverse of wave length) and frequency or frequencies. Since the wave length of these waves are so long for these low frequencies, we are essentially detecting these waves as observers from a frame of reference "*within these waves.*" The approximately 30 Hz waves which we observe always look like well formed sine waves.

For the usual kinematic wave equation $c = \lambda \upsilon$ we can relate the frequency and the time as $\upsilon = 1/t$. For more complex wave equations, this simple relation of time and frequency may not hold.

Analysis of the periodic nature of the observed data waveform amplitude in time (oscilloscope tracing) and power density vs. frequency (spectrum analyzer display) allows us to deduce the relationship of time t and frequency υ. One procedure is to identify the frequency μ with a frequency domain time τ. From this theoretical model we can construct a five dimensional geometry in the coordinates (x,y,z,t,τ) where $\mu \equiv 1/\tau$ and $t \neq \tau$. We have explored in detail elsewhere, the construction and application of five and eight dimensional geometries. See References 40 and 41 and references therein.

Measurement and analysis of the acoustic-seismic modes and magnetic field oscillations in the ELF and the VLF region of the earth and earth-ionosphere cavity, leads us to re-examine issues related to the measurement process. We proceed from a generalized wave equation with coherent, solitary wave solutions to a wave equation with five independent variables, three dimensions of space, the usual time and associated frequency, and an additional time-like variable with a unique additional frequency variable.

Interestingly, the problem of measurement of the ELF phenomena is opposite, in a sense, to that for high energy process--x-rays, gamma rays, elementary particles and quarks. For ELF phenomena, the observer is significantly smaller (internal) than that which is observed, which is external large-scale phenomena. Whereas, for high energy quantum processes where ħ applies, the observer is significantly larger (outside) than that which is observed. Treatment of the problem of an internal or inside observer of an external large-scale phenomena is made in terms of a five dimensional wave equation. A four-space description may well suffice if the observer is larger than the scale of that which is observed.

The recent GUT (Grand Unification Theory) for strong, weak and electromagnetic interaction, involve ten and eleven dimensional spaces. Part of the subspaces of the GUT theory is the five-dimensional space of Kaluza-Klein with four spatial dimensions (one a periodic rotational spatial dimension) and the usual time dimension.

We explore in more detail, the relationship of macroscopic electromagnetic and gravitational interaction. Also examined is the modification of the gauge conditions as applied to electromagnetic interactions in the ELF region. In the conventional view, an electromagnetic wave of about 7.8 Hz has a wavelength of approximately the circumference of the earth. In the conventional view of wave packets, we are dealing with a photon the size of the earth! [5,14]

We can treat the problems of an internal (inside) observer or an external (outside) observer. It may be most useful to treat the wave equation as solvable in five-space for the internal observer whereas our-space may suffice for the external observer, where the relationship of υ and t is less complex. We can define a frequency υ associated with t as $\upsilon = 1/t$, we associate a wave number k with υ as $k = 1/\lambda$ where $c_\omega = \lambda \upsilon$ and a wave $q = 1/\Lambda$ where $c_w = \Lambda \mu$ where c is the velocity of the wave so that c can be equal to c_ω or c_w where w equals $2\pi\mu$.

We can write a general form of the electromagnetic field $F_{\mu\upsilon}$ which depends on the electric and magnetic fields E and B so that (for $\psi \to \mu,\tau$)

$$F(x,\chi,t,\tau) = F_o(x,\chi,t,\tau)\, e^{i(kx - \omega t + q\chi - wt)} \tag{13}$$

which comprises an eight-dimensional representation. Elsewhere, we have examined the symmetry conditions and Lorentz invariance in five and eight dimensional geometries in which the group elements of the five dimensional space is a subset of the group elements of the eight dimensional space. The twister algebra of the eight-space is mappable to the spin or calculus of the five space [41,42,43] (See Appendix II, a brief description of some data.)

One of the outgrowths of this procedure is the formulation of the relationship of the transverse and longitudinal components of the $\underline{E}$ and $\underline{B}$ fields. The existence of actual longitudinal components of $\underline{E}$ and $\underline{B}$, non-Hertzian waves, entails modification of gauge invariance, which we presented in detail elsewhere. [14,44] Extensive evolution of multidimensional models have been made including the application to the design of specific parameters for emission and reception antenna.

Exciting possibilities of a new picture of explaining, understanding and utilizing earth resonance modes may emerge from our five and eight dimensional wave equations. Our theoretical work and experimental data has allowed us to predict earthquake and volcanic activity with approximately 84% accuracy. [4] Future research should lead to more accurate predictions and predictive methods.

Specifically, we can define an orthogonal set of dimensions x,y,z,t,τ. In general, we can express a form of the five dimensional generalized wave equation from the Laplacian form

$$\underline{\underline{\Pi}}^2\,\psi = []^2\psi - 1/c_\upsilon^2\,\partial^2\psi/\partial\tau^2 \tag{14}$$

where we define a new five dimensional operator $\underline{\underline{\Pi}}^2$ and where $[]^2$ is the usual D'Alembertian operator

$$[]^2 \equiv \nabla^2 - 1/c_\omega^2\,\partial^2/\partial t^2 \tag{15}$$

and ∇^2 is the del operator representing the spatial part of the equation as $\nabla^2 = \partial^2/\partial x^2 + \partial^2/\partial y^2 + \partial^2/\partial z^2$. We define the wave amplitude as dependent on the five space independent parameter as $\psi(x,y,z,t,\tau)$. We define a velocity c_w associated with the time variable τ and frequancy μ and c_ω is the velocity associated with the time variable t.

Assuming a functional dependence of w on $w(t,\tau)$ and ω on $\omega(t,\tau)$ and where w and ω are "locked" so that the time frequency domain simultaneously display a similar wave form, where amplitude versus time t and power spectrum versus frequency υ are similar forms.

We express these conditions in a form of a general wave equation as

$$\underline{\underline{\Pi}}^2\,\psi = \frac{1}{c_\omega^2}\frac{\partial^2\psi}{\partial t^2}\frac{1}{c_w^2}\frac{\partial^2\psi}{\partial t^2} - \frac{1}{2\pi}\frac{1}{c_o^2}\int\int\int\int_{-\infty}^{\infty} A\psi e^{i(kx - \omega t + q\chi - wt)}\, d\omega\, d\upsilon\, dt\, d\tau \tag{16}$$

Here we consider ω and w being in the same units (we are not using the usual definition $\omega = \upsilon/2\pi$ but $\omega \neq w$ where ω and w are distinct frequencies). The velocity c_o is associated with wave emission in both time and frequency domain where c_o can be a function of c_ω and c_w and the wave function ψ is a function of the five dimensional space, as $\psi(x,y,z,t,\tau)$. The variable amplitude A has a dependence on variables as $A(\omega,\upsilon,c_\omega,c_w)$ to insure Lorentz invariant conditions are obeyed.

251

The usual relations can hold $c_\omega = \omega/k$ and $c_w = w/q$ for wave numbers k and q. But from the above generalized wave equation, we can consider the possible form of a dispersion relation which is a complex form involving the relationship of ω, w, c_ω, c_w and c_o with the associated wave numbers k and q.

In order for the time and frequency waveforms to appear to be of similar forms, the form of dispersion relation is such that the generalized wave equation is dispersion-free, or of a nonlinear form in terms of the integral term in terms $A(x,y,z,t,\tau)$ which overcomes dispersive terms.

For some general problems, the appropriate dispersion relation can assume a complex statistical form and may take on a nonanalytic form which is unsolvable except by Noval computer analysis techniques. As we suggested in our discussion of the sine-Gordon equation, it may not be most efficient to proceed from a dispersion relation. The determination of the form of the term A depends on the form of the dispersion relation and the insurance of Lorentz invariant conditions for the wave generalized equation in some applications which are relatively invariant.

Let us examine a possible form of the wave equations and solutions in a first approximation as follows. We rewrite our nonlinear equation so that the linear terms appear on the left and the nonlinear terms on the right.

$$\underline{\underline{\Uparrow}}^2\,\psi - \frac{1}{c_\omega^2}\frac{\partial^2\psi}{\partial t^2} = \frac{1}{c_w^2}\frac{\partial^2\psi}{\partial t^2} - \frac{1}{2\pi}\frac{1}{c_o^2}\int\int\int\int_{-\infty}^{\infty} A\psi\, e^{i\omega t}\, e^{iw\tau}\, d\omega\, d\upsilon\, dt\, d\tau \qquad (17)$$

We can treat the term $(1/c^2)(\partial^2\psi/\partial\tau^2)$ as either linear or nonlinear. If we choose a linear form for this term then t and τ are additive and this tensor can be combined linearly with $(1/c_\omega^2)(\partial^2\psi/\partial\tau^2)$ so that the term in τ adds only a coordinate shift. Hence we are reduced to the trivial case where we need only consider a single frequency term ω related in a simple manner to the time t. Since the trivial case is not useful, we will resume our consideration of the term $(1/c_w^2)(\partial^2\psi/\partial\tau^2)$ in terms of its nonlinear form.

We can define a complex form of a coupling constant that defines the relationship of the times t and τ and frequencies ω and w. We denote this term as $g^2(\omega,t,w,\tau)$. The relationship of the quantities in the term $g^2(\omega,t,w,\tau)$ determine, in part, the term $A(\omega,w,t,\tau,c_\omega,c_w)$ and can be derived from the five dimensional Fourier transforms.

We can write the usual Fourier Transforms for frequency and time in four-space as

$$\theta(t) = \frac{1}{\sqrt{2\pi}}\int_{-\pi}^{\pi}\phi(\omega)\, e^{i\omega t} \qquad (18a)$$

and

$$\phi(t) = \frac{1}{\sqrt{2\pi}}\int_{-\pi}^{\pi}\theta(t)\, e^{i\omega t} \qquad (18b)$$

In five dimensions, the Fourier transforms are expressible in terms of a 4x4 matrix array, $\phi(\omega,t,w,\tau)$ and can be written in a form as follows:

$$\theta(t,w,\tau) = \alpha\int\phi(\omega,w,\tau)E\, d\omega\, dw\, d\tau \qquad (19a)$$

$$\theta(\omega,w,\tau) = \beta\int\theta(t,w,\tau)E\, dt\, dw\, d\tau \qquad (19b)$$

$$\xi(w,\omega,t) = \gamma\int\zeta(\tau,\omega,t)E\, d\tau\, d\omega\, dt \qquad (19c)$$

$$\zeta(\tau,\omega,t) = \delta\int\xi(w,\omega,t)E\, dw\, d\omega\, dt \qquad (19d)$$

where α, β, γ and δ are constants including a 2π factor and E is the exponential $E = e^{\pm i(\omega t + w\tau)}$.

The above applies for any relationship of t, τ, ω and w, and simplifies for our particular case. Consider a specific example of our data for the polarity shift phase and amplitude modulated 30 Hz Wave, with rotational vector frequency 1.54 Hz. This waveform can be described as a rotational or a screw wave in an (x,t,τ) coordinate space where the amplitude or power is expressed in the x dimension. We define a wave function solution which is related to ψ for psi (x,y,z,t,τ) in the above equations as $U(x,t,\tau)$ in terms of one spatial dimension only. The nonlinear terms of our wave equation are expressed as

$$-\frac{1}{c_w^2}\frac{\partial^2 \psi}{\partial \tau^2} - \frac{1}{2\pi c_o^2}\int\int\int\int A\,\psi E\, dV = \frac{g^2}{2\pi}\int\int\int\int P(A,\psi)E\, dV \qquad (20)$$

where dV is the form differential $dV = d\omega\, dw\, dt\, d\tau$ and E is the exponential function $E \propto e^{\pm i(\omega t + w\tau)}$. Then we can write $U(x,t,\tau)$ which is a viable solution to the above wave equation

$$U(x,t,\tau) = 4\eta_o\, sech\left(\frac{x - V_\omega t}{l_\omega}\right) sech\left(\frac{x - V_w t}{l_w}\right) \qquad (21)$$

where η is a constant and where the terms in ω and w separate out since the 1.5 Hz rotational wave (as the Foucault pendulum demonstrates) represents a five dimensional rotation of the 30 Hz Wave. If the rotational effect did not exist then the above form would reduce to the usual solitary wave form.

$$U(x,t) = 4\eta_o\, sech^2\eta \qquad (22)$$

where $\eta = (x-vt)/l$ and where v is a simple function of v_ω and v_w and τ represents a linear coordinate transformation of t. The terms in v_w and τ then just add an arbitrary phase shift to the term $\eta = (x-vt)/l$ and represents a unit length normalization so that the argument of the hyperbolic function is dimensionless.

All measurements of the 30 Hz waves and some of the 10 Hz waves couple to the earth's rotational velocity of about 1.5 Hz so that the above simple case does not hold in general.

The integral $P(A,\psi)E$ of the wave equation is derived from the coupling constant expression involving x,ω,t,w,τ. [44]

Detailed computer analysis from our data will better determine the allowable forms of $P(A.\psi)E$. [44] We see that the solutions to these wave equations will be similar to the sine-Gordon equation with soliton solutions. See Appendix II for discussion of geometric conditions on spatial, time and frequency dependence of some of our data.

Possible Activation of Earth and Earth-Ionospheric Resonance States by Solar Wind Activity

We speculate on some additional causes for seismic and volcanic phenomena as well as activation of major storm systems. Seismic activity produces physical movement of the earth in its own steady-state magnetic field and produces charge and acoustic coupling which modifies the earth's field locally. Major field coupling effects can occur due to solar wind effects, particularly during heavy particle interactions which follow major solar flares.

Planetary magnetic field organize ionized matter. The resulting magnetospheres are unique domains or "cells" of plasma that are semi-isolated and considerably different from neighboring plasma regions. The earth's magnetosphere and that of other planets are affected by planetary rotation and ionic flows from the sun and its rotational dynamics and we suspect that the diurnal cosmic bombardment also plays a significant magnetospheric role. Co-rotational plasma flows, production of plasma waves, radio emissions, and ion acceleration of thermal electrons to hundreds of MeV occur within the magnetospheres. Plasma processes can involve the usual plasma instabilities related to ionic interaction with the earth's magnetic lines of force.

Primary driving forces of magnetospheric features involve the solar wind (and its induced changes from solar flare processes) and the earth's rotation (including the steady state magnetic field as well as disturbances within this field due to the earth's physical "adjustments" such as seismic activity and volcanoes). Planetary and stellar magnetic fields are believed to organize ionized matter in stellar and galactic systems.

The deformation of the earth's magnetosphere occurs as the solar wind (an internal expansion of the solar corona) interacts and mixes with the intrinsic magnetic field generated by or intrinsic to the earth. The solar wind at the earth's orbit has an ion density of about 10 ions/cm^3 with an energy density of a few times 10^{-8} dynes/cm^2.

Some General Comments on Theoretical Prediction of Seismic and Volcanic Activity: Research Conducted 1979-1990

Our data and records indicate that earthquakes and volcanic activity seem to be able to be modified by sunspot/solar flare activity and cycles in a manner similar to the effect of tidal action (caused by the gravitational pull of the moon) except that this sunspot/solar flare/earth interaction produces electromagnetic activity in the earth environment. Weather processes also appear to depend on solar cycle patterns. As is known, sunspot magnetic "storm" activity produces changes in the solar wind. The heavy particle interaction with the ionosphere produces charged state changes within the E layer and other charged layers of the atmosphere which creates induced magnetic fields. These fields and current flows interact with, affect, modify and perturb the earth's "steady state" magnetic field. As the lines of force of the earth are affected and "wiggled," the core mantle interface is affected. Extreme effects produce energy releases in the form of earthquakes and volcanic activity. Upper atmospheric effects can drive the jet streams and modify their paths and structures and thus affect the weather. Concurrent with major seismic and volcanic occurrences is the production of lightning which also affects weather. As is well known, increased atmospheric gases and particulate matter can affect global weather patterns.

Large earthquakes can also "wiggle" the lines of force of the earth's magnetic field and hence, in turn, affect ionosphere ionization states, thus affecting weather and other reverberatory seismic modes of excitation. Lightning strokes and piezoelectric releases in rocks in the earth produce electromagnetic field spike-like impulses which are measurable with suitable instrumentation. All these phenomena produce characteristic electromagnetic wave signatures which we can and do record and analyze.

Earthquake tables containing predicted location and estimated magnitude can be constructed in a manner analogous to tide tables. Predictions of possible volcanic activity can also be generated. These and other data might also be utilized to generate long term weather profiles.

Conclusion

The earth and the life forms upon its surface vibrate and resonate in harmony in such a manner that radiant energy from the sun and materials and vibrations of the earth support this life and its evolution. Some of the major normal oscillatory modes of the earth are in the 9-13 Hz range which, interestingly enough, is about the power spectrum peaking for most people's alpha frequency.

It is clear that we depend on "Mother Earth" for our life, but whether, in some sense, the earth itself depends on the life forms on its lands and in its seas and atmosphere, is another matter; ie. is there a symbiosis between the earth and the life it supports?

We know that man can create great changes, some of which have polluted the air, land, streams and seas with chemicals, radioactivity and electromagnetic waves. Some changes wrought by man can be repaired by the earth but others may not be so easily repaired. The question of why man should pollute his life support system continues to go unanswered. Some of the electromagnetic waves generated by man may have global significance for the earth and the life forms upon it.

In this project and in this paper, we have explored the earth's magnetic field emanations, those that are natural and those activated by man. There are a multitude of natural modes, such as the earth's mechanical rotation, seismic activity, volcanoes, solar wind and solar broadband noise activity and many others.

Also impressed upon the environment are many man-made sources disturbing both the atmosphere and the earth. Some of these emissions may be reaping irreparable damage to the ionosphere and earth which, in turn, threatens our very existence. We must examine what we are doing as people, as societies and as nations!. If we do not develop a new consciousness and awareness, destruction of life will inevitably result.

When the earth and the life upon it is in harmony, the system is mutually life enhancing. Man has (or has had) great abilities and potentials and yet, most of his recent technologies have been to strip nature away from us--to shield us from nature, to "conquer" and control her while designing ever more dangerous weapons systems with which to more efficiently strip all life from the planet. We must examine why man has moved toward such insane motivation, toward mutual destruction and whether mutual life enhancement and sanity can again become the noble objective of pursuit which desperately need implementation today.

Acknowledgements

The authors appreciate the assistance of Harold Faretto in helping us with some of the measurements and engineering work on TRL projects and thanks to Hal Treacy for his assistance in providing us some necessary equipment for TRL projects.

Fundamental Excitatory Modes of the Earth and Earth-Ionosphere Resonant Cavity
APPENDIX I
Measurements Of Seismic Precursor Excitation Modes

Earth mechanical or seismic oscillations produce longitudinal excitations in the earth itself. These oscillations produce local and global disturbances which involve local field coupling which perturb the earth's global steady-state magnetic field. The seismic excitations produce pressure and stress waves which have acoustic components as well as "wiggle" the Earth's lines of force producing a magnetic fluctuation component. These acoustic and magnetic components are related, albeit, in a complex manner. The magnetic field oscillations associated with seismic and volcanic activity all lie at the low end of the EMR spectrum (0.3 to 300 Hz). Also, the associated acoustic modes lie in a similar frequency range to that of the ELF activity.

As we have explained elsewhere, ELF magnetic field oscillations have transverse as well as longitudinal modes of excitation and the longitudinal modes appear to be acoustic-like, at least as considered in "four space." Most of the earth-activated ELF modes are non-sine wave-like, having a number of Fourier components.

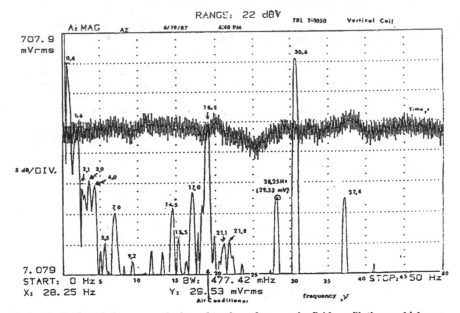

Figure 1. Typical frequency and time domains of magnetic field oscillations which are dominated by the approximately 30 Hz signal. The about 1.56 Hz signal heterodynes with the 30 Hz signal as seen in the time domain. Most of the frequencies below about 11 Hz are natural geologic and atmospheric oscillations.

The 0.4 Hz is the Earth's fundamental magnetic rotational component and the 1.56 Hz is its standard rotational vector and the 3.0 to 3.2 Hz magnetic pulsation component is the seismic precursor signal.

The 5.5 to 5.9 Hz signal is usually associated with excitation in the ionospheric D layer and the 9.2 Hz signal appears associated with heavy particle interactions from solar flare activity bombarding the ionosphere. Frequencies of 12 Hz and above are primarily from man-made sources. The 17 and 18.5 Hz are from the air conditioner compressor near the building where the data was recorded and disappeared when the air conditioner was shut off. The 30 Hz signal with 28.25 Hz side lobes is probably a spurious emission of Project ELF.

256

Precursor frequencies of magnetic field oscillations, observed before the onset of seismic activity, are usually the third and fourth harmonic of the earth's rotational excitation. The difference between the earth's steady state magnetic field and the earth's mechanical rotational axis is 22.5 degrees at the poles. Hence, the first oscillation of the actual rotational vector is 0.4 Hz (= 3.14 x 360/22.5). The second harmonic is 0.8 Hz and the third harmonic is 1.6 (more precisely, 1.56 for the nonlinear progression) and around 3.16 to 3.2 is the fourth harmonic. [45]

As the earth rotates, the Coriolis force is stored in the earth's body. We observe this energy storage as a build-up of a magnetic signal of about 3.16 to 3.2 Hz. From about 24 to 72 hours before an impending event, the approximately 3.2 Hz signal disappears. Triangulation on the maximum magnetic amplitude of the 3.2 Hz signal is used to locate the future or impending event. If the 3.2 signal reappears within the 72-hour time frame, then the time line starts running again for another 24 to 72 hours.

The 1.56 Hz signal is almost always present and, from this signal which we term *rotational vector*, we can estimate the magnitude of the future event. The percentage of deviation from the normal value of the rotational vector gives an approximate magnitude, depending also upon the distance from the measurement instrumentation and the epicenter of the impending event.

We observe a range of "rotational vector" values between 1.26 Hz and 1.80 Hz. The smaller the deviation from the normal value of the rotational vector, the smaller the quake will be. The deviation of the rotational vector is proportional to the impending quake magnitude. For example, 1.26 Hz or 1.80 Hz can be associated with over 6.5 to 7 on the Richter scale depending on the distance from the detection point to the site of the impending event.

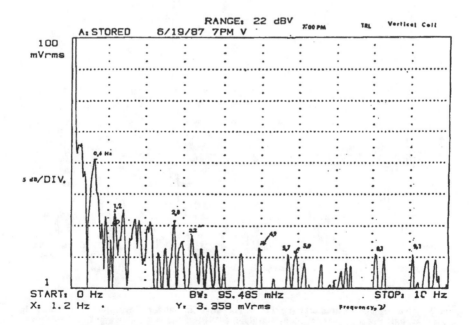

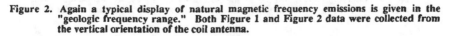

Figure 2. Again a typical display of natural magnetic frequency emissions is given in the "geologic frequency range." Both Figure 1 and Figure 2 data were collected from the vertical orientation of the coil antenna.

The earth and sun have oscillatory modes which we can treat in a manner analogous to that of a ringing bell. These acoustic modes are actually the mechanical motion and they perturb the earth's steady-state magnetic field of 0.5 gauss at the San Francisco Bay area latitude, giving rise to magnetic field oscillations and electromagnetic waves. Hence, magnetic and acoustic modes are related to each other.

The longitudinal modes of ELF waves can be detected by a coil of about 17 miles of AWG #44 wire wound on a spool. These modes travel at about $v_B = 3\,(10^5)$ cm/sec as compared to $c = 3\,(10^{10})$ cm/sec, so that for transverse electromagnetic radiation for a 7.80 Hz signal wavelength, $\lambda = 25{,}000$ miles--approximately the circumference of the earth. For the longitudinal modes of excitation traveling at v_B for a wavelength, λ is about 1/4 mile at a frequency of 7.80 Hz. The coil detection system responds well to the nonlinear ELF waves and the coil containing about 52,800 feet of wire which is very adequate to measure down to the frequency of the thrust waves associated with Love and Rayleigh wave activity of about 0.2 Hz with a wavelength of about 10 miles. We have observed the 0.20 Hz thrust waves associated with on-going seismic events on numerous occasions.

The nonlinear coil response acts like a "giant resonant" (as in nuclear physics) detection (without the circuit amplifying and smoothing elements) and has a peak primary response at 48 Hz for $f_R = 1/(2\pi\sqrt{LC})$ for the coil system currently in use at this frequency range acts as if it has a large cross section (or giant resonance) for ELF detection.

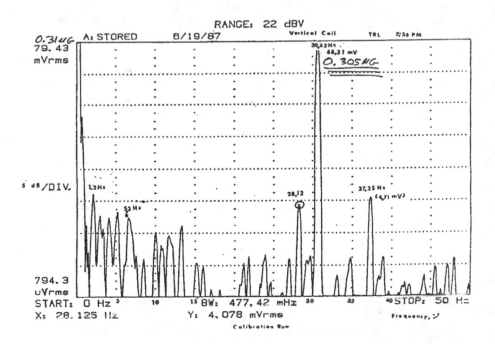

Figure 3. Measurement is made for the 0 to 50 Hz range in which calibration of the 30.625 Hz signal is made as a secondary standard to NBS. Table 1 following this figure gives approximate relative magnitude of the 30 Hz signal measured at various times, as time averages in various locations. All data in these figures was taken prior to 1987.

The T-1050 magnetic field detector has a series of amplifying, smoothing elements and notch filters which make for flat response from 0.1 to 50,000 Hz for magnetic field strengths only detection with a sensitivity of about 10^{-6} Gauss for the high pass circuit; 0.1 to 200 Hz for the low pass circuit with a sensitivity of 0.5×10^{-10} Gauss.

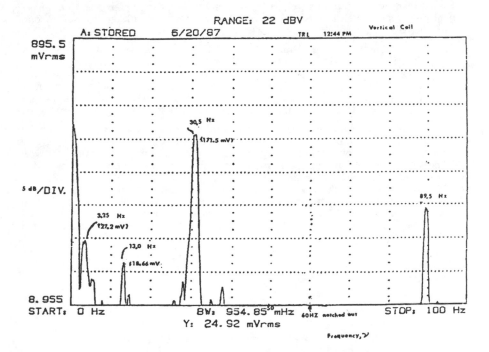

Figure 4. In the 0 to 100 Hz range with 60 Hz notch filters to notch out the 60 Hz, we see the spurious side bands of 30.5 and 89.5 Hz from the center band of 70 to 76 Hz for Project ELF. The 30 Hz signal is nearly always larger than the 89.5 signal since it lies near one of the Earth's rotational resonant frequencies. In earlier measurements there were also some effects of coil roll-off.

Table I.
Relative Magnitude of the Observed 30 Hz in Various Locations

Seattle, WA	90.00 µG
Phoenix, AZ	88.00 µG
Vancouver, BC	37.50 µG
Portland, OR	30.00 µG
San Leandro, CA	03.00 µG
New Orleans, LA	00.99 µG
Kirkland, AZ	00.30 µG

Fundamental Excitatory Modes of the Earth and Earth-Ionosphere Resonant Cavity
APPENDIX II
ELF Data Evidence Implying the Need for a Five Dimensional Geometry

Measurements for ELF frequencies in the region below about 300 Hz with a coil magnetic detection system indicate the need for a mathematical model consisting of greater than four dimensions (of space and time) in order to explain the observation that impulses in the ELF region appear as identical patterns both on the oscilloscope (time domain) and a spectrum analyzer (frequency domain).

Since a wavelength or wave impulse of say 7.8 Hertz (in Maxwell's theory) requires approximately 25 thousand miles for a complete excursion in the conventional electromagnetic theory, any person who is an observer of this frequency and the device employed to display the result are inside of the effect of the wave impulse. Therefore, either the observation of the event and the display of it are erroneous or a higher dimensional treatment of the results are necessary to explain our observations.

It is possible to explain the observation of similarity of space and time displays by resorting to a five dimension geometric description of ELF phenomena. The five dimensions are composed of three spatial amplitude dimensions, one of time associated with the amplitude or usual time or A-t plane where $t = 1/\upsilon$ and a frequency associated with the amplitude frequency or A-μ plane where $\tau = 1/\mu$. We can call this "space" the x,y,z,t,τ space. (We define the amplitude variation as dependent on the usual parameters as $A(x,t,\upsilon)$ and the new amplitude dependent on the new parameters $A'(x,\tau,\mu)$.) Note that this five dimensional space with all macro-dimensions is not like Kaluza-Klein geometry with a compactified spatial dimension. [46]

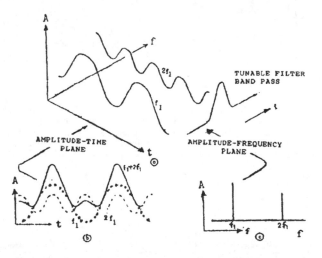

Figure 1. a) Three-dimensional coordinates showing time, frequency and amplitude. The addition of a fundamental and its second harmonic is shown as an example.
b) View seen in the t-A plane. On an oscilloscope, only the composite $f_1 + 2f_1$ would be seen.
c) View seen in the t-A plane. Note how the components of the composite signal are clearly shown.
The nonlinear coil response acts like a "giant resonant" (as in nuclear physics) detection (without the circuit amplifying and smoothing elements) and has a peak primary response at 48 Hz for $f_R = 1/(2\pi\sqrt{LC})$ for the coil system currently in use at this frequency range acts as if it has a large cross section (or giant resonance) for ELF detection.

Figure 2. Five second display from spectrum analyzer set at 100 KHz resolution and zero span gathered in Portland, OR. Top trace 10 Hz electromagnetic variable polarity waves.

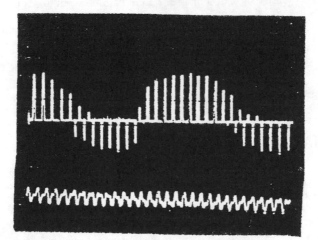

Figure 3. Two 1 second traces of ambient magnetic field impulses gathered in California with a coil detection system and displayed in the time domain on a storage oscilloscope. Top trace railed; bottom trace normalized headroom. Note the variabe polarity serpentine like characteristics similar to Figure 2.

Figure 4. This diagram was taken from the cover of Reference 31, A.C. Scotts text "Active and Non-Linear Wave Propagation in Electronics." This figure is generated as a mechanical model to represent the wave output of a system obeying thes sine-Gordon equation. A series of small pegs or nails of equal length are placed on a wire (which would look like the proverbial white picket fence. To create the figure, the center flexible wire is rotated two and a half twists, creating the display observed and photographed in this figure. Observe the similarity between this figure and Figure 3 which is the time display of the 30 and 1.5Hz signals *hetrodyning*.

261

Maxwell's equations are only approximately valid in the ELF region when coil emitters and sensors are utilized and the results plotted by means of oscilloscopes and spectrum analyzers. Gauge invariant conditions are modified in the frequency range of ELF. [44]

We present examples of frequency and time domain measurements which lend support to the statements we have just delineated. Figure 1 shows the unusual relationship of a typical sine wave and its representation in the frequency and time domain. Figure 2 is an example of the serpentine-like appearance of the "Russian Woodpecker" in the frequency domain as demodulated by a spectrum analysis. Shown are the 10 Hz pulses on a high frequency "carrier."

Figure 3 is a representation of a magnetic field measurement of about a 30 Hz signal gathered with a coil detection system and displayed on an oscilloscope in the time domain. Although the frequencies are 10 and 30 Hz respectively, the ELF region of interest that they occupy are close enough to note that both of these displays are highly similar and serpentine-like, implying a comparative measurement in the frequency domain; and time domain at higher frequencies do not exist at ELF.

The trace in Figure 4 is generated as a mechanical model of soliton waves generated by the sine-Gordon equation which has periodic soliton solutions. Note the similarity to Figure 3 which is a magnetic signal observed with the T-1050 displayed in the time domain at about 31.5 Hz which can be decomposed into a 30 Hz signal *heterodyned* with a 1.5 Hz rotational vector. The amplitude is 2 volts/division for a one second trace. Pulse modulation at 3.33 Hz was observed. Normally one would describe this signal as a Heterodyne of 30 Hz and 1.5 Hz but the rotational wave characteristics are better seen in the nonrailed signal in Figure 3, the second trace. The signal in the top trace in Figure 3 is actually a slice in two dimension from a rotational wave in five dimensions or five dimensional rotor or screw wave which is like a five dimensional twister algebra related to a four dimensional spinor calculus.

In Figure 5 is a five dimensional representation of this screw or rotational wave. To represent it in a two dimensional figure on flat paper, we consider amplitude as a function of x only rather than x,y and z. The other variables of the wave amplitude are time, t, and frequency, μ, where $t \neq 1/\mu$ but $t = 1/\upsilon$ for $\upsilon \neq \mu$. This figure can explain the observation seen in Figure 2 and 3 and a great deal of other similar data with the model of the multidimensional soliton wave. Figure 6 is a schematic representation of the tube-like waveform in x,t,τ space which moves like a "slinky" toy. The "slinky" coils represent the Fourier of the magnetic field and can be seen as the wave with spikes in space and time.

Simple solutions to solitary wave equations can be found in terms of the usual conic section form of the trigonometric functions of sines and cosines built on the relationship to the sphere. Simple second order equations generate parabolic and hyperparabolic forms $x^2 +- y^2 = c$ as contour integration of exponential functions in the complex plane $z = x + iy$ for $i = \sqrt{-1}$.

We will examine the properties of Fourier and Laplace transforms and Lorentz invariants conditions and relate these to Gausses' theorem.

Consider a simple description of the waveforms and their relationship and the frequency and time domain to generate Figure 5. Consider the simple case where the cross-sectional area of the volume generated by and swept out in the frequency and time domain in circular so that more forms observed in the frequency and time domains appear the same. Also consider the envelope on the curve as seen as extended oscillation, in their amplitude dependence or frequency and time, obey sine waves such as $x = \sin(t/t_o)$ and $x = \sin(f/f_o)$ contained under the envelope of the curve. Note that in general, the five dimensional travelling wave can be elliptical, having different "periods" or extensions in time and frequency, for example, the extension in time could be associated with the major axis and the extension in frequency with the minor axis of the ellipse in Figure 5, w represent only one period cycle in the x-t and x-f planes.

In actuality, waveforms in space and time and frequency dependence extend out in the x-t (also termed x-υ) and x-f (also termed x-τ or x-μ) planes so that wave effects extend as amplitude disturbances in the five dimensions of (x,y,z,t,τ) for t association with υ and τ with μ or f in Figure 5. Then the t-f, x-t and x-f planes represent slices through the five dimensional space. Even though the figure looks like these planes are generated by a projective geometry, in fact we consider a mapping procedure which does not produce distortion. We will consider an example of this later. For the wave amplitude extension, we consider one dimension x of space only rather than x,y,z considering the relationship of frequency and time, the simplest case becomes $t = 1/f$ (where we use f and υ interchangeably). The case where $t = 1/f$ is given in Figure 1 in which we observe a sine like wave forms in the time domain and spike wave forms in the frequency domain. In Figure 5, we can associate f with υ (and t) or μ (and τ). This form in general obeys that for a rectangular hyperbola of the form $(t/t_o)^2 - (f/f_o)^2 = 1$ asymmetrically bound to the upper quadrant t and f axis, with a lower image and symmetry exists for $t_o = f_o c_o^2$ so that $t^2 - f^2 = t_o^2 = f_o^2 c_o^4$ is a unit length space of c_o for velocity $c_o = l_n/t$ for l_n dimensionless, gives the gradient of the asymptotes gives $\tan \alpha = \pm 1$ for $\alpha = 45$ degrees. Note also that $t = t_o \cosh \eta$ for $\eta = (fc_n)/(f_o c_o)$ which is dimensionless.

Returning again to Figure 5, the frequency and time relationship of τ and μ act as a rotational frequency in x, τ space. The general relationship x,t,τ can be represented as a circle or ellipse as a slice through the 3 space x,t,τ in the one dimension of space, x approximation in which we use to solve the sine-Gordon equation. Then we can write $(t^2/t_o^2) + (\tau^2/\tau_o^2) = I$ where $|t/(rc_o)| \leq 1$ and $|\tau/(rc_o)| \leq 1$ and let $c_o = 1$ and where r is an average radius of a circle, $t/r = \cos \theta$ and $\tau/r = \sin \theta$ and $t + 1'\tau = r(\cos\theta + i \sin\theta)$ where r is in units of t and τ (in seconds) for $C_o = 1$ and $0 \leq \theta \leq 2\pi$. Consider the case where v=30Hz "heterodynes" with the μ=1.5Hz signal, then t=0.033Hz and τ =0.667 (about 1 second). Then the rotational process in 5D spaces moves 1.5Hz as seen in our data in Figures 2 and 3. The value r is defined as $r = \sqrt{t_o^2 + \tau_o^2}$ with eccentricity $e = \sqrt{(t^2/t_o^2) + (\tau^2/\tau_o^2)}$ or taking $t_o + \tau_o = 1$ then we can write $e = 1/r\sqrt{1 - \tau^2/t^2}$ so that $t = e \cdot 1/\sqrt{1 - \tau^2/t^2}$.

In general, proceeding with the Cauchy Integral theorem in the complex plane, we define an analytic function f(z) in a simple region of space bounded by a region, R of the Argand plane. Then on a simple closed path in R, we have $\oint f(z) \, dz = 0$ where $z = x + iy$ where the vector z makes an angle θ with x. For a circle $\oint dz/z = 2\pi i$ for $i = \sqrt{-1}$ and x, y and θ are real. For this example $z = Re^{i\theta}$ the real and imaginary exponential occupy a fundamental role in describing our sections and the geometry of electro-magnetic waves.

Consider the conformal mapping from a linear space to an elliptic or hyperbolic plane. For W=u+iv, x=a sinμ coshμ and y=a cosμ sinhμ for v= a constant not equal to zero maps to an ellipse and μ= constant not equal to zero maps to a hyperbola in the z plane. If W=sin^{-1}(z/a) then z = a sin w. Note that these procedures generalized to the complex plane, generate the conic forms in terms of exponents such as e^{ix}= cos x + sin x and e^x= sinh x + cosh x and cos ix = cosh x. A number of the trigonomentric and hyperbolic functions can be generated from the real and imaginary exponents. Note that the hyperbolic functions sinh x and cosh x describe th path of a chain suspended at two points.

Turning our attention to what appears as a series of Fourier components, we consider Laplace and Fourier transforms. Consider the Laplace transforms of the Bessel function $J_o(t) = \sum (-1)^n (t/2)^{2n} / (n!)^2$ which gives the transform $\phi(P) = 1/\sqrt{1 - \rho^2}$; where P$\rightarrow$0 and

263

$\int_0^\infty J_o(t)dt = 1$. Returning to the usual Fourier transforms written before for $\theta(t)$ and ϕ (w), we have ϕ (w) as e^{-w} transforming as $\sqrt{2/\pi}(1/(1+t^2))$ or e^w transforming as $\sqrt{2/\pi}(1/(1-t^2))$. Let us briefly turn our attention to invarient conditions.

Note simply that if $x^2/r^2 + y^2/r^2 = 1$ for $z = x + ib$ represents a parabola then the complex conjugate of z or $z' = x - ib$ generates the equation of a hyperbola $x^2/r^2 - y^2/r^2 = 1$ for $|z| = r$. Orthogonality of states can be defined here in terms of z and z'.

Forms such as $\int dx \sqrt{1-x^2} = arc\sin x$ occur. For the proper time $\iota' = \iota\sqrt{1-\beta^2}$ for $\beta \equiv v/c$ which defines the "four vector velocity" v=dx/dt and the time-like component is

$$P_4 = t = mv_4 = \frac{imc}{\sqrt{1-\beta^2}}.$$

An equivalent set of Lorentz conditions are made at a low velocity of propagation and these are the forms that we use in the theoretical soliton model presented in the text.

In the the usual four space representation we use the (+ + + -) signature so that x_4 or time = ict where a four vector space (space and time representation) requires a complex space description. Consider the example of a uniform magnetic field along the z axis when charged particles will move in a circle in this field. We will say a few words about the particle path of motion in the relativistic approximation detailing more in a future paper. In a uniform magnetic field, the particle velocity and momentum goes as $v \propto u \cos kt' + u \sin kt'$ where $k = (e\mu_o B)/mc$ and in a uniform electric field $v' = c \sinh Kt' + b \cosh Kt'$ for $K = eE/mc$ where the proper time, t' define the relativistic and invariant is given as $t = [(1+\mu^2+v^2)/c^2]t'$ and $dt/dt' > 1$ which means that when the energy of the system increases, the orbital particle path becomes smaller which is the resonant principle of the cyclotron. Also this principle applies to the MASE process in ELF phenomena for the magnetic amplification of stimulated eletrons.

Invariance principles are statements of conservation laws. Gauss's theorem defines the conservation of flux or charge within a closed surface S. Applied to electric fields $\iint_S E \cdot dS = 4\pi Q/\varepsilon_o$ where Q is the algebraic sum of all charges inside the surface, S where the integral is taken over the surface, S. This condition expresses the manner in which a Faraday shield works and applies to the conditions on the operation of the shielded coil for the T-1050 detector. We can write Gauss's theorem in terms of the electric potential or $\iint E \cdot \overline{ds} = -\nabla^2 v$ so that we have Poisson's equations as $\nabla \cdot E = -\nabla^2 V = (4\pi Q)/\varepsilon_o$ where units $\varepsilon_o = 1$ are often used. We can apply Gauss's theorem to fields other than electric fields--such as the earth's gravitational field. Application of Gauss's theorem and a Poisson-like formalism can be applied to magnetic flux conservation on a hyper-dimensional surface defined in five dimensional space. This conservation principle leads to the great stability ELF flux phenomena that is well described by Soliton physics.

The usual form of Poisson's equation is $\nabla^2 v = 4\pi\rho$ in three dimensions, which Einstein expanded to four dimensions $[]^2\phi = 4\pi G/C^4$ in general relativity. We write a similar expression in x,y,z,t,τ space as $\underline{[]}^2\phi = 32\pi\rho_\beta/C_o^4$ where ρ_β is the magnetic flux density in five dimensional states. We define the generalized Laplacian form as before as $\underline{[]}^2\psi = []^2\psi - 1/c_o^2 \, \partial^2/\partial\tau^2$.

The magnetic flux field, B in five dimensions acts as a bundle of the magnetic flux lines in a variable diameter cable tube. The magnetic flux potential is conserved in five dimensions forming Gauss's theorem in an analogous manner to that done in three dimensions, we define a surface integral over a four space S_4 and define a magnetic flux vector B_5 in five dimensional space having a flux density ρ_B of so that

264

$$\iiint_{S_4} = B_5 \cdot dS_4 = \frac{16\pi\rho_B}{C_o^2}$$

See reference 9 for a more detailed description of the model and its application to geophysical and other magnetic flux phenomena.

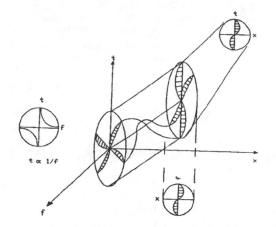

Figure 5. In this figure is displayed a symbolic representation of the rotational or twister wave in five dimensional space.

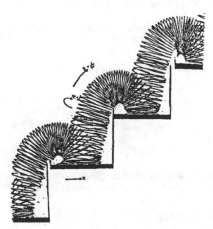

Figure 6. Representation of the Fourier components of an ELF wave in a Five Dimension Space in which the frequency and time domain appear to be similar to each other.

References & Footnotes

[1] W. V. Bise. "New Method of Presensing and Monitoring Volcanic Activity at Mt Saint Helens, Washington," *Planetary Association for Clean Energy* 2:2; page 9, May 1980.

[2] W. V. Bise. "A Week in the Magnetic Life of Mt. St. Helens," *Planetary Association for Clean Energy* 2:3; page 10, July-August 1980.

[3] W. V. Bise. "Magnetic ELF Comparisons with Acoustic Measurements on Mt. St. Helens" *Planetary Association for Clean Energy* 3:1; page 26, June 1981.

[4] W.L. Van Bise and E.A. Rauscher. "Instrumentation and Techniques for Analysis of Extremely Low Frequency (ELF) Magnetic Field Impulses Preceeding Geologic Events," *Bull. American Physics Society* 32,67, 1987; PSRL-I7630AJ, October 1986, TRL Report, San Leandro, CA. 94579

[5] E.A. Rauscher and W.L. Van Bise. "Obaervations of Local and Global Earth Ionospheric Excitations for Earthquake and Volcanic Predictions," *Bull. Am. Phys. Soc.* 32,67, 1987; PSRL 17741, October 1986, TRL Report SAn Leandro, CA. 94579

[6] W.V. Bise and E.A. Rauscher: "Non-superconducting Systems for Detecting and Analyzing Low Intensity Pure Magnetic Fields." Tecnic Reearch Laboratories Report, published in *Bull Am Phys Soc*, Volume 34, p. 109, 1989.

[7] W.V. Bise and E.A. Rauscher: "Geomagnetic Pulsations of the Earth's Magnetosphere." Tecnic Reearch Laboratories Report, PSRL-24125 (1987).

[8] N. Tesla, Patent No. 787,412. April 18, 1905. "Act of Transmitting Electrical Energy Through the Natural Medium" and related patents and papers.

[9] E.A. Rauscher. "Application of Soliton Physics to Plasma-MHD and Super Conducting BCS: Theoretical Implications for Primary Energy." PSRL-3107, TRL Report, November 1982.

[10] E.A. Rauscher. "Electron Interaction and Quantum Plasma Physics." *J. Plasma Physics* 2, 517 (1968).

[11] E.A. Rauscher. "Quantized Plasmas" *Bull. Amer. Phys. Soc.* 15, 1639 (1970).

[12] E.A. Rauscher. "Nonlinear Coherent Modes in MHD-Plasmas," *Bull. Am. Phys. Soc.* 28, 1337 (1983).

[13] N. Tesla, The Colorado Springs Notes. published by J. Ratliff, The Tesla Publishing Company, Milbrae, CA 1980.

[14] E.A. Rauscher and W. Van Bise, "Geologic Significance of the Fundamental Excitatory Modes of the Earth and Earth-Ionosphere Resonant Cavity," PSRL -70253, January 1986.

[15] E.A. Rauscher commuted daily to Lawrence Livermore Laboratory on Tesla Avenue from early 1966 to end of 1969.

[16] E.A. Rauscher, "A Unifying Theory of Fundamental Processes," UCRL 20808, June 1971, University of California LBL Press.

[17] E.A. Rauscher and W. Van Bise, "Multiple Wave Analysis Measurements at Low Frequency (LF) and Extremely Low Frequency (ELF) Artificial and Natural Radiation," Pac. NW Center for the Study of Non-Ionizing Radiation Report. Portland, Oregon, September 1980.

[18] Project ELF, formerly refered to as Project Sanguine and Project Seafarer, is reviewed in several issues of *Aviation Week and Space Technology* and also J.R. Waite "Project Sanguine," *Science* 178, 272 (1972).

[19] Nikola Tesla: Lectures, Patents, Articles, Belgrade, Yugoslavia: Nikola Tesla Museum, 1956.

[20] K.M. Swezey, "Nikola Tesla," *Science* 127, 1147, 1958.

[21] M. Sugiara and J.P. Heppner, "Electric and Magnetic Fields in the Earth's Environment," American Institue of Physics Handbook, McGraw-Hill Book Company, 1972; Section 5 pages 265-303.

[22] N. Tesla. "The Transmission of Electric Energy Without Wires," *Electrical World and Engineer.* New York: March 5, 1904.

[23] M.M. Frocht, "Experimental Stress Analysis," Handbook of Physics, McGraw-Hill Book Company, 2nd edition, Part 3, Chapter 6, page 84.

[24] E.V. Condon "Molecular Optics," Part 6, Chapter 6, page 117.

[25] J.A. Silva and G. Lochak, Quanta, New York: McGraw-Hill Books Company 1969. (page 1969)

[26] F.M. Lowry, Optical Rotary Power, London: Longman's, Green & Co, 1935.

[27] E.V. Condon, "Electromagnetic Waves," Part 6, Chapter 7, page 5.

[28] H.B. Callen, "The Energy Band Theory of Solids, Part 8, Chapter 8, page 24.

[29] J.C. Anderson, Magnetism and Magnetic Materials; London: Chapman and Hall, Ltd. 1968; Page 199.

[30] E.A. Rauscher "Properties of Nonlinear Coherent Modes in MHD Plasmas and a Possible Solution of the Plasma Combinent Problem," *Proceedings of the Tesla Centennial Symposium* (An IEEE Centennial Activity) Editor, E.A. Rauscher, Tesla 1984.

[31] E.A. Rauscher, "Solitary Waves, Coherent Nondispersive Solutions in Complex Minkowski Spaces," *Bull. Amer. Phys. Soc.* 27 135 (1982).

[32] C. Rebbi, "Solitons," *Scientific American;* page 92, February 1979.

[33] *Aviation Week and Space Technology*, page 19, Nov. 8 , 1976.

[34] E.A. Rauscher and W. Van Bise, "Non-Superconducting Apparatus for Detecting Magnetic and Electromagnetic Fields." Patent No. 4,724,390, Issued February 1988.

[35] E.A. Rauscher, W. Van Bise, and H.J. Faretto: "Field survey measurement of the ambient ELF-VLF-LF frequencies and intensities at the Kalish Ranch site and surrounding areas, Kirkland Arizona." Tecnic Reearch Laboratories Report, PSRL-27643, August 1987.

[36] E.A. Rauscher, "Theoretical Examination of Nonlinear Far From Equilibrium Self-Organizing Phenomena," TRL Report PSRL-103764, Abstract March 1985, Presented at the USPA. July 1985, Dayton, Ohio.

[37] A.C. Scott, "Active and Non-Linear Wave Propagation," *Electronics*, 1985.

[38] W. Van Bise, "Extremely Low Frequency (ELF) Radio and Magnetic Signals--New Biologic Clues in the Pacific Northwest Environment," N.W. Center for the Study of Man-Ionizing Radiation Report, Feb 1980, Portland, OR.

[39] J. Beal and E.A. Rauscher, "Modern Science Reads the Earth's Electromagnetic Fields," Invited paper at the University of California at Davis, Symposium co-sponsored by the Institute for the Study of Natural Systems, September 10, 1988.

[40] C. Ramon and E.A. Rauscher, "Super-Luminal Transformations in Complex Minkowski Spaces," *Found of Physics* 10, 661 (1980)

[41] E.A. Rauscher Paper in The Frontiers of Physics (Iceland Conference Proceedings, Reyjavik, Iceland) B.D. Josephson and A. Puharich, Editors, Essentia Research Associates.

[42] E.A. Rauscher, "Complex Coordinate Geometries in General Relativity and Electromagnetism," *Bull. Am. Phys. Soc. 23*, 84 (1978).

[43] E.A. Rauscher, "Electromagnetic and Non-Linear Phenomena in Complex Minkowski Spaces," *Bull. Am. Phys. Soc.* 28, 351 (1983).

[44] Detailed analysis of a variety of wave equations with oscillatory solutions is in progress using an iterative procedure in which multi-frequency and multi-time analysis is made with the computer code CORFAC in FORTRAN IV, IBM Compatible. An update version is in progress.

[45] E.A. Rauscher, "The Earth's Coriolis Force as an Energy Storage and Release Mechanism for Mediating Seismic Activity," *ONR Grant* through the University of California, Berkeley for (1) modeling systems using multidimensional geometries (2) Modeling seismic activity (1970-1974)

[46] After we hypthothesized our five-dimensional modeling a device was developed and marketedthat could measure events in the time-frequency domain called the modulation domain. The Hewlett-Packard HP5371A, frequency and time interval analyzer was introduced in 1988 and is a very important instrument for analysis of seismic activity as a readout from the T-1050 Detector, as well as other uses.

[47] E.A. Rauscher and W. Van Bise, "Relaxation of guge invariant conditions for ELF and VLF phenomena and their implications for magnetic and electromagnetic wave transmission." Tecnic Research Laboratories Report, published in *Bull Am Phys Soc 34*, p. 82, 1989.

About the Authors:

Dr. Elizabeth Rauscher received her BS, MS and PhD from the University of California, Berkley in Nuclear Engineering and Nuclear Science and Cosmological Models. Dr. Rauscher's main fields of research include nuclear physics, astrophysics and cosmological models, plasma physics, biomedical engineering, and geophysical monitoring. In addition to publishing over 100 papers and 4 books, Dr. Rauscher co-founded Technic Research Laboratories

William Van Bise is a biomedical engineer with degrees Tulane University and the Oregon Health Sciences University School of Medicine, including doctoral studies at the Oregon Graduate Center. A member of IEEE, ICWA, and AAAS, he has considerable background in biomedical instrumentation design.

(Together they hold US patents #4,724,390, #4,723,536, and #4,889,526. The first patent '390 is a non-superconducting ELF amplifier that triangulates the location of a pending earthquake, based on the well-known ELF signals that the piezoelectric rock emits while stressed, just before the impending movement we call "an earthquake." It is a masterful work that I highly recommend for those who want to learn more about the earth Schumann cavity, earthquake detection, and ELF waves, while being very readable and comprehensive. The USGS could be using the Rauscher/Van Bise invention to warn Californians, for example, before any major earthquake. Some of the patent content is discussed on p. 233-234 of this book. – Ed. note)

Elizabeth Rauscher and William Van Bise can be reached at Tecnic Research Labs, 3500 S. Tomahawk, Bldg. 188, Apache Junction, AZ 85219 and BVR@uswest.net

SECTION III

Miscellaneous Tesla Articles
and Reference Material

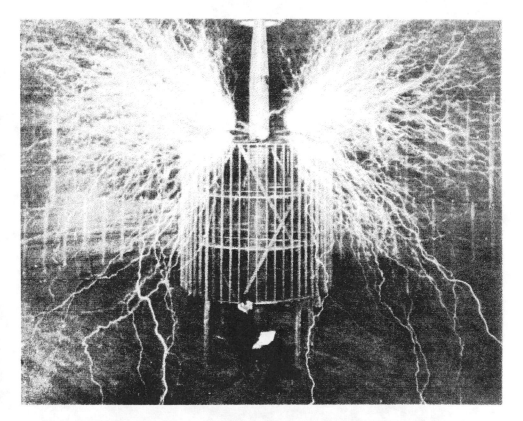

Tesla amazingly seated at the base of a huge coil in action with remarkable symmetry at Colorado Springs without any concern for the lightning he is creating. The fir tree at the top supports his antenna. "By providing a sort of roof, I get on the under side a support on which the electrical density is nothing, and then the support is absolutely safe, even in rain, because it keeps dry, and it is always in contact with a conducting surface of low electric density." – Nikola Tesla, quoted in Anderson, *Nikola Tesla on His Work with Alternating Currents*

269

15   The Homopolar Generator:
Tesla's Contribution

Reprinted from *Proceedings of the International Tesla Symposium.* 1986

Thomas Valone, M.A., P.E.

"The One-Piece Faraday Generator" is now published as the book, *The Homopolar Handbook.* The Forbes 1888 dynamo, with like poles at each end of the rotating armature magnet and opposite pole in the center (U.S. patent #338,169), simplified current extraction in a manner similar to Tesla's two-disk concept described below. – Ed. note

Abstract

With the continued interest in Faraday or homopolar generators, it is good to review Tesla's experiments in this field. Tesla proposed several methods for increasing the output of the generator, including the "current accumulator." In this paper, the range of homopolar generators Tesla experimented, the Forbes Unipolar generator and my work on the one-piece homopolar generator will be discussed. We will look at why Tesla believed that he could build a self-sustaining dynamo.

Introduction

This paper will center on a very simple but very intriguing device that is a model of the planet earth (see Figure 1). It's the unipolar, acyclic, or homopolar generator which is also referred to as a "Faraday disk dynamo" after Michael Faraday who discovered it in 1831. (Hence, the title of my book, The One-Piece Faraday Generator, available through High Energy Enterprises, PO Box 5636, Security, CO 80911.) Note the self-sustaining nature of the earth model that's presented in the following Scientific American article. (For your convenience, the article, "Modeling Magnetism: The earth as a dynamo," by S. Weisburd is reprinted in full.) [Science News, V.128, p.220, 1985] They are describing an unusual mechanism of the earth which pumps current in a spiral manner strengthening the magnetic field, as described by Tesla in the following pages. (This could have great significance for a new, liquid metal Homopolar Generator if someone wants to try a novel approach. The article says, "F.H. Busse proved that by virtue of the dynamo action of the fluid motions, the magnetic field could increase substantially from a small initial value"-p.124). It appears to be a "free energy" device, with no dissipative effects, but the theory, as well as my experiment, has shown that a one-piece homopolar generator should exhibit back torque. Therefore the earth should be slowing down quite noticeably. However, it's not appreciably slowing, so there must be a method nature uses to avoid the back torque.

Modeling Magnetism: The earth as a dynamo

In 1600 William Gilbert, the physician of Queen Elizabeth I published a treatise on magnetism called *De Magnete,* in which he dispelled the notion that lodestones are attracted to heavenly bodies. Instead, he concluded from an experiment with a spherical lodestone that the earth itself is a giant magnet.

Centuries passed before scientists developed any reasonable ideas as to what causes this geomagnetism. The main, dipolar part of the earth's field clearly resembles that produced by a bar magnet. But it has become apparent that the field could not arise from permanently magnetized minerals in the earth. Most of the earth is too hot for such materials to retain their magnetism for long, and in order to create all of the changes observed in the magnetic field, solid magnets would have to scurry around within the earth--an impossible feat that would result in massive upheavals of the planet. Moreover, earthquake data indicate that the outer region of the core is a fluid.

Scientists believe this outer core is a rotating liquid made principally of molten iron and nickel, which conduct electricity. This view of the core has led to the only surviving idea out of many theories (including the notion, once considered and then dropped by Albert Einstein and others, that magnetism is an inherent property of all rotating masses).

Forty years ago Walter M. Elsasser at the University of California at San Diego and Edward C. Bullard of the University of Cambridge in England developed the "self-exciting dynamo" model for the core. The illustration below ... [Figure 1] ... shows a simple example of a dynamo invented by the 19th century British scientist Michael Faraday. When the metal disk spins in the initial presence of a magnetic field, currents are generated in the disk. In a self-exciting dynamo, these currents are fed into a solenoid, or coil, which creates a magnetic field of its own.

If the spinning fluids of the earth's outer core act like the disk in the dynamo, the earth could similarly produce a large magnetic field, provided there was a small magnetic field around at the beginning. (The small field that pervades the galaxy would be a good candidate, according to some scientists.) Another provision would be that the core fluids keep moving, and the unanswered question here is what energy source is responsible for doing just that.

Of course, the actual core movements must be considerably different from and much more complex than a spinning disk. So the present focus of research is to devise complicated flow patterns consistent with the magnetic field's behavior--its reversals, secular variations and now possibly the jerk.

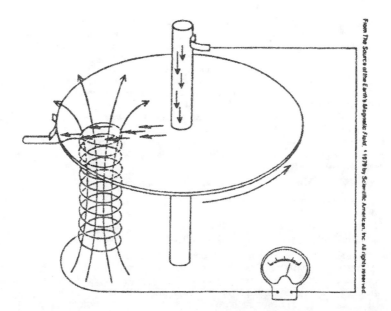

Figure 1. In this self-sustaining Faraday disk dynamo, an electric current (small arrows) in the copper disk reinforces the magnetic field of the coil (from "The Source of the Earth's Magnetic Field" by Scientific American, 1979).

Another interesting aspect of the earth as a unipolar generator (the title of another journal article) is the electromotive voltage that is produced in the rotating armature. Faraday thought it would be measurable in the rotating reference frame. He looked for a voltage in rivers and streams. My experiments with a small LED voltmeter, described in my book, show that no voltage is measurable in the rotating environment. The reason is that (see Figure 2) there is an equal and opposite electrostatic field created when a charge displacement is induced by the Lorentz force. This essentially maintains a neutral environment on the disk, even during sizable current flow, that will act as a voltage regulator. I could measure a voltage across the rotating disk in the stationary lab frame but my LED voltmeter could not measure even a millivolt when one hundred times that was present in the lab frame. In other words, we can't draw power from the earth's homopolar generator while rotating with it.

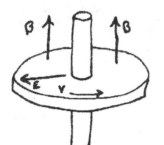

Effective Electric Field set up by $\vec{E} = \vec{V} \times \vec{B}$

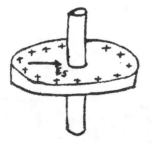

Results in a charge displacement that causes an external electrostatic field in opposite direction. (E_s)

Figure 2. Charges displace until equilibrium is established. Fields cancel within a disk. This explains why a constant voltage will be maintained across the disk even during high current output.

To calculate the voltage generated with an homopolar generator (Figure 3), we find that the equation depends upon the magnetic flux density, rotational speed, and the radius of the disk squared. The internal resistance is the only thing that limits the power output of the device. It is important to note also, before going on with the aspects that relate to Tesla's article, that all of the recent experiments that have reported anomalous effects have all been done with the one-piece Faraday generator, the one with the magnet rotating with the disk. (Dr. Stephan Marinov, on p. A-73 of my book, has a couple of published articles to this effect. Also, Dr. P. Tewari just published "Generation of Electrical Power from Absolute Vacuum by High Speed Rotation of Conducting Magnetic Cylinder" in <u>Magnets</u>, August 1986, p. 16, based upon his experiments, not to mention Trombly and DePalma.) We notice that the one-piece is closer the a model of the earth as well.

Calculation of the HOMOPOLAR GENERATOR Voltage

As seen in <u>Electromechanical Devices</u> by Woodson and Melcher, p.287 & 289, we use,

$$V_{oc} = \frac{\omega B_o}{2}(R_o^2 - R_i^2) \tag{1}$$

where V_{oc} is the open circuit voltage, R_o is the outer radius of the disk, R_i is the inner radius, and ω is the angular frequency (or the frequency of rotation in Hertz multiplied by 2π).

Now if we have a disk generator with the following characteristics:

R_i = 1cm = 0.01 m
R_o = 10 cm = 0.1 m
ω = 3820 rpm = 400 rad/sec
B_o = 10,000 gauss = 1 Tesla = 1 Wb/m²

then we can calculate the open circuit voltage. It should come out to be:

V_{oc} = 2.0 volts

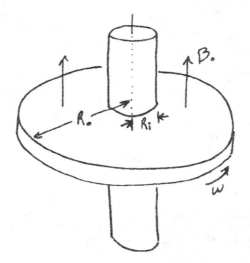

Figure 3. Calculation of the homopolar generator voltage.

Back to Nikola Tesla's article, we see that he performed a few experiments with models of the device and published the results and theory in an article entitled, "Notes on a Unipolar Dynamo" (<u>Electrical Engineer</u>, Sept. 2, 1891, p.258). {Note:

This article of Tesla's is reprinted in its entirety in the Appendix of my book, The One-Piece Faraday Generator.} For your convenience, key portions are excerpted for reference throughout this paper.

The general design, shown in Figure 4, is discussed in the first portions of Tesla's article, (also reprinted in Inventions, Researches and Writings of Nikola Tesla, by Thomas C. Martin). "... such a machine (the homopolar generator) differs from ordinary dynamos in that there is no reaction between armature and field." It is a key sentence which Tesla qualifies by limiting the circumstance to magnets that are weakly energized.

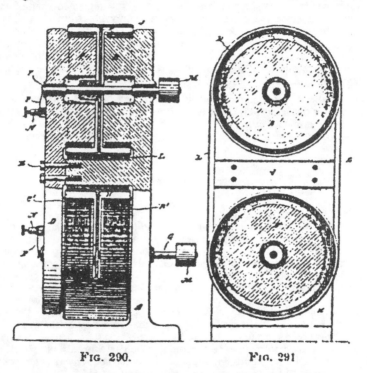

FIG. 290.   FIG. 291

Figure 4. Tesla's Design: Coupling two homopolar generators together for higher output and better brush conductivity.

In my experimentation with strong (almost 1 Tesla field strength) ceramic magnets mounted on a copper disk, I was able to measure the reaction between the disk (armature) and field which is technically labelled "back torque" (the force which slows down the spinning disk when current is drawn from it). We can ask, "How does the one-piece homopolar generator experience back torque when there is no stator, only a rotor?" The best explanation that I could come up with when I measured it is the following: The electrons in the armature (disk) current push against the magnetic field, not the magnet, causing the reaction force. By the way, it's just the radial component of the current that contributes to the back torque, according to the traditional methods for applying the torque equation.

Tesla, however describes it as a reaction between the magnetic field set up by the armature current and the electromagnet's coil current. This is probably equivalent to the equation of torque (T) being equal to the current density crossed with the magnetic flux density (JxB). He also notes another aspect of the generator that is a key to reducing back torque, the symmetry of the external circuit:

> ... Considered as a dynamo machine, the disc is an equally interesting object of study. In addition to its peculiarity of giving currents of one direction without the employment of commutating devices, such a machine differs from ordinary dynamos in that there is no reaction between armature and field. The armature current tends to set up a magnetization at right angles to that of the field current, but since the current is taken off uniformly from all points of the periphery, and since, to be exact, the external circuit may also be arranged perfectly symmetrical to the field magnet, no reaction can occur. This, however, is true only as long as the magnets are weakly energized, for when the magnets are more or less saturated, both magnetizations at right angles seemingly interfere with each other.

> For the above reason alone it would appear that the output of such a machine should, for the same weight, be much greater than that of any other machine in which the armature current tends to demagnetize the field. The extraordinary output of the Forbes unipolar dynamo and the experience of the writer confirm this view ...

Symmetrical External Circuit

We note that Tesla refers to the "external circuit" which is made to be "perfectly symmetrical" to reduce the reaction to zero. This was a popular notion, which still may have profound significance. Adam Trombly, the builder of the most successful "over-unity" homopolar generator in recent history, also emphasized to me the symmetry of the external circuit in his design in order to reduce back torque). It is believed, according to the theory noted by G.W. Howe in The Electrician, (Nov. 5, 1915, p.169), and others, that the torque or "reaction" in a unipolar generator, that tends to slow down or retard its motion (and thus keep its efficiency less than 100%) is due to the interaction between the magnetic flux and the current-carrying conductors in the external circuit. Our present theory only looks at the armature current and the magnetic field but this aspect of the force may be the neglected part. The next section refers to the eddy currents that are set up in the disk with external symmetry. They tend to magnetize the field, which is a beneficial effect. A disk without external symmetry pulling current off from one spot (like my generator, to a great extent) will tend not to contribute to reinforcing the field.

It's interesting to note that Howe also published an article 37 years later entitled "A Novel Form of D.C. Motor" (Wireless Engineer, Nov. 1952, p.285) in which a spiral path homopolar generator is described. It was subsequently built by Ku and Kamal a short time later (see J. Franklin Inst., v.258, 1954, p.7) and tested.

Tesla also notes in the next paragraph that in all other motors and generators "the armature current tends to demagnetize the field" which may be greatly reduced in his design of a unique Faraday generator.

Beneficial Eddy Currents

The next illustration (Figure 5) of a unipolar dynamo with relatively small magnets demonstrates a principle that Tesla wishes to exploit. He points out that path "n" will tend to predominate because the current will choose the path "which offers the least opposition." He believes that the currents in such a generator tend to reinforce the magnetic field and may even "continue to flow" when the field magnet is turned off (assuming electromagnets).

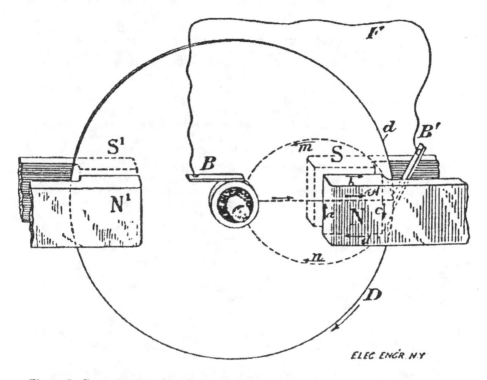

Figure 5. General schematic of a homopolar generator.

... In consequence of this there will be a constant tendency to reduce the current flow in the path A B' m B, while on the other hand no such opposition will exist in path A B' n B, and the effect of the latter branch or path will be more or less preponderating over that of the former. The joint effect of both the assumed branch currents might be represented by that of one single current of the same direction as that energizing the field. In other words, the eddy currents circulating in the disc will energize the field magnet. This is a result quite contrary to what we might be led to suppose at first, for we would naturally expect that the resulting effect of the armature currents would be such as to oppose the field current, as generally occurs when a primary and secondary conductor are placed in inductive relations to each other. But it must be remembered that this results from the peculiar disposition in this case, namely, two paths being afforded to the current, and the latter selecting that path which offers the least opposition to its flow. From this we see that the eddy currents flowing in the disc partly energize the field, and for this reason when the field current is interrupted the currents in the disc will continue to flow , and the field magnet will lose its strength with comparative slowness and may even retain a certain strength as long as the rotation of the disc is continued ...

... If the latter [disc] were rotated as before in the direction of the arrow D, the field would be dragged in the same direction with a torque, which, up to a certain point, would go on increasing with the speed of rotation, then fall off, and, passing through zero, finally become negative; that is, the field would begin to rotate in opposite direction to the disc. In experiments with alternate current motors in which the field was shifted by currents of differing phase, this interesting result was observed. For very low speeds of rotation of the field the motor would show a torque of 900 lbs. or more, measured on a pulley 12 inches in diameter. When the speed of rotation of the poles was increased, the torque would diminish, would finally go down to zero, become negative, and then the armature would begin to rotate in opposite direction to the field.

Tesla notes further on that this effect depends upon the "resistance, speed of rotation, and the geometrical dimensions of the resulting eddy currents." He then suggests that "at a certain speed there would be a maximum energizing effect," presenting the intriguing notion that the field is being dragged in the same direction as the rotation of the disk until a maximum is reached where the field would tend to reverse as the rotation speed is increased. He is proposing here that there is a phase relationship between the field concentric to a conducting disk, as illustrated by a split phase AC motor analogy. We will see shortly an illustration with solid and dotted spiral lines on a disk which demonstrates Tesla's "phase" relationship. Depending upon the direction of the spiraling eddy currents, clockwise (path "n") or counter-clockwise (path "m"), the magnetic field will tend to be in the same direction as the external magnetic field or opposing it. Thus it is reasonable to assume that as the disk increases speed, the current may start out spiraling, say, in a clockwise manner reinforcing the external field, and then reverse to a counterclockwise spiral as the speed increases. A good computer simulation of the variables involved would reveal this relationship and may suggest, as Tesla does, an optimum speed of operation for self-generation.

Another article that reinforces Tesla's ideas is "A Laboratory Self-Exciting Dynamo" by Lowes and Wilkinson, reprinted in Magnetism and the Cosmos, in 1965, by NATO Advanced Study Institute on Planetary and Stellar Magnetism in the Departments of Physics and Mathematics at the University of Newcastle upon Tyne. On page 124, they mention that "a more efficient geometry was found, so efficient that the dynamo would self-excite in a completely homogeneous state (i.e. with no insulation) at a much lower rotor speed than was believed possible." Their design is based upon conducting spheres or cylinders rotating like eddy currents in a conducting medium (also see page 126 of Sci. Amer., "The Source of the Earth's Magnetic Field," 1979).

My Experiment With Field Rotation

I'd like to mention that I have tried the experiment of rotating an 8" disk magnet on a non-conductive (wood) disk within one centimeter of a copper disk (and vice versa), along with the help of Dan Winter in Buffalo, NY (see Figure 6). We were unable to find an effect of rotating a symmetric magnetic field on the output voltage though we didn't look at the output current which is what Dr. Tesla is referring to.

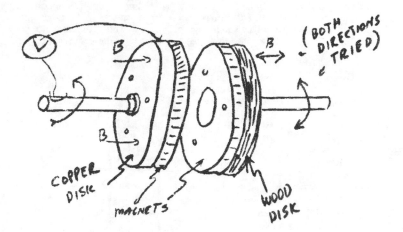

Figure 6. Dragged rotating field experiment.

Here, it is important to also note experiments (described in my book) by Cramp and Norgrove (1936) and Das Gupta (1963) which have failed to find any torques on a non-conductive magnet adjacent to a conductive disk carrying current. In fact, several scientists have proposed, as Mr. Klicker mentioned at the Tesla 1986 Conference, that no experiment can resolve whether the magnetic field is rotating if it is symmetric. However, this may not have a bearing on the phenomena that Tesla is talking about concerning the "dragging" of the field, since he concentrates on the spiraling currents.

Removing The External Magnetic Field

The next part of Dr. Tesla's article proposes that in a solid disk, as described above, we may be able to find that the field magnet may be removed while the generator disk is kept rotating. Tesla suggests that, due to favorable eddy currents, the entire generator may continue to function and even increase in output when the speed is increased, forming a fascinating "current accumulator."

> To return to the principal subject; assume the conditions to be such that the eddy currents generated by the rotation of the disc strengthen the field, and suppose the latter gradually removed while the disc is kept rotating at an increased rate. The current, once started, may then be sufficient to maintain itself and even increase in strength, and then we have the case of Sir William Thomson's "current accumulator." But from the above considerations it would seem that for the success of the experiment the employment of a disc *not subdivided* would be essential, for if there should be a radial subdivision, the eddy currents could not form and the self-exciting action would cease. If such a radially subdivided disc were used it would be necessary to connect the spokes by a conducting rim or in any proper manner so as to form a symmetrical system of closed circuits ...

279

In the next illustration (Figure 7), we see a suggestion for giving the disk an additional push from the generated currents by leading them through conductors that pass into coils. Here, the coils then encounter a reverse polarity magnetic field which tends to give the coils a small amount of push. This effect may not work at all but tends to lead the reader into thinking about curving the generated currents to an advantage.

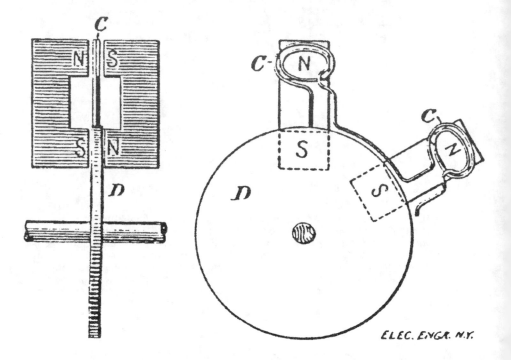

Figure 7. Possible enhancement to give discs an additional "push."

Sub-Dividing The Disk

The suggestion of sub-dividing the disk is now discussed, in order to "do away with the field coils" entirely! As illustrated in Figure 8, sub-dividing the disk spirally (actually cutting the disk in radial directions that spiral outward) tends to create a self-generated magnetic field.

... But a unipolar dynamo or motor, such as shown in Fig 292, may be excited in an efficient manner by simply properly subdividing the disc or cylinder in which the currents are set up, and it is practicable to do away with the field coils which are usually employed. Such a plan is illustrated in Fig. 295. The disc or cylinder D is supposed to be arranged to rotate between the two poles N and S of a magnet, which completely cover it on both sides, the contours of the disc and poles being represented by the circles d and d' respectively, the upper pole being omitted for the sake of clearness. The cores of the magnet are supposed to be hollow, the shaft C of the disc passing through them If the unmarked pole be below, and the disc be rotated screw fashion, the current will be, as before, from the centre to the periphery and may be taken off by suitable sliding contacts, B B', on the shaft and periphery respectively. In this arrangement the current flowing through the disc and external circuit will have no appreciable effect on the field magnet.

But let us now suppose the disc to be subdivided spirally, as indicated by the full or dotted lines, Fig. 295. [Fig 8; ed. note] The difference of potential between a point on the shaft and a point on the periphery will remain unchanged, in sign as well as in amount ...

We note here that in AC induction motors, eddy currents have to be controlled by laminating the core to obtain a reasonable efficiency, which demonstrates the same principle. As the illustration shows, the dotted path of radial current generation is the preferred path of current that suffers the standard "back torque" and goes in the opposite direction to the rotation of the disk in a usually successful attempt to slow it sown. (My generator slowed down from back torque rather well, to my disappointment.) However, if the disk is subdivided in the solid line manner, the current generated will now enhance the magnetic field (since rotating currents generate magnetic fields).

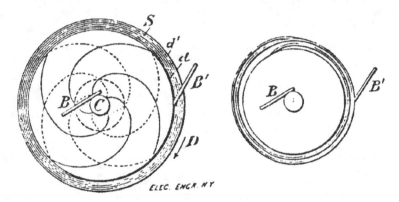

Figure 8. Spiral Disc Detailed.

At this point I would like to propose that the etched disk for sale by Borderland Research Foundation, in the pattern of a golden mean spiral, may be an interesting unipolar generator for the Tesla experiment proposed.

Of course, as Figure 8 shows, we may use an external spiral or coil encircling the disk to obtain a similar effect. Note the similarity between this drawing and the first one (from <u>Scientific American</u>) which is a model for the earth's core.

Forbes Dynamo

The Forbes dynamo is now discussed, from Figure 4, which seems to be simply a very efficient homopolar generator modified with two disks for higher voltage. For that example, Tesla proposes using the external coil but also a conductive belt, in what turns out to be a very innovative idea for increasing conductivity and decreasing resistance of the dynamo. The current is thus extracted only from the shafts of both generators.

> ... Instead of subdividing the disc or cylinder spirally, as indicated in Fig. 295, it is more convenient to interpose one or more turns between the disc and the contact ring on the periphery, as illustrated in Fig. 296. [Fig 8; ed. note]

> A Forbes dynamo may, for instance, be excited in such a manner. In the experience of the writer it has been found that instead of taking the current from two such discs by sliding contacts, as usual, a flexible conducting belt may be employed to advantage. The discs are in such case provided with large flanges, affording a very great contact surface. The belt should be made to bear on the flanges with spring pressure to take up the expansion. Several machines with belt contact were constructed by the writer two years ago, and worked satisfactorily; but for want of time the work in that direction has been temporarily suspended. A number of features pointed out above have also been used by the writer in connection with some types of alternating current motors.

Noting that his work on these generators has been suspended in the recent past, Tesla abruptly ends a most entertaining article unequalled in all of the homopolar literature. [Note: More information on Forbes and his dynamo can be found in Robert Belfield's article, Jour. IEEE, Sept. 1976, p.344.]

RELATIVITY COMPARISONS	
Rectilinear Motion	**Circular Motion**
- No voltage developed when bar and meter move together	- Voltage not developed when disk and meter move together, but electric field is generated
- No difference between motion of observer and charge: $M_1 = V \times P_1$	- Difference between rotating charged sphere or rotating observer (Schiff, 1939) B Field vs. no field - Ring currents developed causing magnetic field for sphere rotation
- No absolute motion detectable	- Absolute Rotation measured (wrt inertial frame) Sagnac, Marinov; see Marinov, Foundations of Physics Vol. 8, 1978 p.137
- Special Relativity applies	- Special relativity doesn't apply
- No volume charge by special relativity transformation laws	- Volume charge: $E = V \times B$ $D = \varepsilon_o E_{mak}$ $\rho = \nabla \cdot D$ $\rho = -2\varepsilon_o \omega B$
- No forces for uniform, constant velocity	- Centrifugal and coriolis forces generated

Relative Motion

In order to help some researchers distinguish between what they believe is true for the linear motion and what really happens on a rotating disk, I have included a couple of charts, not published previously. In Figure 9, we see the left hand column contains some major aspects of rectilinear motion taken from a classical physics text. On the right is my rotational analog to each of the same experiments, i.e., rotate the magnet (+) but not the meter (-) nor the disk(-), etc. Notice that the results are the same whether we rotate the magnet or not (to the best of our knowledge).

RELATIVITY COMPARISONS

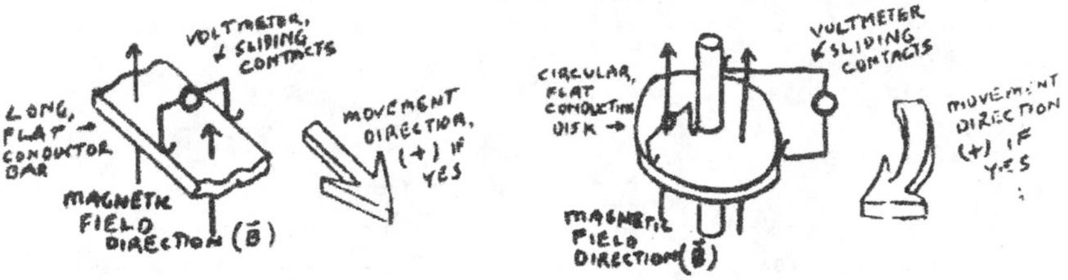

MOVEMENT			VOLTAGE		MOVEMENT			VOLTAGE
B-SOURCE	METER	BAR	(YES OR NO)		B-SOURCE	METER	DISK	(YES OR NO)
—	—	—	NO		—	—	—	NO
—	—	+	YES		—	—	+	YES
—	+	—	YES		—	+	—	YES
—	+	+	NO ρ=0		—	+	+	YES/NO ρ≠0
+	—	—	NO		+	—	—	NO
+	—	+	YES		+	—	+	YES
+	+	—	YES		+	+	—	YES
+	+	+	NO ρ=0		+	+	+	YES/NO ρ≠0

Figure 9. Comparison between linear and rotational motion.

The only debatable part comes when we rotate the meter and the disk with or without the magnet rotating. Here if we ask about an "emf" or electromotive force, we know that here is one present (yes), or if we ask if there is a nonuniform charge density, we answer yes, but if we ask about "voltage", the reaction of a meter, in the rotating frame, we have to answer no. (I placed a small, specially designed voltmeter on the disc to test this unusual effect.)

The Relativity Comparison Table is a comparison between linear and circular motion in a more theoretical fashion. Notice the many differences present for any rotating object.

My $1000 Homopolar Generator

When I came back from California in 1980, after a trip exploring the $25,000 generator that Bruce DePalma had built at the Sunburst Community (now called The Builders); I was determined to build one myself. This evolved into a Master's Degree project for the Physics Department at Buffalo's State University. Thanks to Erie Community College, I was able to test it at the school, in one of the labs.

In Figure 10, we see the results of a typical run. At the top, the trial was performed with the circuit closed, generating about 380 Amps and the DC motor demanding 266 Watts. Note the slowdown time here was about 0.57 minutes. Next, at the bottom, we see the test run with the circuit open, generating just voltage. Here the motor demand lessened to 249 Watts. Note the slowdown time is now longer (0.64 minutes) showing less "resistance" with the lack of back torque. The last verification of this analysis is the comparison of generated power (about 25 Watts) and the difference in the motor demand (about 17 Watts in this case). They are about the same.

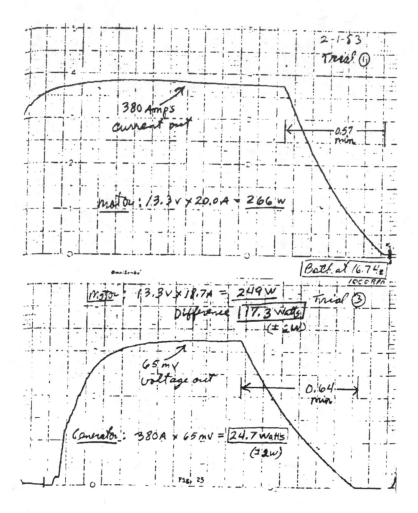

Figure 10. Current run charts for the $1000 generator.

In Figure 11, we see the results of another trial. Here the generated power and the difference in the motor demand was even closer -- less than 1 Watt between them. Quantitatively and qualitatively we see evidence for the existence of back torque in a one-piece Faraday generator. The output of the generator tended to be compensated by the increase in the motor demand for power from the batteries.

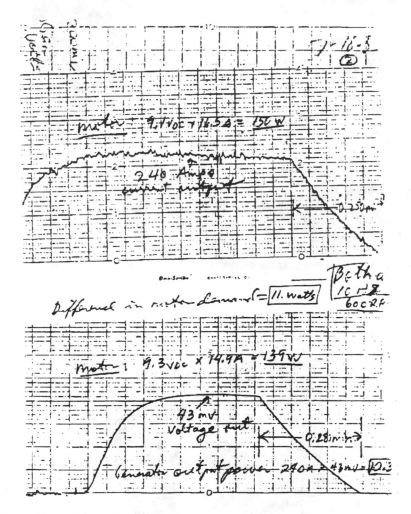

Figure 11. Additional run charts.

We know that a decrease in resistance of the system, from 1 milliohm down to about 1 microhm (recommended by Adam Trombly) would improve the performance. Also some of the design ideas of Tesla's would also contribute to a more self-sustaining generator.

The last few figures show various pictures of my generator. They are the best photos of my large generator with 8" ceramic magnets. For more information, including a complete copy of the Trombly-Kahn patent application, I would recommend my book, The One-Piece Faraday Generator.

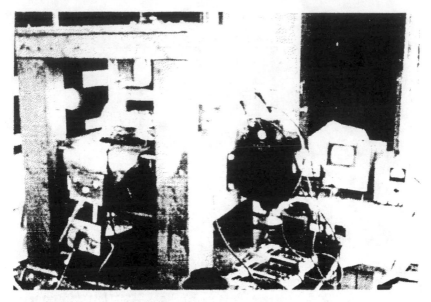

Figure 12. Homopolar generator - nonrotating.
  Background: Digital Frequency Counter, Oscilloscope, Power Supply for Circuit.
  Foreground: DC Motor, 2 Variacs for Heaters, 3 Digital Voltmeters.

Figure 13. Homopolar generator - front view exposing the General Electric 2500 Ampere 50 millivolt current shunt.

Figure 14. Homopolar generator with strip chart recorder shown on left.

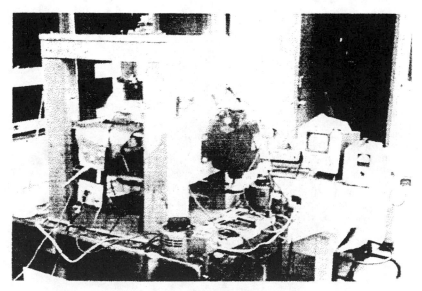

Figure 15. Homopolar generator - rotating.
Square wave on Oscilloscope is the photocell circuit output.

Figure 16. Homopolar generator - Metal arm at top holds photocell close to rotating magnet assembly. A strip of aluminum foil has been attached at the top of the magnets, covering half of the circumference for reflection of light into photocell.

About The Author:

Thomas F. Valone holds a MA in Physics, 1984; a BS in Physics and BSEE, State University of NY at Buffalo, 1974. Tom is a Professor of Physics and Electronics at Erie Community College in Buffalo, NY, a Professional Engineer, and President of his own company, Integrity Electronics & Research. His company is engaged in computer sales and specialized instrument manufacture. Tom has been involved in non-conventional energy research for the past ten years. He established his prominence in this field by beginning an investigation into a non-conventional energy generator in 1979, as part of a Master's Degree program at the State University

This information and picture was accurate back in 1986. The author can be reached at iri@erols.com and www.integrityresearchinstitute.org

16   Tesla's Ionizer and Ozonator

Thomas Valone
Reprinted from *Tesla: A Journal of Modern Science*, 1997

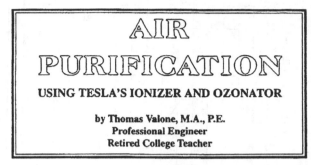

AIR

PURIFICATION

USING TESLA'S IONIZER AND OZONATOR

by **Thomas Valone, M.A., P.E.**
Professional Engineer
Retired College Teacher

OVERVIEW

Indoor and outdoor air quality has deteriorated due to airtight homes, more artificial materials and excessive air particulates in every major city. Oxygenating and ionizing the air has been shown to be a superior technique for purification, even surpassing the best air filtration methods. Nikola Tesla's patent on a method for generating ozone was issued in 1896. With a better invention than one would believe, we acknowledge Tesla for inventing the medically-approved method for producing ozone during its centennial. His method has been incorporated into Alpine air purifiers along with a mysterious invention of Dr. Pat Flanagan's that produces ionization at a distance. The merging of ion and ozone production in one machine, both based on Tesla technology, is a remarkable outgrowth of two separate inventions.

TESLA'S OZONE PATENT

At the turn of the century, there was a great deal of high voltage in Tesla's laboratory, as is seen in several photographs of Tesla seated in the midst of high voltage discharge which generated a good amount of ozone. Tesla was exposed to ozone ("triatomic oxygen") for years throughout his work.

The circuit that Tesla used to produce ozone is very simple. The diagram (from Tesla's 1896 patent #568,177) in Fig. 1 shows two parallel plates connected to a high voltage transformer, along with a fan to move the air through the parallel plates. The patent examiner reviewing this case cited three patents for synthesis of ozone to purify liquids from as early as 1882. It is worth noting that ozone was discovered in 1785 by Van Marum who passed air through an electric discharge.

Since 1896, three classes of ozone generators have become available: electric spark discharge; ultraviolet (UV) light; and "cold plasma discharge." As noted in **Explore More!** (No.7, 1994, p.30) only the cold plasma discharge method (which is the technical name for Tesla's invention) produces the purest ozone, with virtually no nitrogen oxides, and "is the type used by German medical doctors and health clinics worldwide for the treatment of disease."

OZONE/OXYGEN THERAPY

The book <u>Oxygen Therapies</u> by

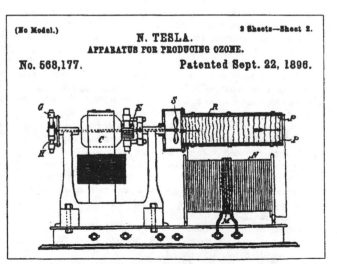

(No Model.)   2 Sheets—Sheet 2.

N. TESLA.
APPARATUS FOR PRODUCING OZONE.

No. 568,177.   Patented Sept. 22, 1896.

Fig. 1 - Tesla's Original Patent Application for the Ozone Machine

Ed McCabe (available from **The Tesla Resource Center**) is an excellent review on the subject. Journal articles such *as "Ozone Selectively Inhibits Growth of Human Cancer Cells"* (Sweet et al., Science, Vol. 209, 1980, p.931) and *"Do Oxygen Therapies Work?"* (East West Jour., Sept. 1989, p.70) have attracted much attention to the therapeutic effects of ozone, even for blood dialysis treatment of AIDS (more information available on the Internet). Dr. Andrija Puharich (a speaker at the 1984 ITS Symposium) presented a paper at the Sixth World Ozone Conference in 1983 on the successful treatment of neoplasms (cancer) with ozone. He stated, *"if gaseous ozone is administered directly into cancerous tissue in mice, the tumor would dissolve in a matter of seconds to minutes leaving the surrounding tissue unaffected."*

Unfortunately, the FDA has classified ozone as a drug, even though it is a naturally occurring gas at about 0.03 parts-per-million (ppm) concentration worldwide in country fresh air. FDA approval for ozone therapy as a treatment modality has still been withheld. In fact, a few years ago I saw an issue of the FDA newsletter which featured an illustrated cover story about an ozone machine which they had seized. (DISCLAIMER: We are legally only allowed to mention anecdotal information and emphasize that ozone can be used only on an experimental basis, without recommending or prescribing.) Companies such as OZ-TECH (POB 730, Alton, NH 03809) or Excalibur (314 W. 53rd St., NY, NY 10019) sell water treatment ozone generators which some people have configured

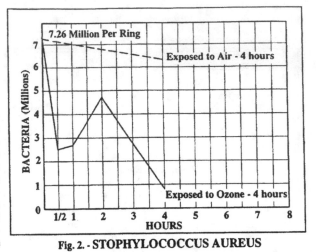

Fig. 2. - STOPHYLOCOCCUS AUREUS

for rectal insufflation or body bag exposure for skin absorption. The Bradford Institute (800-227-4473) is a good source of quality information on ozone therapy, operating their own hospital in Mexico, where it is legal. Canada and Europe are also countries where ozone therapy is legal and effective for a wide range of diseases. Dr. Michael Prytula, with a clinic in Niagara Falls, Canada, states in his *article "Ozone Therapy: Using Oxygen to Heal"* (**Holistic Health Journal**, Vol.2, No.4, 1995), *"The only problem ozone has in being fully accepted is that, like all naturally occurring products, it cannot be patented. Imagine how the manufacturers of pharmaceutical products and manufacturers of radiation equipment would feel about ozone getting equal press coverage."*

EPA FINDS INDOOR AIR CAN BE WORSE THAN OUTDOOR AIR

Nature only produces ozone with thunderstorms and alternatively with sunlight. Indoor air, however, contains no ozone whatsoever. Laboratory tests have shown (Fig. 2) that only 0.05 ppm of ozone will kill mold, E. coli, salmonella, and staph germs in 4 to 6 hours. At the same time, trace amounts of ozone (0.01 to 0.03 ppm) reduce odors, and increase oxygen absorption by hemoglobin. Consequently, the use of ozone generators in homes, offices, beauty parlors, hotels, and even in the President's limousine, is becoming more popular.

Since the average person spends over 90% of their time indoors, he/she should be treating the problem more seriously. In 1987, an EPA report ranked indoor pollution at the top of the list of environmental risks that Americans face. In one of the most exhaustive studies ever, the EPA fitted 800 people around the country with battery-operated sensors, in 1980, to measure levels of 20 chemicals in the air around them. In a few cities, the levels of 11 of those chemicals were higher in the family den (up to 70 times higher!) than were found outside the huge petrochemical plants and refineries nearby.

Indoor air is filled with pollutants such as benzene, carbon monoxide, sulfur dioxide, trichlorethylene, carbon tetrachloride, mold spores, pollens, fungi, etc. However, during the energy crisis a few decades ago, we started to build houses and office building that are airtight. Airtight buildings do not breathe and must rely upon a certain number of air exchanges per day. It is distressing to note that most hotels are not designed for any air exchange or circulation at all! Further compounding the problem is the finding reported in **Science** **News** (March 27, 1993) for sick buildings, labeled "*The Ventilation Conundrum.*" In a double-blind study which doubled the flow of outdoor air, from 30 to 64 cubic feet per minute (cfm), into a building already reporting complaints of "sick building syndrome", there was no decrease in worker symptoms. The concentration of volatile chemicals had decreased but the workers "perceived absolutely no difference."

Clothing, furnishings, construction products, paint, plywood, and particle board all "outgas" chemicals. Also, heating and cooling systems "can grow microbial products such as bacteria, fungi, and protozoa" (**Indoor Air Review**, Oct., 1992, p. 11). These colonies of microbes grow more often in the ventilation systems or air ducts which return air to the furnace, because they are not subject to filtering until just before the furnace.

"DIRTY AIR CAN SHORTEN YOUR LIFE," STUDY SAYS

A front page article from the **Washington Post** (Mar. 10, 1995, p.1) used the above title for the largest study ever conducted on the health effects of airborne particles which found that people in the nation's most polluted cities are 15 to 17 percent more likely to die prematurely than those in cities with the cleanest air, like Topeka, Kansas, thus losing approximately one year of life expectancy. The study cited particles 2.5 microns (millionths of

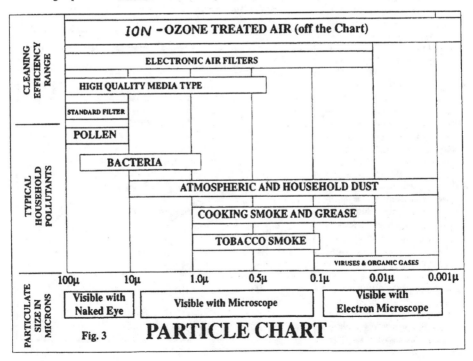

Fig. 3 PARTICLE CHART

291

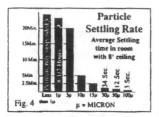

Fig. 4   μ = MICRON

a meter) and smaller as the culprits, because they stick in the deep recesses of the lungs and limit oxygen exchange.

In July, 1995, <u>Science News</u> (Vol. 148, p.5) reported that "*a spate of studies show that daily hospital admissions and deaths from respiratory disease tend to fluctuate nearly lockstep with variations in airborne dust—even when particulate levels fall within federal limits.*" Their findings indicate that the increase in hospital admissions for congestive heart failure matched the increase in small particulate levels of 10 microns and smaller. A revealing sideline was the fact that the same effect was seen from an increase in carbon monoxide levels in the air. Therefore, it may be possible that very small particulates affect the respiratory system in a way similar to carbon monoxide poisoning.

In a report in the November 1, 1996 *Washington Post*, it was written that the EPA is trying to change the standards for the industry from 10 Parts per Million (ppm) to 2.5 ppm so that finer particulates will be captured before they are emitted to the atmosphere. However, the National Association of Manufacturers (NAM) has been fighting against the new standards to control "particulate matter." Joining this opposition to tighten controls are the American Petroleum Institute, Geneva Steel, Chevron,

DuPont, Xerox, American Automobile Manufacturers Association and others.

What are these particles and how do they affect us indoors? Most studies, including those from local environmental testing firms, confirm that indoor air can be more polluted than outdoor air. One report, from the testing of the Takoma Park Library (in Maryland) by Aerosol Monitoring & Analysis, Inc., found on the average, 7000 to 10,000 particles per cubic centimeter (cc) throughout the library. Our nose and lungs have to process all of these particles which can consist of dead skin, bacteria, pollen, dander, dust mite droppings, viruses, etc. The lungs and the cilia of the cells can expel particles that are bigger than one micron.

Many people rely upon air filtration, and even high efficiency "HEPA" filters, to grab those nasty particles. However, the best filters will only arrest particles that are bigger than one micron, with varying retention rates. Clients inform me that even with the best electronic or electrostatic filters on their furnaces, they find very fine dust on everything. In Fig. 3, we notice that most dust particle concentration is in the sub-micron (below one micron) range, where the majority of the 7000 to 10,000 particles per cc reside. <u>Consumers Digest</u> (Oct. 1992, p.40) reports that "sunbeams in a home wouldn't be visible if it weren't for the suspended airborne particles that scatter the light in the beam's path." In Fig. 4, we see that it is the sub-micron particles which

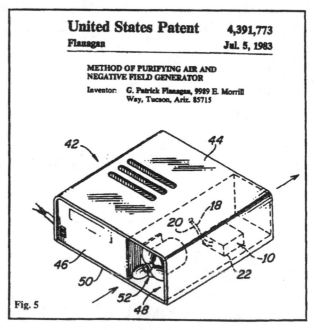

Fig. 5

are permanently suspended in the air, mostly from the "Brownian motion" of such microscopic particles.

IONIZATION SOLVES SUB-MICRON PROBLEM

Ironically, it is the smallest particles which can be ionized most easily. They quickly move under the influence of an electric field set up by an ionizer. Ions (from the Greek for "traveler") are electrically charged air molecules that have gained or lost an electron. The Ion Effect by Fred Soyka (Ballantine Books, 1991) is the only book that I am aware of on the subject of ions and health. It describes the medical experiments that have successfully treated a variety of respiratory diseases, mostly with negative ions. Negative ions tend to lower serotonin in the brain, making people more alert. Russian studies have indicated that "atmospheric ozone and ions are the vehicles of freshness." The results of two Russian doctors (Priroda, No.9, p.26, 1976) testing indoor air with only 0.02 ppm of ozone and negative ions is reported in Explore More! (No. 16, 1996, p.21).*

THE MYSTERIOUS ELEC-TRON FIELD GENERATOR

In the early 1980's, Dr. Patrick Flanagan made a discovery with a Tesla coil and a few dielectrics, producing an ionization field which even affected adjacent rooms. He states, "I found that electrons could be released from electrical insulators and semiconductors by means of a Tesla coil" (Acres USA, Mar., 1992, p.20). Shortly afterwards, he found how to miniaturize the package, using a flyback transformer operating at about 35 kHz and at least 5000 volts (see Fig. 5). When I spoke to him at that time, he described his laboratory electrometer which pegged at its limit, far from the device, when he reached the right parameters for the circuit. Tests revealed that the electron field released foam from the walls of pillow factories, in most of the rooms in the building. Dr. Flanagan received two patents on the device (#4,391,773 and #4,743,275) and markets the machine as model ECP-1000 (Vortex Indus., 1109 S. Plaza Way #399, Flagstaff, AZ 86001).

By the mid-1980's Mr. Bill Converse was given the opportunity to test Dr. Flanagan's device and formed his own company shortly thereafter to market a very similar machine with an added ozone generator. Surviving a court test of patent infringement, Mr. Converse formed what is now known as Alpine Industries and has sold almost 1 million of the machines worldwide. Upon testing the flyback transformer in the Alpine XL-15, which includes a "radio-wave ionizer," I found that its peak resonant frequency was at 35 kHz and it operated at over 5000 volts. The mysterious ceramic or dielectric which Flanagan determined was the "radiator" of the ionization field, may actually be the removable ozone plate. The Alpine machine's ionization field, reaching out to 60 feet from the machine, does seem to reduce airborne dust. Tests with black-mirrored surfaces (daily dusting), halogen flashlights (sunbeam effect test at night) and air vent tissue paper (dust collection) all seem to confirm the effect.

Alpine's Model 880 or XL-15 incorporate Tesla's cold plasma discharge method for producing ozone and the Tesla/Flanagan method for producing ionization at a distance. While I find it gratifying that such Tesla technology has penetrated the market so successfully, it would be great if acknowledgment were given to Tesla and Flanagan in Alpine's sales literature.

*A free ion/ozone booklet and a copy of "Fresh Air Curative Effect Related to Ions and Traces of Ozone" from Explore More! is available from the author by calling 800-295-7674.

This article is an abridged version of the presentation at the 1996 International Tesla Society Symposium entitled, "Tesla's Contribution to Indoor Air Purification." The complete transcript will be available in the 1996 Proceedings of the ITS Symposium. A video of the lecture and slide presentation is available from ITS for $29.95.

Ed. note: Alpine is now called "EcoQuest International" but still has Bill Converse as president. He has never settled with Pat Flanagan for making millions with Patrick's invention. It may also be true that a lawsuit was never initiated by Flanagan for patent infringement but certainly could be. The video presentation shows the spectrum graph of the Alpine transformer and is also distributed by Integrity Research Institute.

17   FBI Documents on Tesla

Declassified and finally made available to the public through Freedom of Information Act.

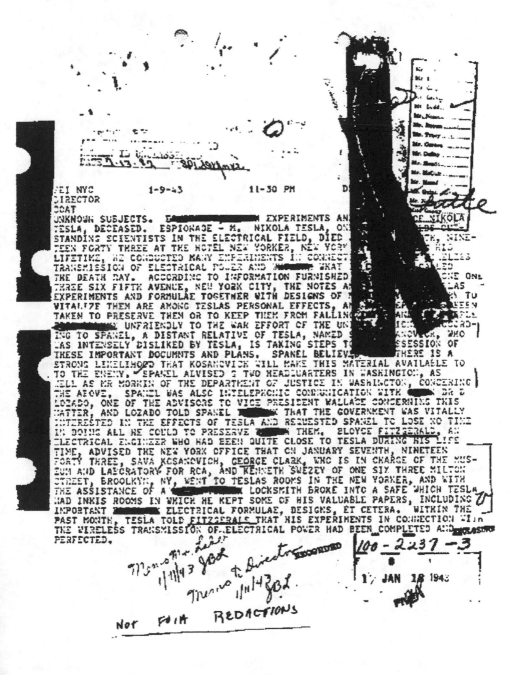

FBI NYC   1-9-43   11-30 PM   D

DIRECTOR
COAT
UNKNOWN SUBJECTS.   EXPERIMENTS AN   CE NIKOLA
TESLA, DECEASED. ESPIONAGE - M. NIKOLA TESLA, ON
STANDING SCIENTISTS IN THE ELECTRICAL FIELD, DIED   TH, NINE-
TEEN FORTY THREE AT THE HOTEL NEW YORKER, NEW YORK
LIFETIME, HE CONDUCTED MANY EXPERIMENTS IN CONNECT
TRANSMISSION OF ELECTRICAL POWER AND   WHAT
THE DEATH RAY. ACCORDING TO INFORMATION FURNISHED   ONE ONE
THREE SIX FIFTH AVENUE, NEW YORK CITY, THE NOTES A   LAS
EXPERIMENTS AND FORMULAE TOGETHER WITH DESIGNS OF   TO
VITALIZE THEM ARE AMONG TESLAS PERSONAL EFFECTS, AN
TAKEN TO PRESERVE THEM OR TO KEEP THEM FROM FALLING
UNFRIENDLY TO THE WAR EFFORT OF THE UN   CCORD-
ING TO SPANEL, A DISTANT RELATIVE OF TESLA, NAMED   WHO
WAS INTENSELY DISLIKED BY TESLA, IS TAKING STEPS T   SESSION OF
THESE IMPORTANT DOCUMENTS AND PLANS. SPANEL BELIEV   THERE IS A
STRONG LIKELIHOOD THAT KOSANOVICH WILL MAKE THIS MATERIAL AVAILABLE TO
TO THE ENEMY. SPANEL ALVISED 2 TWO HEADQUARTERS IN WASHINGTON, AS
WELL AS MR MORKIN OF THE DEPARTMENT OF JUSTICE IN WASHINGTON, CONCERNING
THE ABOVE. SPANEL WAS ALSO INTELEPHONIC COMMUNICATION WITH   DR D
LOZADO, ONE OF THE ADVISORS TO VICE PRESIDENT WALLACE CONCERNING THIS
MATTER, AND LOZADO TOLD SPANEL   THAT THE GOVERNMENT WAS VITALLY
INTERESTED IN THE EFFECTS OF TESLA AND REQUESTED SPANEL TO LOSE NO TIME
IN DOING ALL HE COULD TO PRESERVE   THEM. BLOYCE FITZGERALD, AN
ELECTRICAL ENGINEER WHO HAD BEEN QUITE CLOSE TO TESLA DURING HIS LIFE
TIME, ADVISED THE NEW YORK OFFICE THAT ON JANUARY SEVENTH, NINETEEN
FORTY THREE, SAVA KOSANOVICH, GEORGE CLARK, WHO IS IN CHARGE OF THE MUS-
EUM AND LABORATORY FOR RCA, AND KENNETH SWEZEY OF ONE SIX THREE MILTON
STREET, BROOKLYN, NY, WENT TO TESLAS ROOMS IN THE NEW YORKER, AND WITH
THE ASSISTANCE OF A   LOCKSMITH BROKE INTO A SAFE WHICH TESLA
HAD INKIS ROOMS IN WHICH HE KEPT SOME OF HIS VALUABLE PAPERS, INCLUDING
IMPORTANT   ELECTRICAL FORMULAE, DESIGNS, ET CETERA. WITHIN THE
PAST MONTH, TESLA TOLD FITZGERALD THAT HIS EXPERIMENTS IN CONNECTION WITH
THE WIRELESS TRANSMISSION OF ELECTRICAL POWER HAD BEEN COMPLETED AND
PERFECTED.

100-2237-3

17 JAN 18 1943

Not FOIA REDACTIONS

NOT FOIA REDACTIONS

JAN 12 1943

FBI NYC   1-12-43   11-06 PM   WXS
DIRECTOR
WEST
UNSUBS --- EQUIPMENT, EXPERIMENTS AND RESEARCH OF NIKOLA TESLA, DE
CEASED, ESPIONAGE - M. RETEL UNDER ABOVE HEADING TO THE BUREAU
FROM THIS OFFICE DATED JAN NINE LAST. INQUIRY DEVELOPS THAT TESLA DIED
JAN EIGHT, RATHER THAN THURSDAY, JAN SEVEN, AS STATED IN
REFERENCE TELETYPE. ON THE NIGHT OF JAN EIGHT, SAVA KOSANOVICH, GEORGE
CLARK, AND KENNETH SWEEZEY VISITED TESLA-S HOTEL WITH A REPRESENTATIVE
OF SHAW WALKER CO. IN ORDER TO OPEN THE SAFE IN THE ROOM OF TESLA.
KOSANOVICH LATER REPORTED TO WALTER GORSUCH, OFFICE OF ALIEN PROPERTY
CUSTODIAN, NYC, THAT HE WENT INTO THE ROOM IN ORDER TO SEARCH FOR A
WILL OF TESLA. KOSANOVICH AND THE OTHERS MADE THE SEARCH OF THE SAFE
IN THE PRESENCE OF THREE ASST MANAGERS OF HOTEL NEW YORKER AS WELL AS
REPRESENTATIVES OF THE YUGOSLAVIAN CONSULATE, IDENTITIES OF LATTER NOT
YET KNOWN. AFTER THE SAFE WAS OPENED, SWEEZEY, TOOK FROM THE SAFE A
BOOK CONTAINING TESTIMONIALS SENT TO TESLA ON THE OCCASION OF HIS SEVEN-
TY FIFTH BIRTHDAY. THIS BOOK WAS ARRANGED FOR TESLA BY SWEEZEY.
KOSANOVICH TOOK FROM THE ROOM THREE PICTURES OF TESLA, TWO BEING EN-
LARGED NEWSPAPER PICTUREX. ACCORDING TO MANAGERS OF HOTEL AND KOSANO-
VICH HIMSELF, NOTHING ELSE WAS REMOVED FROM THE ROOM OR SAFE. THE
SAFE WAS THEN CLOSED UNDER A NEW COMBINATION, WHICH COMBINATION IS
NOW IN POSSESSION OF KOSANOVICH. ON SATURDAY AFTERNOON, JAN
NINE, GORSUCH AND FTIZGERALD OF ALIEN PROPERTY CONTROL WENT TO
HOTEL AND SEIZED ALL THE PROPERTY OF TESLA, CONSISTING OF ABOUT TWO
TRUCKLOANDS OF MATERIAL, SEALED ALL ARTICLES AND TRANSFERRED THEM TO
THE MANHATTAN STORAGE AND WAREHOUSE CO. NY, WHERE THEY ARE NOW LOCATED.
AT THAT TIME THERE WERE ALSO IN THIS WAREHOUSE APPROXIMATELY
THIRTY BARRELS AND BUNDLES BELONGING TO TESLA WHICH HAD BEEN THERE SINC
ABOUT NINETEEN THIRTY FOUR. THESE HAVE ALSO BEEN SEALED AND
ARE NOW UNDER ORDERS OF ALIEN PROPERTY CUSTODIAN. IN VIEW OF FACT TESLA
IS A US NATURALIZED CITIZEN, ALIEN PROPERTY CUSTODIAN FEELS THAT ITS
JURISDICTION OVER PROPERTY IS DOUBTFUL BUT FEELS THAT NO OTHER
AGENCY WILL BE ABLE TO GET TO THIS PROPERTY FOR AT LEAST TWO DAYS.
COPIES DESTROYED ALL INFORMATION CONTAINED
HEREIN IS UNCLASSIFIED

NOT FOIA REDACTION

REDACTIONS ON THIS PAGE ARE
NOT FOIA DELETIONS

Office Memorandum • UNITED STATES GOVERNMENT

TO : MR. TOLSON DATE: Jan. 30, 1951

FROM : L. P. NICHOLS

SUBJECT:

called yesterday and said he would
be in town for the next two or three days and wanted to see me.
I told him I would be glad to see him late yesterday or today
at his convenience. He is coming in today at 2:30 p.m.

I asked him if there was anything special which would
require my doing any checking. He stated there were two things he
wanted to discuss.

1. The case of Nicola Tesla and Abraham N. Spanel,
President of International Later Corporation

Our files reflect that Nicola Tesla was one of the
world's outstanding scientists and in fact designed the generators
installed at Niagara Falls. He died in New York on January 7,
1943, and is supposed to have left details and plans for a so-
called death ray.

Our files also reflect that Colonel Erskine of
Military Intelligence called us on January 9, 1943, advising that
Tesla had died, that A. Spanel had communicated with the War Depart-
ment regarding this death, that Tesla had a nephew named Sava
Kosanovich who had taken possession of Tesla's papers and Spanel
thought the papers might be used against our Government.

We made an immediate inquiry in New York and the first
report was that Kosanovich and others entered Tesla's room with the
aid of a locksmith, broke into a safe containing some of Tesla's
valuable papers including formula.

Coincident with this, on January 6, L. M. C. Smith called
Mr. Tamm regarding the death of Tesla and Smith stated he was
talking to the Alien Property Custodian about seizing these items.

We interviewed Spanel who expressed concern over Tesla's
effects and Spanel stated that Kosanovich had turned over the
effects of Tesla to the Alien Property Custodian. Spanel further
stated the day before Tesla died, he tried to get in touch with
the War Department to make available certain patents

RECORDED - 15   MAR 2 1951

COPIES DESTROYED   FX 8   INDEXED - 18

R94

297

April 20, 1976

Mr. Clarence Kelly
Director
F.B.I.
Washington, DC

Dear Mr. Kelly:

Mr. Allen and Mr. Ruchlehaus,   former acting Director of the
FBI, contacted me in 1973 regarding the unavailability of
American microfilm records of Nikola Tesla's unpublished diary
(now in the Belgrade museum, arranged by month per folder).

At the time I discounted the possibility that these unpublished
discoveries had military significance. But because of experiments
now under way at Hill AFB, I now suspect such military
applications exist and feel it imperative that you be notified,
particularly in view of the fact that the Soviets have primary
access to the entire collection.

Two photos of each page exist.

After Tesla's death, scientists from the Navy and OSS performed
a cursory examination of the diary and notes, which if my
memory serves me correctly, was one month long, hardly enough
time to decipher Tesla's torturous handwriting. Though Tesla
wrote in English, his penmanship was small, blurred, and as
difficult to translate as a foreign language.

According to the museum director (1971), the Soviets had made
copies of some portions, but not the Colorado Springs diary,
which numbers 500 pages, 20 that directly pertain to ball
lightning, and 20 or so relevant to the equipment construction.
(We copied the most significant portions, but feel more exists)

EX-115   REC-52 /00 - 2237-

_____ an article _____ magazine, EDN (an electrical
engineering magazine), but only with the very recent receipt
of an unpublished manuscript from John J. O'Neill's book
(PRODIGAL GENIUS) did I place credence on Tesla's later claim
to military applications. Incidentally, some of O'Neill's
descriptions were inaccurate and exagerated, as we have exceeded
Tesla's results and are familiar with the experiments. At any
rate, there are three possible military applications.

MAY 8 197

CON<s>FIDENTIAL</s>

OFFICE THE UNDER SECRETARY OF DEFENSE

WASHINGTON, D.C. 20301

RESEARCH AND
ENGINEERING

FEDERAL GOVERNMENT

9 FEB 1981

/

MEMORANDUM FOR THE DIRECTOR, FEDERAL BUREAU OF INVESTIGATION

SUBJECT: Papers Recovered on the Death of Nicola Tesla (U)

(U) We understand that the FBI may have possession of a number of papers
found after the death of Nicola Tesla in 1943. Nicola Tesla was a brilliant
electrical engineer (i.e. the Tesla Coil) who was a pioneer in various
aspects of electrical transmission phenomena.

(C) We believe that certain of Tesla's papers may contain basic principles
which would be of considerable value to certain ongoing research within the
DoD. It would be very helpful to have access to his papers.

(U) Since we have really no idea of the possible volume of these papers,
we would be happy to provide a researcher who could assist you in reducing
the magnitude of the search. If there are further questions, I am the
point of contact within the DoD and can be reached at 695-6364 or 695-7417.

Allan J. MacLaren
LtColonel, USAF
Military Assistant
Strategic and Space Systems

REGISTERED
1059638

FBI
ALL INFORMATION CONTAINED
HEREIN IS UNCLASSIFIED
DATE 4-8-93 BY 9803   b7C
FOIPA No. 356,608
        362,001 100-2237-33

Per DoD letter dated 3-29-93
with enclosures 1 and 2
All DoD info is unclassified
DoD referenced FOIPA #356,608 and
        362,001
4/8/93 9803   b7C

ALL INFORMATION CONTAINED
HEREIN IS UNCLASSIFIED
DATE   BY

2 FEB 12 1981

b7C

Classifed by: DUSDRE/S&SS
Declassify on: February 1987

DECLASSIFIED BY 1049
ON 1-29-93
Per Army letter 5-15-89
        291960

SO MAR 30 1981

CON<s>FIDENTIAL</s>

FD-36 (Rev. 8-26-82)

FBI

TRANSMIT VIA:
☐ Teletype
☒ Facsimile
☒ AIRTEL

PRECEDENCE:
☐ Immediate
☐ Priority
☐ Routine

CLASSIFICATION:
☐ TOP SECRET
☐ SECRET
☐ CONFIDENTIAL
☐ UNCLAS E F T O
☐ UNCLAS

Date   8/18/83

TO: DIRECTOR, FBI   SECRET
           ATTN: INTD. SUPERVISOR

FROM: SAC, CINCINNATI   (P)

NIKOLA TESLA

(OO: CI)

This communication is classified "Secret" in its
entirety.

          Re telephone call of SA   , Cincinnati
Division, to Supervisor   FBIHQ, on 8/11/83.

          Enclosed for the Bureau and New York is one copy
each of pertinent pages from the 1981 book titled "Tesla:
Man Out of Time" by Margaret Cheney, with important passages
underlined.

          For information of Bureau and New York,

                            at Wright-Patterson Air Force Base (WPAFB) and

          also at WPAFB, have both been in contact with SA
          at the Dayton, Ohio RA regarding possible FBI

          SECRET   Classified by
                                              Declassify on: OADR
          Classified by: 8262
          Declassify on: OADR

          ENCLOSURE

2 - Bureau (Enc. 1)
2 - New York (Enc. 1)
2 - Cincinnati

(6)   ALL INFORMATION CONTAINED

                                             17 AUG 22 1983

Approved: Transmitted
                                  (Number)   (Time)

(Sample of available FBI documents on Nikola Tesla. See www.pbs.org/tesla – Ed. note)

300

18 Selected Tesla Patents

Tesla is reported to have held about 700 patents including the foreign ones. Many books, such as Keith Tutt's book, include a list of all of the US, British, and Canadian patents. www.pbs.org/tesla also lists Tesla patents. Here are three of the ones that are referred to in the wireless transmission of energy research papers included in this book.

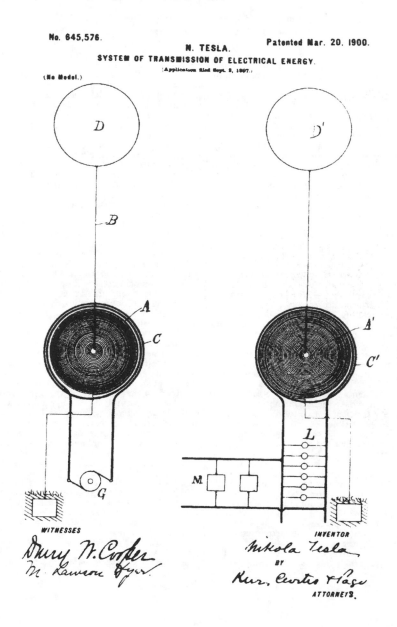

No. 645,576.

N. TESLA.

Patented Mar. 20. 1900.

SYSTEM OF TRANSMISSION OF ELECTRICAL ENERGY.

(Application filed Sept. 2, 1897.)

(No Model.)

UNITED STATES PATENT OFFICE.

NIKOLA TESLA, OF NEW YORK, N. Y.

SYSTEM OF TRANSMISSION OF ELECTRICAL ENERGY.

SPECIFICATION forming part of Letters Patent No. 645,576, dated March 20, 1900.

Application filed September 2, 1897. Serial No. 650,343. (No model.)

To all whom it may concern:

Be it known that I, NIKOLA TESLA, a citizen of the United States, residing at New York, in the county and State of New York, have in-
5 vented certain new and useful Improvements in Systems of Transmission of Electrical Energy, of which the following is a specification, reference being had to the drawing accompanying and forming a part of the same.
10 It has been well known heretofore that by rarefying the air inclosed in a vessel its insulating properties are impaired to such an extent that it becomes what may be considered as a true conductor, although one of ad-
15 mittedly very high resistance. The practical information in this regard has been derived from observations necessarily limited in their scope by the character of the apparatus or means heretofore known and the quality of
20 the electrical effects producible thereby. Thus it has been shown by William Crookes in his classical researches, which have so far served as the chief source of knowledge of this subject, that all gases behave as excellent
25 insulators until rarefied to a point corresponding to a barometric pressure of about seventy-five millimeters, and even at this very low pressure the discharge of a high-tension induction-coil passes through only a part of the
30 attenuated gas in the form of a luminous thread or arc, a still further and considerable diminution of the pressure being required to render the entire mass of the gas inclosed in a vessel conducting. While this is true in
35 every particular so long as electromotive or current impulses such as are obtainable with ordinary forms of apparatus are employed, I have found that neither the general behavior of the gases nor the known relations between
40 electrical conductivity and barometric pressure are in conformity with these observations when impulses are used such as are producible by methods and apparatus devised by me and which have peculiar and hitherto
45 unobserved properties and are of effective electromotive forces, measuring many hundred thousands or millions of volts. Through the continuous perfection of these methods and apparatus and the investigation of the
50 actions of these current impulses I have been led to the discovery of certain highly-important and useful facts which have hitherto been unknown. Among these and bearing directly upon the subject of my present application
55 are the following: First, that atmospheric or other gases, even under normal pressure, when they are known to behave as perfect insulators, are in a large measure deprived of their dielectric properties by being subjected to the
60 influence of electromotive impulses of the character and magnitude I have referred to and assume conducting and other qualities which have been so far observed only in gases greatly attenuated or heated to a high tem-
65 perature, and, second, that the conductivity imparted to the air or gases increases very rapidly both with the augmentation of the applied electrical pressure and with the degree of rarefaction, the law in this latter respect be-
70 ing, however, quite different from that heretofore established. In illustration of these facts a few observations, which I have made with apparatus devised for the purposes here contemplated, may be cited. For example, a con-
75 ductor or terminal, to which impulses such as those here considered are supplied, but which is otherwise insulated in space and is remote from any conducting-bodies, is surrounded by a luminous flame-like brush or discharge
80 often covering many hundreds or even as much as several thousands of square feet of surface, this striking phenomenon clearly attesting the high degree of conductivity which the atmosphere attains under the influence
85 of the immense electrical stresses to which it is subjected. This influence is, however, not confined to that portion of the atmosphere which is discernible by the eye as luminous and which, as has been the case in some in-
90 stances actually observed, may fill the space within a spherical or cylindrical envelop of a diameter of sixty feet or more, but reaches out to far remote regions, the insulating qualities of the air being, as I have ascertained,
95 still sensibly impaired at a distance many hundred times that through which the luminous discharge projects from the terminal and in all probability much farther. The distance extends with the increase of the elec-
100 tromotive force of the impulses, with the diminution of the density of the atmosphere, with the elevation of the active terminal above the ground, and also, apparently, in a slight measure, with the degree of moisture contained in

the air. I have likewise observed that this region of decidedly-noticeable influence continuously enlarges as time goes on, and the discharge is allowed to pass not unlike a conflagration which slowly spreads, this being possibly due to the gradual electrification or ionization of the air or to the formation of less insulating gaseous compounds. It is, furthermore, a fact that such discharges of extreme tensions, approximating those of lightning, manifest a marked tendency to pass upward away from the ground, which may be due to electrostatic repulsion, or possibly to slight heating and consequent rising of the electrified or ionized air. These latter observations make it appear probable that a discharge of this character allowed to escape into the atmosphere from a terminal maintained at a great height will gradually leak through and establish a good conducting-path to more elevated and better conducting air strata, a process which possibly takes place in silent lightning discharges frequently witnessed on hot and sultry days. It will be apparent to what an extent the conductivity imparted to the air is enhanced by the increase of the electromotive force of the impulses when it is stated that in some instances the area covered by the flame discharge mentioned was enlarged more than sixfold by an augmentation of the electrical pressure, amounting scarcely to more than fifty per cent. As to the influence of rarefaction upon the electric conductivity imparted to the gases it is noteworthy that, whereas the atmospheric or other gases begin ordinarily to manifest this quality at something like seventy-five millimeters barometric pressure with the impulses of excessive electromotive force to which I have referred, the conductivity, as already pointed out, begins even at normal pressure and continuously increases with the degree of tenuity of the gas, so that at, say, one hundred and thirty millimeters pressure, when the gases are known to be still nearly perfect insulators for ordinary electromotive forces, they behave toward electromotive impulses of several millions of volts like excellent conductors, as though they were rarefied to a much higher degree. By the discovery of these facts and the perfection of means for producing in a safe, economical, and thoroughly-practicable manner current impulses of the character described it becomes possible to transmit through easily-accessible and only moderately-rarefied strata of the atmosphere electrical energy not merely in insignificant quantities, such as are suitable for the operation of delicate instruments and like purposes, but also in quantities suitable for industrial uses on a large scale up to practically any amount and, according to all the experimental evidence I have obtained, to any terrestrial distance. To conduce to a better understanding of this method of transmission of energy and to distinguish it clearly, both in its theoretical aspect and in its practical

bearing, from other known modes of transmission, it is useful to state that all previous efforts made by myself and others for transmitting electrical energy to a distance without the use of metallic conductors, chiefly with the object of actuating sensitive receivers, have been based, in so far as the atmosphere is concerned, upon those qualities which it possesses by virtue of its being an excellent insulator, and all these attempts would have been obviously recognized as ineffective if not entirely futile in the presence of a conducting atmosphere or medium. The utilization of any conducting properties of the air for purposes of transmission of energy has been hitherto out of the question in the absence of apparatus suitable for meeting the many and difficult requirements, although it has long been known or surmised that atmospheric strata at great altitudes—say fifteen or more miles above sea-level—are, or should be, in a measure, conducting; but assuming even that the indispensable means should have been produced then still a difficulty, which in the present state of the mechanical arts must be considered as insuperable, would remain—namely, that of maintaining terminals at elevations of fifteen miles or more above the level of the sea. Through my discoveries before mentioned and the production of adequate means the necessity of maintaining terminals at such inaccessible altitudes is obviated and a practical method and system of transmission of energy through the natural media is afforded essentially different from all those available up to the present time and possessing, moreover, this important practical advantage, that whereas in all such methods and systems heretofore used or proposed but a minute fraction of the total energy expended by the generator or transmitter was recoverable in a distant receiving apparatus by my method and appliances it is possible to utilize by far the greater portion of the energy of the source and in any locality however remote from the same.

Expressed briefly, my present invention, based upon these discoveries, consists then in producing at one point an electrical pressure of such character and magnitude as to cause thereby a current to traverse elevated strata of the air between the point of generation and a distant point at which the energy is to be received and utilized.

In the accompanying drawing a general arrangement of apparatus is diagrammatically illustrated such as I contemplate employing in the carrying out of my invention on an industrial scale—as, for instance, for lighting distant cities or districts from places where cheap power is obtainable.

Referring to the drawing, A is a coil, generally of many turns and of a very large diameter, wound in spiral form either about a magnetic core or not, as may be found necessary. C is a second coil, formed of a conductor of much larger section and smaller length, wound around and in proximity to the coil A. In the transmitting apparatus the coil A constitutes the high-tension secondary and the coil C the primary of much lower tension of a transformer. In the circuit of the primary C is included a suitable source of current G. One terminal of the secondary A is at the center of the spiral coil, and from this terminal the current is led by a conductor B to a terminal D, preferably of large surface, formed or maintained by such means as a balloon at an elevation suitable for the purposes of transmission, as before described. The other terminal of the secondary A is connected to earth and, if desired, also to the primary in order that the latter may be at substantially the same potential as the adjacent portions of the secondary, thus insuring safety. At the receiving-station a transformer of similar construction is employed; but in this case the coil A', of relatively-thin wire, constitutes the primary and the coil C', of thick wire or cable, the secondary of the transformer. In the circuit of the latter are included lamps L, motors M, or other devices for utilizing the current. The elevated terminal D' is connected with the center of the coil A', and the other terminal of said coil is connected to earth and preferably, also, to the coil C' for the reasons above stated.

It will be observed that in coils of the character described the potential gradually increases with the number of turns toward the center, and the difference of potential between the adjacent turns being comparatively small a very high potential, impracticable with ordinary coils, may be successfully obtained. It will be, furthermore, noted that no matter to what an extent the coils may be modified in design and construction, owing to their general arrangement and manner of connection, as illustrated, those portions of the wire or apparatus which are highly charged will be out of reach, while those parts of the same which are liable to be approached, touched, or handled will be at or nearly the same potential as the adjacent portions of the ground, this insuring, both in the transmitting and receiving apparatus and regardless of the magnitude of the electrical pressure used, perfect personal safety, which is best evidenced by the fact that although such extreme pressures of many millions of volts have been for a number of years continuously experimented with no injury has been sustained neither by myself or any of my assistants.

The length of the thin-wire coil in each transformer should be approximately one-quarter of the wave length of the electric disturbance in the circuit, this estimate being based on the velocity of propagation of the disturbance through the coil itself and the circuit with which it is designed to be used. By way of illustration if the rate at which the current traverses the circuit, including the coil, be one hundred and eighty-five thousand miles per second then a frequency of nine hundred and twenty-five per second would maintain nine hundred and twenty-five stationary waves in a circuit one hundred and eighty-five thousand miles long and each wave would be two hundred miles in length. For such a low frequency, to which I shall resort only when it is indispensable to operate motors of the ordinary kind under the conditions above assumed, I would use a secondary of fifty miles in length. By such an adjustment or porportioning of the length of wire in the secondary coil or coils the points of highest potential are made to coincide with the elevated terminals D D', and it should be understood that whatever length be given to the wires this condition should be complied with in order to attain the best results.

As the main requirement in carrying out my invention is to produce currents of an excessively-high potential, this object will be facilitated by using a primary current of very considerable frequency, since the electromotive force obtainable with a given length of conductor is proportionate to the frequency; but the frequency of the current is in a large measure arbitrary, for if the potential be sufficiently high and if the terminals of the coils be maintained at the proper altitudes the action described will take place, and a current will be transmitted through the elevated air strata, which will encounter little and possibly even less resistance than if conveyed through a copper wire of a practicable size. Accordingly the construction of the apparatus may be in many details greatly varied; but in order to enable any person skilled in the mechanical and electrical arts to utilize to advantage in the practical applications of my system the experience I have so far gained the following particulars of a model plant which has been long in use and which was constructed for the purpose of obtaining further data to be used in the carrying out of my invention on a large scale are given. The transmitting apparatus was in this case one of my electrical oscillators, which are transformers of a special type, now well known and characterized by the passage of oscillatory discharges of a condenser through the primary. The source G, forming one of the elements of the transmitter, was a condenser of a capacity of about four one-hundredths of a microfarad and was charged from a generator of alternating currents of fifty thousand volts pressure and discharged by means of a mechanically-operated break five thousand times per second through the primary C. The latter consisted of a single turn of stout stranded cable of inappreciable resistance and of an inductance of about eight thousand centimeters, the diameter of the loop being very nearly two hundred and forty-four centimeters. The total inductance of the primary circuit was approximately ten thousand centimeters, so that the primary circuit vibrated generally according to adjustment, from two hundred and thirty thousand to two hundred and fifty thousand times per second. The high-tension coil A in the form of a flat spiral was composed of fifty turns of heavily-insulated cable No. 8 wound in one single layer, the turns beginning close to the primary loop and ending near its center. The outer end of the secondary or high-tension coil A was connected to the ground, as illustrated, while the free end was led to a terminal placed in the rarefied air stratum through which the energy was to be transmitted, which was contained in an insulating-tube of a length of fifty feet or more, within which a barometric pressure varying from about one hundred and twenty to one hundred and fifty millimeters was maintained by means of a mechanical suction-pump. The receiving-transformer was similarly proportioned, the ratio of conversion being the reciprocal of that of the transmitter, and the primary high-tension coil A' was connected, as illustrated, with the end near the low-tension coil C' to the ground and with the free end to a wire or plate likewise placed in the rarefied air stratum and at the distance named from the transmitting-terminal. The primary and secondary circuits in the transmitting apparatus being carefully synchronized, an electromotive force from two to four million volts and more was obtainable at the terminals of the secondary coil A, the discharge passing freely through the attenuated air stratum maintained at the above barometric pressures, and it was easy under these conditions to transmit with fair economy considerable amounts of energy, such as are of industrial moment, to the receiving apparatus for supplying from the secondary coil C' lamps L or kindred devices. The results were particularly satisfactory when the primary coil or system A', with its secondary C', was carefully adjusted, so as to vibrate in synchronism with the transmitting coil or system A C. I have, however, found no difficulty in producing with apparatus of substantially the same design and construction electromotive forces exceeding three or four times those before mentioned and have ascertained that by their means current impulses can be transmitted through much-denser air strata. By the use of these I have also found it practicable to transmit notable amounts of energy through air strata not in direct contact with the transmitting and receiving terminals, but remote from them, the action of the impulses, in rendering conducting air of a density at which it normally behaves as an insulator, extending, as before remarked, to a considerable distance. The high electromotive force obtained at the terminals of coil or conductor A was, as will be seen, in the preceding instance, not so much due to a large ratio of transformation as to the joint effect of the capacities and inductances in the synchronized circuits, which effect is enhanced by a high frequency, and it will be obviously un-

derstood that if the latter be reduced a greater ratio of transformation should be resorted to, especially in cases in which it may be deemed of advantage to suppress as much as possible, and particularly in the transmitting-coil A, the rise of pressure due to the above effect and to obtain the necessary electromotive force solely by a large transformation ratio. While electromotive forces such as are produced by the apparatus just described may be sufficient for many purposes to which my system will or may be applied, I wish to state that I contemplate using in an industrial undertaking of this kind forces greatly in excess of these, and with my present knowledge and experience in this novel field I would estimate them to range from twenty to fifty million volts and possibly more. By the use of these much greater forces larger amounts of energy may be conveyed through the atmosphere to remote places or regions, and the distance of transmission may be thus extended practically without limit. As to the elevation of the terminals D D' it is obvious that it will be determined by a number of things, as by the amount and quality of the work to be performed, by the local density and other conditions of the atmosphere, by the character of the surrounding country, and such considerations as may present themselves in individual instances. Thus if there be high mountains in the vicinity the terminals should be at a greater height, and generally they should always be, if practicable, at altitudes much greater than those of the highest objects near them in order to avoid as much as possible the loss by leakage. In some cases when small amounts of energy are required the high elevation of the terminals, and more particularly of the receiving-terminal D', may not be necessary, since, especially when the frequency of the currents is very high, a sufficient amount of energy may be collected at that terminal by electrostatic induction from the upper air strata, which are rendered conducting by the active terminal of the transmitter or through which the currents from the same are conveyed. With reference to the facts which have been pointed out above it will be seen that the altitudes required for the transmission of considerable amounts of electrical energy in accordance with this method are such as are easily accessible and at which terminals can be safely maintained, as by the aid of captive balloons supplied continuously with gas from reservoirs and held in position securely by steel wires or by any other means, devices, or expedients, such as may be contrived and perfected by ingenious and skilled engineers. From my experiments and observations I conclude that with electromotive impulses not greatly exceeding fifteen or twenty million volts the energy of many thousands of horse-power may be transmitted over vast distances, measured by many hundreds and even thousands of miles, with terminals not more than thirty to thirty-five thousand feet above the level of the sea, and even this comparatively-small elevation will be required chiefly for reasons of economy, and, if desired, it may be considerably reduced, since by such means as have been described practically any potential that is desired may be obtained, the currents through the air strata may be rendered very small, whereby the loss in the transmission may be reduced.

It will be understood that the transmitting as well as the receiving coils, transformers, or other apparatus may be in some cases movable—as, for example, when they are carried by vessels floating in the air or by ships at sea. In such a case, or generally, the connection of one of the terminals of the high-tension coil or coils to the ground may not be permanent, but may be intermittently or inductively established, and any such or similar modifications I shall consider as within the scope of my invention.

While the description here given contemplates chiefly a method and system of energy transmission to a distance through the natural media for industrial purposes, the principles which I have herein disclosed and the apparatus which I have shown will obviously have many other valuable uses—as, for instance, when it is desired to transmit intelligible messages to great distances, or to illuminate upper strata of the air, or to produce, designedly, any useful changes in the condition of the atmosphere, or to manufacture from the gases of the same products, as nitric acid, fertilizing compounds, or the like, by the action of such current impulses, for all of which and for many other valuable purposes they are eminently suitable, and I do not wish to limit myself in this respect. Obviously, also, certain features of my invention here disclosed will be useful as disconnected from the method itself—as, for example, in other systems of energy transmission, for whatever purpose they may be intended, the transmitting and receiving transformers arranged and connected as illustrated, the feature of a transmitting and receiving coil or conductor, both connected to the ground and to an elevated terminal and adjusted so as to vibrate in synchronism, the proportioning of such conductors or coils, as above specified, the feature of a receiving-transformer with its primary connected to earth and to an elevated terminal and having the operative devices in its secondary, and other features or particulars, such as have been described in this specification or will readily suggest themselves by a perusal of the same.

I do not claim in this application a transformer for developing or converting currents of high potential in the form herewith shown and described and with the two coils connected together, as and for the purpose set forth, having made these improvements the subject of a patent granted to me November

2, 1897, No. 593,138, nor do I claim herein the apparatus employed in carrying out the method of this application when such apparatus is specially constructed and arranged for securing the particular object sought in the present invention, as these last-named features are made the subject of an application filed as a division of this application on February 19, 1900, Serial No. 5,780.

What I now claim is—

1. The method hereinbefore described of transmitting electrical energy through the natural media, which consists in producing at a generating-station a very high electrical pressure, causing thereby a propagation or flow of electrical energy, by conduction, through the earth and the air strata, and collecting or receiving at a distant point the electrical energy so propagated or caused to flow.

2. The method hereinbefore described of transmitting electrical energy, which consists in producing at a generating-station a very high electrical pressure, conducting the current caused thereby to earth and to a terminal at an elevation at which the atmosphere serves as a conductor therefor, and collecting the current by a second elevated terminal at a distance from the first.

3. The method hereinbefore described of transmitting electrical energy through the natural media, which consists in producing between the earth and a generator-terminal elevated above the same, at a generating-station, a sufficiently-high electromotive force to render elevated air strata conducting, causing thereby a propagation or flow of electrical energy, by conduction, through the air strata, and collecting or receiving at a point distant from the generating-station the electrical energy so propagated or caused to flow.

4. The method hereinbefore described of transmitting electrical energy through the natural media, which consists in producing between the earth and a generator-terminal elevated above the same, at a generating-station, a sufficiently-high electromotive force to render the air strata at or near the elevated terminal conducting, causing thereby a propagation or flow of electrical energy, by conduction, through the air strata, and collecting or receiving at a point distant from the generating-station the electrical energy so propagated or caused to flow.

5. The method hereinbefore described of transmitting electrical energy through the natural media, which consists in producing between the earth and a generator-terminal elevated above the same, at a generating-station, electrical impulses of a sufficiently high electromotive force to render elevated air strata conducting, causing thereby current impulses to pass, by conduction, through the air strata, and collecting or receiving at a point distant from the generating-station, the energy of the current impulses by means of a circuit synchronized with the impulses.

6. The method hereinbefore described of

transmitting electrical energy through the natural media, which consists in producing between the earth and a generator-terminal elevated above the same, at a generating-sta-
5 tion, electrical impulses of a sufficiently-high electromotive force to render the air strata at or near the elevated terminal conducting, causing thereby current impulses to pass through the air strata, and collecting or re-
10 ceiving at a point distant from the generating-station the energy of the current impulses by means of a circuit synchronized with the impulses.

7. The method hereinbefore described of
15 transmitting electrical energy through the natural media, which consists in producing between the earth and a generator-terminal elevated above the same, at a generating-station, electrical impulses of a wave length
20 so related to the length of the generating circuit or conductor as to produce the maximum potential at the elevated terminal, and of sufficiently-high electromotive force to render elevated air strata conducting, causing
25 thereby a propagation of electrical impulses through the air strata, and collecting or receiving at a point distant from the generating-station the energy of such impulses by means of a receiving-circuit having a length
30 of conductor similarly related to the wave length of the impulses.

8. The method hereinbefore described of transmitting electrical energy through the natural media, which consists in producing
35 between the earth and a generator-terminal elevated above the same, at a generating-sta-

tion, a sufficiently-high electromotive force to render elevated air strata conducting, causing thereby a propagation or flow of electrical energy through the air strata, by conduction, 40 collecting or receiving the energy so transmitted by means of a receiving-circuit at a point distant from the generating-station, using the receiving-circuit to energize a secondary circuit, and operating translating de- 45 vices by means of the energy so obtained in the secondary circuit.

9. The method hereinbefore described of transmitting electrical energy through the natural media, which consists in generating 50 current impulses of relatively-low electromotive force at a generating-station, utilizing such impulses to energize the primary of a transformer, generating by means of such primary circuit impulses in a secondary sur- 55 rounding by the primary and connected to the earth and to an elevated terminal, of sufficiently-high electromotive force to render elevated air strata conducting, causing thereby impulses to be propagated through the 60 air strata, collecting or receiving the energy of such impulses, at a point distant from the generating-station, by means of a receiving-circuit connected to the earth and to an elevated terminal, and utilizing the energy so 65 received to energize a secondary circuit of low potential surrounding the receiving-circuit.

                    NIKOLA TESLA.

Witnesses:
    M. LAWSON DYER,
    G. W. MARTLING.

No. 787,412.

PATENTED APR. 18, 1905.

N. TESLA.

ART OF TRANSMITTING ELECTRICAL ENERGY THROUGH THE NATURAL
MEDIUMS.

APPLICATION FILED MAY 16, 1900. RENEWED JUNE 17, 1902.

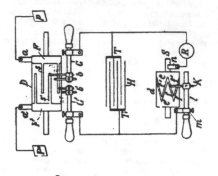

Fig.2

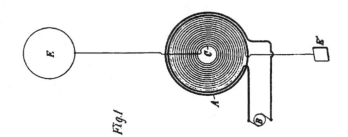

Fig.1

Witnesses:
Raphael Netter
M. Lawson Dyer.

Nikola Tesla *Inventor*
by Ken, Page & Cooper *Att'ys*

308

No. 787,412.   Patented April 18, 1905.

UNITED STATES PATENT OFFICE.

NIKOLA TESLA, OF NEW YORK, N. Y.

ART OF TRANSMITTING ELECTRICAL ENERGY THROUGH THE NATURAL MEDIUMS.

SPECIFICATION forming part of Letters Patent No. 787,412, dated April 18, 1905.

Application filed May 16, 1900. Renewed June 17, 1902. Serial No. 112,034.

To all whom it may concern:

Be it known that I, NIKOLA TESLA, a citizen of the United States, residing in the borough of Manhattan, in the city, county, and State
5 of New York, have discovered a new and useful Improvement in the Art of Transmitting Electrical Energy Through the Natural Media, of which the following is a specification, reference being had to the drawings accompanying
10 and forming a part of the same.

It is known since a long time that electric currents may be propagated through the earth, and this knowledge has been utilized in many ways in the transmission of signals
15 and the operation of a variety of receiving devices remote from the source of energy, mainly with the object of dispensing with a return conducting-wire. It is also known that electrical disturbances may be transmitted
20 through portions of the earth by grounding only one of the poles of the source, and this fact I have made use of in systems which I have devised for the purposes of transmitting through the natural media intelligible signals
25 or power and which are now familiar; but all experiments and observations heretofore made have tended to confirm the opinion held by the majority of scientific men that the earth, owing to its immense extent, although pos-
30 sessing conducting properties, does not behave in the manner of a conductor of limited dimensions with respect to the disturbances produced, but, on the contrary, much like a vast reservoir or ocean, which while it may be
35 locally disturbed by a commotion of some kind remains unresponsive and quiescent in a large part or as a whole. Still another fact now of common knowledge is that when electrical waves or oscillations are impressed upon
40 such a conducting-path as a metallic wire reflection takes place under certain conditions from the ends of the wire, and in consequence of the interference of the impressed and reflected oscillations the phenomenon of "sta-
45 tionary waves" with maxima and minima in definite fixed positions is produced. In any case the existence of these waves indicates that some of the outgoing waves have reached the boundaries of the conducting-path and have
50 been reflected from the same. Now I have discovered that notwithstanding its vast dimensions and contrary to all observations heretofore made the terrestrial globe may in a large part or as a whole behave toward disturbances impressed upon it in the same man- 55 ner as a conductor of limited size, this fact being demonstrated by novel phenomena, which I shall hereinafter describe.

In the course of certain investigations which I carried on for the purpose of studying 60 the effects of lightning discharges upon the electrical condition of the earth I observed that sensitive receiving instruments arranged so as to be capable of responding to electrical disturbances created by the discharges at 65 times failed to respond when they should have done so, and upon inquiring into the causes of this unexpected behavior I discovered it to be due to the character of the electrical waves which were produced in the earth by the 70 lightning discharges and which had nodal regions following at definite distances the shifting source of the disturbances. From data obtained in a large number of observations of the maxima and minima of these waves I 75 found their length to vary approximately from twenty-five to seventy kilometers, and these results and certain theoretical deductions led me to the conclusion that waves of this kind may be propagated in all directions 80 over the globe and that they may be of still more widely differing lengths, the extreme limits being imposed by the physical dimensions and properties of the earth. Recognizing in the existence of these waves an unmistakable evi- 85 dence that the disturbances created had been conducted from their origin to the most remote portions of the globe and had been thence reflected, I conceived the idea of producing such waves in the earth by artificial 90 means with the object of utilizing them for many useful purposes for which they are or might be found applicable. This problem was rendered extremely difficult owing to the immense dimensions of the planet, and conse- 95 quently enormous movement of electricity or rate at which electrical energy had to be delivered in order to approximate, even in a remote degree, movements or rates which are manifestly attained in the displays of elec- 100

309

trical forces in nature and which seemed at first unrealizable by any human agencies; but by gradual and continuous improvements of a generator of electrical oscillations, which I have described in my Patents Nos. 645,576 and 649,621, I finally succeeded in reaching electrical movements or rates of delivery of electrical energy not only approximating, but, as shown in many comparative tests and measurements, actually surpassing those of lightning discharges, and by means of this apparatus I have found it possible to reproduce whenever desired phenomena in the earth the same as or similar to those due to such discharges. With the knowledge of the phenomena discovered by me and the means at command for accomplishing these results I am enabled not only to carry out many operations by the use of known instruments, but also to offer a solution for many important problems involving the operation or control of remote devices which for want of this knowledge and the absence of these means have heretofore been entirely impossible. For example, by the use of such a generator of stationary waves and receiving apparatus properly placed and adjusted in any other locality, however remote, it is practicable to transmit intelligible signals or to control or actuate at will any one or all of such apparatus for many other important and valuable purposes, as for indicating wherever desired the correct time of an observatory or for ascertaining the relative position of a body or distance of the same with reference to a given point or for determining the course of a moving object, such as a vessel at sea, the distance traversed by the same or its speed, or for producing many other useful effects at a distance dependent on the intensity, wave length, direction or velocity of movement, or other feature or property of disturbances of this character.

I shall typically illustrate the manner of applying my discovery by describing one of the specific uses of the same—namely, the transmission of intelligible signals or messages between distant points—and with this object reference is now made to the accompanying drawings, in which—

Figure 1 represents diagrammatically the generator which produces stationary waves in the earth, and Fig. 2 an apparatus situated in a remote locality for recording the effects of these waves.

In Fig. 1, A designates a primary coil forming part of a transformer and consisting generally of a few turns of a stout cable of inappreciable resistance, the ends of which are connected to the terminals of a source of powerful electrical oscillations, diagrammatically represented by B. This source is usually a condenser charged to a high potential and discharged in rapid succession through the primary, as in a type of transformer invented by me and not well known; but when it is desired to produce stationary waves of great lengths an alternating dynamo of suitable construction may be used to energize the primary A. C is a spirally-wound secondary coil within the primary having the end nearer to the latter connected to the ground E′ and the other end to an elevated terminal E. The physical constants of coil C, determining its period of vibration, are so chosen and adjusted that the secondary system E′ C E is in the closest possible resonance with the oscillations impressed upon it by the primary A. It is, moreover, of the greatest importance in order to still further enhance the rise of pressure and to increase the electrical movement in the secondary system that its resistance be as small as practicable and its self-induction as large as possible under the conditions imposed. The ground should be made with great care, with the object of reducing its resistance. Instead of being directly grounded, as indicated, the coil C may be joined in series or otherwise to the primary A, in which case the latter will be connected to the plate E′; but be it that none or a part or all of the primary or exciting turns are included in the coil C the total length of the conductor from the ground-plate E′ to the elevated terminal E should be equal to one-quarter of the wave length of the electrical disturbance in the system E′ C E or else equal to that length multiplied by an odd number. This relation being observed, the terminal E will be made to coincide with the points of maximum pressure in the secondary or excited circuit, and the greatest flow of electricity will take place in the same. In order to magnify the electrical movement in the secondary as much as possible, it is essential that its inductive connection with the primary A should not be very intimate, as in ordinary transformers, but loose, so as to permit free oscillation—that is to say, their mutual induction should be small. The spiral form of coil C secures this advantage, while the turns near the primary A are subjected to a strong inductive action and develop a high initial electromotive force. These adjustments and relations being carefully completed and other constructive features indicated rigorously observed, the electrical movement produced in the secondary system by the inductive action of the primary A will be enormously magnified, the increase being directly proportionate to the inductance and frequency and inversely to the resistance of the secondary system. I have found it practicable to produce in this manner an electrical movement thousands of times greater than the initial—that is, the one impressed upon the secondary by the primary A—and I have thus reached activities or rates of flow of electrical energy in the system E′ C E measured by many tens of thousands of horsepower. Such immense movements of electricity give rise to a variety of novel and striking phenomena, among which are those already described. The powerful electrical oscillations in the system E′ C E being communicated to the ground cause corresponding vibrations to be propagated to distant parts of the globe, whence they are reflected and by interference with the outgoing vibrations produce stationary waves the crests and hollows of which lie in parallel circles relatively to which the ground-plate E′ may be considered to be the pole. Stated otherwise, the terrestrial conductor is thrown into resonance with the oscillations impressed upon it just like a wire. More than this, a number of facts ascertained by me clearly show that the movement of electricity through it follows certain laws with nearly mathematical rigor. For the present it will be sufficient to state that the planet behaves like a perfectly smooth or polished conductor of inappreciable resistance with capacity and self induction uniformly distributed along the axis of symmetry of wave propagation and transmitting slow electrical oscillations without sensible distortion and attenuation.

Besides the above three requirements seem to be essential to the establishment of the resonating condition.

First. The earth's diameter passing through the pole should be an odd multiple of the quarter wave length—that is, of the ratio between the velocity of light—and four times the frequency of the currents.

Second. It is necessary to employ oscillations in which the rate of radiation of energy into space in the form of hertzian or electromagnetic waves is very small. To give an idea, I would say that the frequency should be smaller than twenty thousand per second, though shorter waves might be practicable. The lowest frequency would appear to be six per second, in which case there will be but one node, at or near the ground-plate, and, paradoxical as it may seem, the effect will increase with the distance and will be greatest in a region diametrically opposite the transmitter. With oscillations still slower the earth, strictly speaking, will not resonate, but simply act as a capacity, and the variation of potential will be more or less uniform over its entire surface.

Third. The most essential requirement is, however, that irrespective of frequency the wave or wave-train should continue for a certain interval of time, which I have estimated to be not less than one-twelfth or probably 0.08484 of a second and which is taken in passing to and returning from the region diametrically opposite the pole over the earth's surface with a mean velocity of about four hundred and seventy-one thousand two hundred and forty kilometers per second.

The presence of the stationary waves may be detected in many ways. For instance, a circuit may be connected directly or inductively to the ground and to an elevated terminal and tuned to respond more effectively to the oscillations. Another way is to connect a tuned circuit to the ground at two points lying more or less in a meridian passing through the pole E′ or, generally stated, to any two points of a different potential.

In Fig. 2 I have shown a device for detecting the presence of the waves such as I have used in a novel method of magnifying feeble effects which I have described in my Patents Nos. 685,953 and 685,955. It consists of a cylinder D, of insulating material, which is moved at a uniform rate of speed by clockwork or other suitable motive power and is provided with two metal rings F F′, upon which bear brushes a and a′, connected, respectively, to the terminal plates P and P′. From the rings F F′ extend narrow metallic segments s and s′, which by the rotation of the cylinder D are brought alternately into contact with double brushes b and b′, carried by and in contact with conducting-holders h and h′, supported in metallic bearings G and G′, as shown. The latter are connected to the terminals T and T′ of a condenser H, and it should be understood that they are capable of angular displacement as ordinary brush-supports. The object of using two brushes, as b and b′, in each of the holders h and h′ is to vary at will the duration of the electric contact of the plates P and P′ with the terminals T and T′, to which is connected a receiving-circuit including a receiver R and a device d, performing the duty of closing the receiving-circuit at predetermined intervals of time and discharging the stored energy through the receiver. In the present case this device consists of a cylinder made partly of conducting and partly of insulating material e and e′, respectively, which is rotated at the desired rate of speed by any suitable means. The conducting part e is in good electrical connection with the shaft S and is provided with tapering segments f f′, upon which slides a brush k, supported on a conducting-rod l, capable of longitudinal adjustment in a metallic support m. Another brush, n, is arranged to bear upon the shaft S, and it will be seen that whenever one of the segments f′ comes in contact with the brush k the circuit including the receiver R is completed and the condenser discharged through the same. By an adjustment of the speed or rotation of the cylinder d and a displacement of the brush k along the cylinder the circuit may be made to open and close in as rapid succession and remain open or closed during such intervals of time as may be desired. The plates P and P′, through which the electrical energy is conveyed to the brushes a and a′, may be at a considerable distance from each other in the ground or one in the ground and the other in the air, preferably at some height. If but one plate is connected to earth and the other maintained at an elevation, the location of the apparatus must be determined with reference to the position of the stationary waves established by the generator, the effect evidently being greatest in a maximum and zero in a nodal region. On the other hand, if both plates be connected to earth the points of connection must be selected with reference to the difference of potential which it is desired to secure, the strongest effect being of course obtained when the plates are at a distance equal to half the wave length.

In illustration of the operation of the system let it be assumed that alternating electrical impulses from the generator are caused to produce stationary waves in the earth, as above described, and that the receiving apparatus is properly located with reference to the position of the nodal and ventral regions of the waves. The speed of rotation of the cylinder D is varied until it is made to turn in synchronism with the alternate impulses of the generator, and the position of the brushes *b* and *b'* is adjusted by angular displacement or otherwise, so that they are in contact with the segments S and S' during the periods when the impulses are at or near the maximum of their intensity. These requirements being fulfilled, electrical charges of the same sign will be conveyed to each of the terminals of the condenser, and with each fresh impulse it will be charged to a higher potential. The speed of rotation of the cylinder *d* being adjustable at will, the energy of any number of separate impulses may thus be accumulated in potential form and discharged through the receiver R upon the brush *k* coming in contact with one of the segments *f'*. It will be understood that the capacity of the condenser should be such as to allow the storing of a much greater amount of energy than is required for the ordinary operation of the receiver. Since by this method a relatively great amount of energy and in a suitable form may be made available for the operation of a receiver, the latter need not be very sensitive; but when the impulses are very weak or when it is desired to operate a receiver very rapidly any of the well-known sensitive devices capable of responding to very feeble influences may be used in the manner indicated or in other ways. Under the conditions described it is evident that during the continuance of the stationary waves the receiver will be acted upon by current impulses more or less intense, according to its location with reference to the maxima and minima of said waves; but upon interrupting or reducing the flow of the current the stationary waves will disappear or diminish in intensity. Hence a great variety of effects may be produced in a receiver, according to the mode in which the waves are controlled. It is practicable, however, to shift the nodal and ventral regions of the waves at will from the sending-station, as by

varying the length of the waves under observance of the above requirements. In this manner the regions of maximum and minimum effect may be made to coincide with any receiving station or stations. By impressing upon the earth two or more oscillations of different wave length a resultant stationary wave may be made to travel slowly over the globe, and thus a great variety of useful effects may be produced. Evidently the course of a vessel may be easily determined without the use of a compass, as by a circuit connected to the earth at two points, for the effect exerted upon the circuit will be greatest when the plates P P' are lying on a meridian passing through ground-plate E' and will be *nil* when the plates are located at a parallel circle. If the nodal and ventral regions are maintained in fixed positions, the speed of a vessel carrying a receiving apparatus may be exactly computed from observations of the maxima and minima regions successively traversed. This will be understood when it is stated that the projections of all the nodes and loops on the earth's diameter passing through the pole or axis of symmetry of the wave movement are all equal. Hence in any region at the surface the wave length can be ascertained from simple rules of geometry. Conversely, knowing the wave length, the distance from the source can be readily calculated. In like ways the distance of one point from another, the latitude and longitude, the hour, &c., may be determined from the observation of such stationary waves. If several such generators of stationary waves, preferably of different length, were installed in judiciously-selected localities, the entire globe could be subdivided in definite zones of electric activity, and such and other important data could be at once obtained by simple calculation or readings from suitably-graduated instruments. Many other useful applications of my discovery will suggest themselves, and in this respect I do not wish to limit myself. Thus the specific plan herein described of producing the stationary waves might be departed from. For example, the circuit which impresses the powerful oscillations upon the earth might be connected to the latter at two points. In this application I have advanced various improvements in means and methods of producing and utilizing electrical effects which either in connection with my present discovery or independently of the same may be usefully applied.

I desire it to be understood that such novel features as are not herein specifically claimed will form the subjects of subsequent applications.

What I now claim is—

1. The improvement in the art of transmitting electrical energy to a distance which consists in establishing stationary electrical waves in the earth, as set forth.

2. The improvement in the art of transmit-

787,412

ting electrical energy to a distance which consists in impressing upon the earth electrical oscillations of such character as to produce stationary electrical waves therein, as set
5 forth.

3. The improvement in the art of transmitting and utilizing electrical energy which consists in establishing stationary electrical waves in the natural conducting media, and operat-
10 ing thereby one or more receiving devices remote from the source of energy, as set forth.

4. The improvement in the art of transmitting and utilizing electrical energy which consists in establishing in the natural conducting
15 media, stationary electrical waves of predetermined length and operating thereby one or more receiving devices remote from the source of energy and properly located with respect

to the position of such waves, as herein set
forth.   20

5. The improvement in the art of transmitting and utilizing electrical energy, which consists in establishing in the natural conducting media, stationary electrical waves, and varying the length of such waves, as herein set 25
forth.

6. The improvement in the art of transmitting and utilizing electrical energy, which consists in establishing in the natural conducting media stationary electrical waves and shifting 30
the nodal and ventral regions of these waves, as described.

NIKOLA TESLA.

Witnesses:
  M. Lawson Dyer,
  Benjamin Miller.

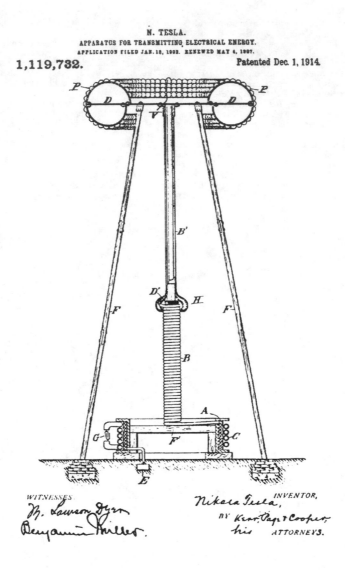

N. TESLA.
APPARATUS FOR TRANSMITTING ELECTRICAL ENERGY.
APPLICATION FILED JAN. 18, 1902. RENEWED MAY 4, 1907.

1,119,732.

Patented Dec. 1, 1914.

UNITED STATES PATENT OFFICE.

NIKOLA TESLA, OF NEW YORK, N. Y.

APPARATUS FOR TRANSMITTING ELECTRICAL ENERGY.

1,119,732.   Specification of Letters Patent.   **Patented Dec. 1, 1914.**

Application filed January 18, 1902. Serial No. 90,245. Renewed May 4, 1907. Serial No. 371,817.

To all whom it may concern:

Be it known that I, NIKOLA TESLA, a citizen of the United States, residing in the borough of Manhattan, in the city, county, and State of New York, have invented certain new and useful Improvements in Apparatus for Transmitting Electrical Energy, of which the following is a specification, reference being had to the drawing accompanying and forming a part of the same.

In endeavoring to adapt currents or discharges of very high tension to various valuable uses, as the distribution of energy through wires from central plants to distant places of consumption, or the transmission of powerful disturbances to great distances, through the natural or non-artificial media, I have encountered difficulties in confining considerable amounts of electricity to the conductors and preventing its leakage over their supports, or its escape into the ambient air, which always takes place when the electric surface density reaches a certain value. The intensity of the effect of a transmitting circuit with a free or elevated terminal is proportionate to the quantity of electricity displaced, which is determined by the product of the capacity of the circuit, the pressure, and the frequency of the currents employed. To produce an electrical movement of the required magnitude it is desirable to charge the terminal as highly as possible, for while a great quantity of electricity may also be displaced by a large capacity charged to low pressure, there are disadvantages met with in many cases when the former is made too large. The chief of these are due to the fact that an increase of the capacity entails a lowering of the frequency of the impulses or discharges and a diminution of the energy of vibration. This will be understood when it is borne in mind, that a circuit with a large capacity behaves as a slackspring, whereas one with a small capacity acts like a stiff spring, vibrating more vigorously. Therefore, in order to attain the highest possible frequency, which for certain purposes is advantageous and, apart from that, to develop the greatest energy in such a transmitting circuit, I employ a terminal of relatively small capacity, which I charge to as high a pressure as practicable. To accomplish this result I have found it imperative to so construct the elevated conductor, that its outer surface, on which the electrical charge chiefly accumulates, has itself a large radius of curvature, or is composed of separate elements which, irrespective of their own radius of curvature, are arranged in close proximity to each other and so, that the outside ideal surface enveloping them is of a large radius. Evidently, the smaller the radius of curvature the greater, for a given electric displacement, will be the surface-density and, consequently, the lower the limiting pressure to which the terminal may be charged without electricity escaping into the air. Such a terminal I secure to an insulating support entering more or less into its interior, and I likewise connect the circuit to it inside or, generally, at points where the electric density is small. This plan of constructing and supporting a highly charged conductor I have found to be of great practical importance, and it may be usefully applied in many ways.

Referring to the accompanying drawing, the figure is a view in elevation and part section of an improved free terminal and circuit of large surface with supporting structure and generating apparatus.

The terminal D consists of a suitably shaped metallic frame, in this case a ring of nearly circular cross section, which is covered with half spherical metal plates P P, thus constituting a very large conducting surface, smooth on all places where the electric charge principally accumulates. The frame is carried by a strong platform expressly provided for safety appliances, instruments of observation, etc., which in turn rests on insulating supports F F. These should penetrate far into the hollow space formed by the terminal, and if the electric density at the points where they are bolted to the frame is still considerable, they may be specially protected by conducting hoods as H.

A part of the improvements which form the subject of this specification, the transmitting circuit, in its general features, is identical with that described and claimed in my original Patents Nos. 645,576 and 649,621. The circuit comprises a coil A which is in close inductive relation with a primary C, and one end of which is connected to a ground-plate E, while its other end is led through a separate self-induction coil B and a metallic cylinder B' to the terminal D.

The connection to the latter should always be made at, or near the center, in order to secure a symmetrical distribution of the current. as otherwise. when the frequency is
5 very high and the flow of large volume, the performance of the apparatus might be impaired. The primary C may be excited in any desired manner, from a suitable source of currents G, which may be an alternator
10 or condenser, the important requirement being that the resonant condition is established, that is to say, that the terminal D is charged to the maximum pressure developed in the circuit, as I have specified in my
15 original patents before referred to. The adjustments should be made with particular care when the transmitter is one of great power, not only on account of economy, but also in order to avoid danger. I have shown
20 that it is practicable to produce in a resonating circuit as E A B B' D immense electrical activities, measured by tens and even hundreds of thousands of horse-power, and in such a case, if the points of maximum
25 pressure should be shifted below the terminal D, along coil B, a ball of fire might break out and destroy the support F or anything else in the way. For the better appreciation of the nature of this danger it
30 should be stated, that the destructive action may take place with inconceivable violence. This will cease to be surprising when it is borne in mind, that the entire energy accumulated in the excited circuit, instead of re-
35 quiring, as under normal working conditions, one quarter of the period or more for its transformation from static to kinetic form, may spend itself in an incomparably smaller interval of time, at a rate of many
40 millions of horse power. The accident is apt to occur when, the transmitting circuit being strongly excited, the impressed oscillations upon it are caused, in any manner more or less sudden, to be more rapid than
45 the free oscillations. It is therefore advisable to begin the adjustments with feeble and somewhat slower impressed oscillations, strengthening and quickening them gradually, until the apparatus has been brought
50 under perfect control. To increase the safety, I provide on a convenient place, preferably on terminal D, one or more elements or plates either of somewhat smaller radius of curvature or protruding more or less be-
55 yond the others (in which case they may be of larger radius of curvature) so that, should the pressure rise to a value, beyond which it is not desired to go, the powerful discharge may dart out there and lose itself harmlessly
60 in the air. Such a plate, performing a function similar to that of a safety valve on a high pressure reservoir, is indicated at V.
Still further extending the principles underlying my invention, special reference
65 is made to coil B and conductor B'. The

latter is in the form of a cylinder with smooth or polished surface of a radius much larger than that of the half spherical elements P P, and widens out at the bottom
70 into a hood H, which should be slotted to avoid loss by eddy currents and the purpose of which will be clear from the foregoing. The coil B is wound on a frame or drum D¹ of insulating material, with its
75 turns close together. I have discovered that when so wound the effect of the small radius of curvature of the wire itself is overcome and the coil behaves as a conductor of large radius of curvature, corresponding to that
80 of the drum. This feature is of considerable practical importance and is applicable not only in this special instance, but generally. For example, such plates at P P of terminal D, though preferably of large
85 radius of curvature, need not be necessarily so, for provided only that the individual plates or elements of a high potential conductor or terminal are arranged in proximity to each other and with their outer
90 boundaries along an ideal symmetrical enveloping surface of a large radius of curvature, the advantages of the invention will be more or less fully realized. The lower end of the coil B—which, if desired, may
95 be extended up to the terminal D—should be somewhat below the uppermost turn of coil A. This, I find, lessens the tendency of the charge to break out from the wire connecting both and to pass along the sup-
100 port F'.
Having described my invention, I claim:
1. As a means for producing great electrical activities a resonant circuit having its outer conducting boundaries, which are
105 charged to a high potential, arranged in surfaces of large radii of curvature so as to prevent leakage of the oscillating charge, substantially as set forth.
2. In apparatus for the transmission of
110 electrical energy a circuit connected to ground and to an elevated terminal and having its outer conducting boundaries, which are subject to high tension, arranged in surfaces of large radii of curvature sub-
115 stantially as, and for the purpose described.
3. In a plant for the transmission of electrical energy without wires, in combination with a primary or exciting circuit a secondary connected to ground and to an elevated
120 terminal and having its outer conducting boundaries, which are charged to a high potential, arranged in surfaces of large radii of curvature for the purpose of preventing leakage and loss of energy, substantially as
125 set forth.
4. As a means for transmitting electrical energy to a distance through the natural media a grounded resonant circuit, comprising a part upon which oscillations are
130 impressed and another for raising the ten-

sion, having its outer conducting boundaries on which a high tension charge accumulates arranged in surfaces of large radii of curvature, substantially as described.

5   5. The means for producing excessive electric potentials consisting of a primary exciting circuit and a resonant secondary having its outer conducting elements which are subject to high tension arranged in prox-
10 imity to each other and in surfaces of large radii of curvature so as to prevent leakage of the charge and attendant lowering of potential, substantially as described.

  6. A circuit comprising a part upon which
15 oscillations are impressed and another part for raising the tension by resonance, the latter part being supported on places of low electric density and having its outermost conducting boundaries arranged in surfaces
20 of large radii of curvature, as set forth.

  7. In apparatus for the transmission of electrical energy without wires a grounded circuit the outer conducting elements of which have a great aggregate area and are
25 arranged in surfaces of large radii of curvature so as to permit the storing of a high charge at a small electric density and prevent loss through leakage, substantially as described.

  8. A wireless transmitter comprising in 30 combination a source of oscillations as a condenser, a primary exciting circuit and a secondary grounded and elevated conductor the outer conducting boundaries of which are in proximity to each other and arranged 35 in surfaces of large radii of curvature, substantially as described.

  9. In apparatus for the transmission of electrical energy without wires an elevated conductor or antenna having its outer high 40 potential conducting or capacity elements arranged in proximity to each other and in surfaces of large radii of curvature so as to overcome the effect of the small radius of curvature of the individual elements and 45 leakage of the charge, as set forth.

  10. A grounded resonant transmitting circuit having its outer conducting boundaries arranged in surfaces of large radii of curvature in combination with an ele- 50 vated terminal of great surface supported at points of low electric density, substantially as described.

NIKOLA TESLA.

Witnesses:
  M. Lamson Dyer,
  Richard Donovan.

19 Bibliography of Tesla Resources

Wizard: The Life and Times of Nikola Tesla
Dr. Marc Siefer

Lightning in His Hand: The Life Story of Nikola Tesla
Inez Hunt & W. Draper

My Inventions : The Autobiography of Nikola Tesla
Nikola Tesla; Ben Johnston, Editor

Tesla
Tad Wise

Inventions, Researches and Writings of Nikola Tesla (hardcover)
Nikola Tesla; Thomas Commerford Martin, Editor

Nikola Tesla: Lecture Before The New York Academy of Sciences, April 6, 1897
Tesla Presents Series Part 2
Nikola Tesla; Leland I. Anderson, Editor

High Frequency Oscillators for Electro-therapeutic and Other Purposes
Nikola Tesla

Nikola Tesla On His Work With Alternating Currents and Their Application to Wireless
    Telegraphy, Telephony, and Transmission of Power
Tesla Presents Series Part 1
Nikola Tesla; Leland I. Anderson, Editor

Nikola Tesla: Guided Weapons & Computer Technology (hardcover & paperback)
Tesla Presents Series Part 3
Nikola Tesla; Leland I. Anderson, Editor

Nikola Tesla's Teleforce & Telegeodynamics Proposals
Limited Edition
Tesla Presents Series Part 4

Nikola Tesla; Leland I. Anderson, Editor

Colorado Springs Notes, 1899-1900 (hardcover)
Nikola Tesla; Commentary by Aleksandar Marincic

Dr. Nikola Tesla - Complete Patents
Nikola Tesla; Compiled by John T. Ratzlaff

Dr. Nikola Tesla - Selected Patent Wrappers, 4 volume set (spiral)
Nikola Tesla; Compiled by John T. Ratzlaff

Tesla Said
Nikola Tesla; Compiled by John T. Ratzlaff

Solutions to Tesla's Secrets
Nikola Tesla; Compiled by John T. Ratzlaff
Thomas Bearden

The Transmission of Electric Energy Without Wires
Nikola Tesla

Fantastic Inventions of Nikola Tesla
Nikola Tesla & David Childress

Nikola Tesla On His Work With Alternating Currents and Their Application
to Wireless Telegraphy, Telephony, and Transmission of Power
Tesla Presents Series, Part 1.
Leland I. Anderson

Nikola Tesla: Lecture Before the New York Academy of Sciences - April 6,
1897 : The Streams of Lenard and Roentgen and Novel Apparatus for Their
Production
Tesla Presents Series, Part 2
Leland I. Anderson
xix, 123 pages, illustrated, indexed.
ISBN 1-893817-01-6 $12.95

Nikola Tesla: Guided Weapons & Computer Technology
Tesla Presents Series, Part 3
Leland I. Anderson, ed.
xv, 241 pages, illustrated, indexed, hardcover & paperback.
ISBN 0-9636012-5-3, 0-9636012-9-6

Nikola Tesla's Teleforce & Telegeodynamics Proposals
Tesla Presents Series, Part 4
viii, 119 pages, illustrated, indexed.
ISBN 0-9636012-8-8
Tesla Coil Builder's Guide to the Colorado Springs Notes of Nikola Tesla
Second Edition

Richard Hull
xxvii, 203 pages, well illustrated, spiral.
ISBN 0-9636012-2-9

Dr. Nikola Tesla Bibliography
John T. Ratzlaff and Leland I. Anderson
250 pages, spiral.
ISBN 0-9636012-6-1

PRIORITY IN THE INVENTION OF RADIO -- TESLA VS. MARCONI. Leland I. Anderson, 9 pages, 8 illustrations, 17 references, 5" X 8".

JOHN STONE STONE ON NIKOLA TESLA'S PRIORITY IN RADIO AND CONTINUOUS-WAVE RADIOFREQUENCY APPARATUS. Leland Anderson, 24 pages, illustrated, 5" X 8".

BALL LIGHTNING & TESLA'S ELECTRIC FIREBALLS. Leland I. Anderson, 31 pages, illustrated, 8" X 11".

THE TESLA BLADELESS TURBINE. Nikola Tesla, C.R. Possell, vi, 24 pages, illustrated, 8" X 11".

TESLA OSCILLATOR AND FLUORESCENT TUBE DRIVER. Gary Peterson, 17 pages, illustrated, 8" X 11".

TESLA'S FUELLESS GENERATOR AND WIRELESS METHOD. Oliver Nichelson, v, 48 pages, illustrated, 8" X 11", ISBN 0-9636012-0-2.

HIGH FREQUENCY OSCILLATORS FOR ELECTRO-THERAPEUTIC AND OTHER PURPOSES. Nikola Tesla, 12 pages

THE TRANSMISSION OF ELECTRIC ENERGY WITHOUT WIRES (reprinted article/brochure). Nikola Tesla, 7 pages

NIKOLA TESLA'S RESIDENCES, LABORATORIES, AND OFFICES, (list). Leland Anderson, 11 pages, illustrated,

TESLA TURBINE REFERENCES, (bibliographic list). Leland Anderson & Warren Rice, 23 pages,

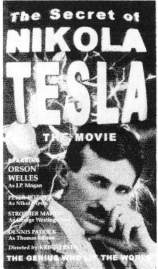

Tesla videos

The Secret of Nikola Tesla: The Movie

Tesla: Master of Lightning

Tesla websites for books, information and tapes

http://www.tfcbooks.com

http://www.teslamemorialsociety.org/videos.htm

http://www.netsense.net/tesla/

http://www.yurope.com/org/tesla/

http://www.nickf.com/tesla.htm

http://www.neuronet.pitt.edu/~bogdan/tesla/index.htm

http://www.mall-usa.com/BPCS/tesla.html

http://www.pbs.org/tesla

20   Glossary of Tesla Terms

Excerpt adapted from "Project Insight: A Study of Tesla's Advanced Concepts," H. W. Jones, *Proceedings of the Tesla Centennial Symposium*, 1984

Advanced Concepts – Those ideas which Tesla was known to have conceived and developed to some extent, but did not pursue to fruition because of lack of funding and laboratory facilities. The more dramatic of these concepts were: free energy, wireless transmission of energy, employment of scalar technology, non-Hertzian waves, the Tesla shield, the Tesla ray.

Ball Lightning – A form of lightning in which a slow-moving, extremely high temperature sphere forms. Only rarely seen in nature, but producible by artificial means. Currently being used in the study of harnessing fusion energy for commercial use. See Robert Golka articles.

DeBroglie Wave – The quantum mechanics wave associated with a particle of matter which can theoretically give rise to intra-atomic interference effects. In his speech accepting the Nobel Prize, DeBroglie emphasized that these waves are real and must not be regarded simply as mathematical oddities or conveniences.

Electromagnetic Pulse (EMP) – A sharp pulse of energy and electromagnetic radiation occurring when an explosion occurs in an unsymmetrical envrionment. Tesla theorized that an EMP would result when two longitudinal scalar electrostatic potential wave patterns met and coupled into a flash of vector electromagnetic energy. See early Tom Bearden articles.

Electromagnetic Theory (EM) – Conventional electrical theory currently taught in our educational system, mainly giving credit only to Hertz, Maxwell, and Faraday. Tesla's work challenged the adequacy of existing EM theory, as do many physicists today. EM theory is only good for "far field" EM waves, as only electrical engineering EM textbooks (e.g. Magid) will admit. The recommended physical perspective, is to ask whether we are within the "near field," i.e., <u>within the first couple of wavelengths</u>. In this region, a capacitively-created EM wave will still retain mainly electrical characteristics and can be stopped by a Faraday cage, however, a inductively-created EM wave still still retain mainly magnetic characteristics and go right through even the most expensive Faraday cage (made with Mu Metal) such as the quarter-million dollar one at Wright-Patterson AFB. Especially when dealing with extremely low frequency (ELF), most staunch EM theorists are stymied because we are always within the near field with ELF waves. See Thomas Phipps, *Heretical Verities* book.

Electromagnetic Wave – A Hertzian wave. A wave that oscillates transversely rather than longitudinally, having electric (E) field and magnetic (B) field effects (each may be detected). If two sine waves are pictured, perpendicular to each other, one on the x-y plane (vertical) and the other on the x-z plane (horizontal), both traveling in the x-direction, the E-field will be designated by the x-y plane wave (if it is polarized light) and the B-field will be the x-z plane wave. Polaroid® sunglasses work because 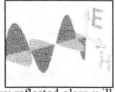 they only let the E-field light through if it is in the x-y plane, whereas any reflected glare will have the E-field oscillating in the x-z plane (which is horizontal). Non-Hertzian waves are not transverse and often occur because near field, distorted waves are created in the experiment.

Energy – The capacity to do work, which is the result of a force moving a mass through a distance. Measured in "joules" it is the timeless version of power times time, such as kilowatt-hours (kWh). A energy conversion example is: 1 watt-hour = 3.4 British thermal unit (Btu), which is used as a measure of heat energy. See zero-point energy.

Ether – (also aether) Simply stated, it is the same as the physical vacuum. This differs from the common understanding of empty space, since theorists regard the ether (and the physical vacuum) as having substance (and particles in negative energy states). With the Silvertooth experiments, now showing a preference of direction for the old Michelson-Morley type of experiment, the ether is coming back into vogue. Very compatible with Eastern mysticism.

Free Energy – Energy which is free. Often confused with perpetual motion, free energy has three aspects: 1) no cost for input; 2) plentiful and inexhaustible; 3) one-time capital expenditure. Renewable energy is free energy. Zero point energy (ZPE) is also free and equated with the ether. A prophetic endorsement for free energy comes from Tesla's comment that "it is a mere question of time when men will succeed in attaching their machinery to the very wheelwork of nature."[1] He implies, what was discovered years later, that in open systems, it appears that energy is not conserved (e.g. ZPE vacuum fluctuations). However, in closed systems, we know that the second law of thermodynamics and energy conservation laws apply. Compare with the physics definition $G = H - TS$ where G is "Gibb's free energy," H is enthalpy, T is absolute temperature, and S is entropy. In words, free energy is the internal energy of a system minus the product of its temperature and its entropy.

Gravity – The phenomenon characterized by the physical attraction of any two material bodies, defined as the product of the masses divided by the square of the distance between them. Today, physicists are surprised to find evidence of antigravity in the accelerated expansion of the distant galaxies (*Science*, Dec. 18, 1998) which was called "the breakthrough of the year." However, Tesla talked about controlling gravity many times. The fact that gravity is always attractive and never repulsive is a curiosity that physicists have always wondered about. The fact that gravity has to travel many times faster than light speed to prevent aberrations has caused a lot of commotion.[2] NASA recently tested the Allais & Saxl experiments during a solar eclipse, showing that there is a shielding effect from the moon when aligned.[3] There are many accepted modalities for creating artificial gravity and antigravity with high energy electromagnetism.[4] See book, *Causality, Electromagnetic Induction and Gravity*, by Dr. Oleg Jefimenko; *Hunt for Zero Point* by N. Cook

---

[1] Nikola Tesla, addressing the American Institute of Electrical Engineers, 1891

[2] Tom Van Flandern, "The Speed of Gravity – What the Experiments Say," *Phys. Lett. A*, 250, 1998, 1-11; also in Future Energy: Proc. of COFE, 1999, IRI

[3] Saxl & Allais, "Observation of Periodic Phenomena with a Massive Torsion Pendulum," Report #702, Integrity Research Institute (IRI)

Longitudinal Wave – A pressure type of wave, similar to sound, in which the vibrations are along the direction of travel, a sequence of compressions and rarefactions. E and B fields are misaligned. Scalar waves are longitudinal, as contrasted with EM "Hertzian" waves which have transverse oscillations. Longitudinal waves are non-Hertzian as a result, as Tesla said many times, regarding his magnifying transmitter. The current density (or any vector field) can be split into transverse J_t and longitudinal J_l components.[5] The transverse or solenoidal current has $\blacktriangledown \cdot J_t = 0$ while longitudinal or irrotational current has $\blacktriangledown \times J_l = 0$. See Dr. Thomas Phipps book, *Heretical Verities*.

Radiant Energy – Term used by Tesla in his two 1901 patents, #685,957 and #685,958, indicating radiation of any kind. In this case, it indicated Roentgen rays or x-rays as they are called today. He intended the single end x-ray tube to operate at the top of a Tesla coil or preferably, the magnifying transmitter.

Scalar Field – In physics, each point in space for a particular potential is assigned a magnitude but no direction. The scalar potential is just the Coulomb potential due to a charge density ρ (x,t). This is the origin of the name "Coulomb gauge." Compare with the vector potential A created by a toroidal magnetic field that satisfies the inhomogeneous wave equation. While EM waves travel at light speed, "the scalar potential 'propagates' instantaneously everywhere in space."[6]

Scalar Wave – (see Longitudinal Wave.) Also Tesla Wave. An oscillating field of pure potential without E and B fields. Starting with $B = 0 = \blacktriangledown \times A$ and $E = 0 = -\blacktriangledown \varphi - 1/c\ (\partial A/\partial t)$ the solutions are $A = \blacktriangledown X$ and $\varphi = -1/c\ (\partial X/\partial t)$ where X is a scalar and obeys the wave equation. X is a scalar wave varying harmonically in time but only longitudinal fields exist because $A = iKX$ which shows that A points in the direction of travel. Normally, this is regarded as a gauge transformation but in quantum mechanics (e.g., Aharonov-Bohm experiment) it has real effects on the electron wave function. Because no energy or momentum transfer occurs, X fields can penetrate all objects and in fact can traverse the whole universe. Scalar waves thus may in fact, travel faster than light speed c, since no c-limited fields are involved.[7] See "Scalar Potentials Fields and Waves," Report #303, Integrity Research Institute.

Tesla Fireball – (see Ball Lightning.) A self-sustaining globe of radiant EM energy, exhibiting soliton behavior.

Tesla Ray – Forerunner of the laser, it was an EM device demonstrated by Tesla and offered to the British government in 1937 as a defense against the Luftwaffe threat. It was ridiculed and rejected. There is a variation is called the Tesla Death Ray, which was a particle beam weapon. See "Tesla, Man of Lightning" video www.pbs.org.

---

[4] "Antigravity Report: Collection of Seminal Articles for Futurists," Report #707, IRI
[5] J.D. Jackson, *Classical Electrodynamics*, Second edition, J. Wiley Pub., 1975. P.222
[6] Ibid., p. 223
[7] Dr. Jack Dea, "Instantaneous Interactions," *Proc. ITS, 1986*, p.4-34 and Raymond Gelinas, "Curl-Free Vector Potential Effects," p. 4-43

Transverse Wave – A standard Hertzian EM vector wave which oscillates laterally, as contrasted with a Tesla electrostatic scalar wave which vibrates longitudinally.

Vacuum – (See Ether.) A plenum which is filled with particles in negative energy states. Dr. Paul Dirac became a Nobel Prize winner for predicting the existence of the positron (antimatter electron with positive charge) after theorizing that under high voltage circumstances, and electron-positron pair can emerge, like magic, from the vacuum and go off in opposite directions. Such experiments (shown here with cloud chamber picture) have verified the vacuum is teaming with activity. See zero-point energy.

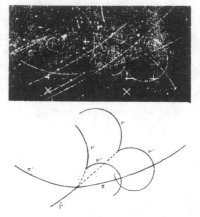

Vector – A force or field that has magnitude and direction, compared with scalar fields. EM waves are vector fields and contain momentum.

Wardenclyffe – Name of the first transmission tower in the world, erected 1901-3 in Shoreham, NY by Tesla which rose to a height of 187 feet. The Tesla Wardenclyffe Project, Inc. is a firm dedicated to recovering the property for a commemorative site. Contact CEO, Gary Petersen, POB 2001, Breckenridge, CO 80424

Zero-Point Energy – The energy of the vacuum that is sustained even at zero absolute temperature and no air (complete vacuum). This is the "very wheelwork of nature" and even implicated in the antigravity effect seen on the acceleration of distant galaxies (*Science*, Dec, 1998) The Casimir effect, experimentally verified, shows that virtual particles, as they emerge from the vacuum, also exert a measurable force. Zero-point energy (ZPE) has so many unusual characteristics that it forms the most intriguing field of study in physics today. It is predicted that, since the ZPE Casimir force already exhibits perpetual wavy motion already in nanotechnology (endless oscillations of nanostructures under tensile stress),[8] we may soon see a revolution in energy production that is fuelless, as Tesla predicted. Recently a vacuum energy transducer was theoretically designed to contain a complete engine cycle with electrical output. "Free energy" was also a phrase used by the JPL scientist, Dr. Pinto, in the abstract of his corresponding journal article.[9] See M. King books, *Tapping the Zero Point Energy* and *Quest for Zero Point Energy* or "Zero Point Energy and the Future" Report #822, IRI.

Further research information on all of the above-mentioned topics is available from the free *Future Energy* newsletter and catalog from Integrity Research Institute, 202-452-7674 or 800-295-7674

---

[8] Gu Hai-Cheng, et al., "Influence of Combination of Casimir Force and Residual Stress on the Behaviour of Micro- and Nano-Electromechanical Systems," *Chinese Phys. Lett.* June, 2002, p.832

[9] F. Pinto, "Engine cycle of an optically controlled vacuum energy transducer," *Physical Review B*, 60 (21) 14740, Dec. 1, 1999

21   **Index**

HARNESSING THE WHEELWORK OF NATURE
Tesla's Science of Energy
by Thomas Valone, Ph.D., P.E.
A compilation of essays, papers and technical briefings on the emerging Tesla Technology and Zero Point Energy engineering that will soon change the entire way we live. Chapters include: Tesla: Scientific Superman who Launched the Westinghouse Industrial Firm by John Shatlan; Nikola Tesla—Electricity's Hidden Genius, excerpt from The Search for Free Energy; Tesla's History at Niagara Falls; Non-Hertzian Waves: True Meaning of the Wireless Transmission of Power by Toby Grotz; On the Transmission of Electricity Without Wires by Nikola Tesla; Tesla's Magnifying Transmitter by Andrija Puharich; Tesla's Self-Sustaining Electrical Generator and the Ether by Oliver Nichelson; Self-Sustaining Non-Hertzian Longitudinal Waves by Dr. Robert Bass; Modification of Maxwell's Equations in Free Space; Scalar Electromagnetic Waves; Disclosures Concerning Tesla's Operation of an ELF Oscillator; A Study of Tesla's Advanced Concepts & Glossary of Tesla Technology Terms; Electric Weather Forces: Tesla's Vision by Charles Yost; The New Art of Projecting Concentrated Non-Dispersive Energy Through Natural Media; The Homopolar Generator: Tesla's Contribution by Thomas Valone; Tesla's Ionizer and Ozonator: Implications for Indoor Air Pollution by Thomas Valone; How Cosmic Forces Shape Our Destiny by Nikola Tesla; Tesla's Death Ray plus Selected Tesla Patents; more.
288 PAGES. 6X9 PAPERBACK. ILLUSTRATED. $16.95. CODE: HWWN

TAPPING THE ZERO POINT ENERGY
Free Energy & Anti-Gravity in Today's Physics
by Moray B. King
King explains how free energy and anti-gravity are possible. The theories of the zero point energy maintain there are tremendous fluctuations of electrical field energy imbedded within the fabric of space. This book tells how, in the 1930s, inventor T. Henry Moray could produce a fifty kilowatt "free energy" machine; how an electrified plasma vortex creates anti-gravity; how the Pons/Fleischmann "cold fusion" experiment could produce tremendous heat without fusion; and how certain experiments might produce a gravitational anomaly.
180 PAGES. 5X8 PAPERBACK. ILLUSTRATED. $12.95. CODE: TAP

QUEST FOR ZERO-POINT ENERGY
Engineering Principles for "Free Energy"
by Moray B. King
King expands, with diagrams, on how free energy and anti-gravity are possible. The theories of zero point energy maintain there are tremendous fluctuations of electrical field energy embedded within the fabric of space. King explains the following topics: Tapping the Zero-Point Energy as an Energy Source; Fundamentals of a Zero-Point Energy Technology; Vacuum Energy Vortices; The Super Tube; Charge Clusters: The Basis of Zero-Point Energy Inventions; Vortex Filaments, Torsion Fields and the Zero-Point Energy; Transforming the Planet with a Zero-Point Energy Experiment; Dual Vortex Forms: The Key to a Large Zero-Point Energy Coherence. Packed with diagrams, patents and photos. With power shortages now a daily reality in many parts of the world, this book offers a fresh approach very rarely mentioned in the mainstream media.
224 PAGES. 6X9 PAPERBACK. ILLUSTRATED. $14.95. CODE: QZPE

THE TESLA PAPERS
Nikola Tesla on Free Energy & Wireless Transmission of Power
by Nikola Tesla, edited by David Hatcher Childress
David Hatcher Childress takes us into the incredible world of Nikola Tesla and his amazing inventions. Tesla's rare article "The Problem of Increasing Human Energy with Special Reference to the Harnessing of the Sun's Energy" is included. This lengthy article was originally published in the June 1900 issue of *The Century Illustrated Monthly Magazine* and it was the outline for Tesla's master blueprint for the world. Tesla's fantastic vision of the future, including wireless power, anti-gravity, free energy and highly advanced solar power. Also included are some of the papers, patents and material collected on Tesla at the Colorado Springs Tesla Symposiums, including papers on: •The Secret History of Wireless Transmission •Tesla and the Magnifying Transmitter •Design and Construction of a Half-Wave Tesla Coil •Electrostatics: A Key to Free Energy •Progress in Zero-Point Energy Research •Electromagnetic Energy from Antennas to Atoms •Tesla's Particle Beam Technology •Fundamental Excitatory Modes of the Earth-Ionosphere Cavity
325 PAGES. 8X10 PAPERBACK. ILLUSTRATED. $16.95. CODE: TTP

THE FANTASTIC INVENTIONS OF NIKOLA TESLA
by Nikola Tesla with additional material by David Hatcher Childress
This book is a readable compendium of patents, diagrams, photos and explanations of the many incredible inventions of the originator of the modern era of electrification. In Tesla's own words are such topics as wireless transmission of power, death rays, and radio-controlled airships. In addition, rare material on German bases in Antarctica and South America, and a secret city built at a remote jungle site in South America by one of Tesla's students, Guglielmo Marconi. Marconi's secret group claims to have built flying saucers in the 1940s and to have gone to Mars in the early 1950s! Incredible photos of these Tesla craft are included. The Ancient Atlantean system of broadcasting energy through a grid system of obelisks and pyramids is discussed, and a fascinating concept comes out of one chapter: that Egyptian engineers had to wear protective metal head-shields while in these power plants, hence the Egyptian Pharoah's head covering as well as the Face on Mars! •His plan to transmit free electricity into the atmosphere. •How electrical devices would work using only small antennas. •Why unlimited power could be utilized anywhere on earth. •How radio and radar technology can be used as death-ray weapons in Star Wars.
342 PAGES. 6X9 PAPERBACK. ILLUSTRATED. $16.95. CODE: FINT

ELECTROGRAVITICS SYSTEMS
Reports on a New Propulsion Methodology
edited by Thomas Valone

An anthology of two rare, unearthed reports of the secret work of T. Townsend Brown. The first report, *Electrogravitics Systems*, was classified until recently, and the second report, *The Gravitics Situation*, is a fascinating update on Brown's anti-gravity experiments in the early 50s. Also included, Dr. Paul LaViolette's research paper on the B-2 as a modern-day version of an eletrogravitics aircraft—a literal U.S. anti-gravity squadron!
116 PAGES. 6x9 PAPERBACK. ILLUSTRATED. $15.00. CODE: EGS

CONFERENCE ON FUTURE ENERGY CD
compiled by Tom Valone

Electrogravitics author Tom Valone's jam-packed CD-ROM with material from the First International Conference on Future Energy (COFE). Comes with Adobe Acrobat Reader, all plenary and contributed COFE papers in .pdf format, activates user's browser like Netscape or Explorer, 15 audiotracks of COFE workshops in .wav format, 14 audiotracks of COFE plenary lectures in .wav format, slide shows of several COFE presentations, a video clip of the Intora Noncombustive Helicopter, a video clip from the *Cold Fusion: Fire From Water* video, the DOE-EIA Energy Forecast Study & Radwaste Study, the Kyoto Protocol for reducing carbon emissions world-wide, more.
CD-ROM AUDIO/VIDEO DISK. $49.00. CODE: CFCD

THE HOMOPOLAR HANDBOOK
A Definitive Guide to Faraday Disk & N-Machine Technologies
by Thomas Valone, M.A., P.E.

The second book from Tom Valone, author of *Electrogravitics Systems* and well-known free energy/anti-gravity scientist, is a milestone work on permanent magnet free energy devices. This book is packed with technical information with chapters on the Faraday Disc Dynamo, Unipolar Induction, the "Field Rotation Paradox," the Stelle Homopolar Machine, the Trombly-Khan Closed-Path Homopolar Generator, the Sunburst Machine, Experimental Results with Various Devices, more.
180 PAGES. 6x9 PAPERBACK. ILLUSTRATED. REFERENCES, APPENDIX & INDEX. $20.00. CODE: HPH

LOST SCIENCE
by Gerry Vassilatos

Rediscover the legendary names of suppressed scientific revolution—remarkable lives, astounding discoveries, and incredible inventions which would have produced a world of wonder. How did the aura research of Baron Karl von Reichenbach prove the vitalistic theory and frighten the greatest minds of Germany? How did the physiophone and wireless of Antonio Meucci predate both Bell and Marconi by decades? How does the earth battery technology of Nathan Stubblefield portend an unsuspected energy revolution? How did the geoaetheric engines of Nikola Tesla threaten the establishment of a fuel-dependent America? The microscopes and virus-destroying ray machines of Dr. Royal Rife provided the solution for every world-threatening disease. Why did the FDA and AMA together condemn this great man to Federal Prison? The static crashes on telephone lines enabled Dr. T. Henry Moray to discover the reality of radiant space energy. Was the mysterious "Swedish stone," the powerful mineral which Dr. Moray discovered, the very first historical instance in which stellar power was recognized and secured on earth? Why did the Air Force initially fund the gravitational warp research and warp-cloaking devices of T. Townsend Brown and then reject it? When the controlled fusion devices of Philo Farnsworth achieved the "break-even" point in 1967 the FUSOR project was abruptly cancelled by ITT.
304 PAGES. 6x9 PAPERBACK. ILLUSTRATED. BIBLIOGRAPHY. $16.95. CODE: LOS

SECRETS OF COLD WAR TECHNOLOGY
Project HAARP and Beyond
by Gerry Vassilatos

Vassilatos reveals that "Death Ray" technology has been secretly researched and developed since the turn of the century. Included are chapters on such inventors and their devices as H.C. Vion, the developer of auroral energy receivers; Dr. Selim Lemstrom's pre-Tesla experiments; the early beam weapons of Grindell-Mathews, Ulivi, Turpain and others; John Hettenger and his early beam power systems. Learn about Project Argus, Project Teak and Project Orange; EMP experiments in the 60s; why the Air Force directed the construction of a huge Ionospheric "backscatter" telemetry system across the Pacific just after WWII; why Raytheon has collected every patent relevant to HAARP over the past few years; more.
250 PAGES. 6x9 PAPERBACK. ILLUSTRATED. $15.95. CODE: SCWT

THE SEARCH FOR A NEW ENERGY SOURCE

THE SEARCH FOR A NEW ENERGY SOURCE
edited by Dr. Gary L. Johnson

Johnson examines sacred texts and modern physics and constructs a free energy device. Material on unexplained phenomena, ball lightning, tornadoes, the earth's magnetic field, dowsing, UFOs, and gravitational anomalies. Also discussed are Tesla, Moray, Newman, Bearden and others. Chapters on Vapor Canopy; Ice Shell; Interplanetary Ice; Heat Balance; Fire from Heaven; Current Loops in the Earth's Core; Field, Aether, or Action-at-a-Distance; Patents; more.
263 PAGES. 6x9 PAPERBACK. ILLUSTRATED. $20.00. CODE: SNES

THE ANTI-GRAVITY HANDBOOK

edited by David Hatcher Childress, with Nikola Tesla, T.B. Paulicki, Bruce Cathie, Albert Einstein and others

The new expanded compilation of material on Anti-Gravity, Free Energy, Flying Saucer Propulsion, UFOs, Suppressed Technology, NASA Cover-ups and more. Highly illustrated with patents, technical illustrations and photos. This revised and expanded edition has more material, including photos of Area 51, Nevada, the government's secret testing facility. This classic on weird science is back in a 90s format!
• **How to build a flying saucer.**
•**Arthur C. Clarke on Anti-Gravity.**
• **Crystals and their role in levitation.**
• **Secret government research and development.**
• **Nikola Tesla on how anti-gravity airships could draw power from the atmosphere.**
• **Bruce Cathie's Anti-Gravity Equation.**
• **NASA, the Moon and Anti-Gravity.**
230 PAGES. 7x10 PAPERBACK. ILLUSTRATED. $14.95. CODE: **AGH**

ANTI–GRAVITY & THE WORLD GRID

Is the earth surrounded by an intricate electromagnetic grid network offering free energy? This compilation of material on ley lines and world power points contains chapters on the geography, mathematics, and light harmonics of the earth grid. Learn the purpose of ley lines and ancient megalithic structures located on the grid. Discover how the grid made the Philadelphia Experiment possible. Explore the Coral Castle and many other mysteries, including acoustic levitation, Tesla Shields and scalar wave weaponry. Browse through the section on anti-gravity patents, and research resources.
274 PAGES. 7x10 PAPERBACK. ILLUSTRATED. $14.95. CODE: **AGW**

ANTI–GRAVITY & THE UNIFIED FIELD

edited by David Hatcher Childress

Is Einstein's Unified Field Theory the answer to all of our energy problems? Explored in this compilation of material is how gravity, electricity and magnetism manifest from a unified field around us. Why artificial gravity is possible; secrets of UFO propulsion; free energy; Nikola Tesla and anti-gravity airships of the 20s and 30s; flying saucers as superconducting whirls of plasma; anti-mass generators; vortex propulsion; suppressed technology; government cover-ups; gravitational pulse drive; spacecraft & more.
240 PAGES. 7x10 PAPERBACK. ILLUSTRATED. $14.95. CODE: **AGU**

THE FREE-ENERGY DEVICE HANDBOOK

A Compilation of Patents and Reports
by David Hatcher Childress

A large-format compilation of various patents, papers, descriptions and diagrams concerning free-energy devices and systems. *The Free-Energy Device Handbook* is a visual tool for experimenters and researchers into magnetic motors and other "over-unity" devices. With chapters on the Adams Motor, the Hans Coler Generator, cold fusion, superconductors, "N" machines, space-energy generators, Nikola Tesla, T. Townsend Brown, and the latest in free-energy devices. Packed with photos, technical diagrams, patents and fascinating information, this book belongs on every science shelf. With energy and profit being a major political reason for fighting various wars, free-energy devices, if ever allowed to be mass distributed to consumers, could change the world! Get your copy now before the Department of Energy bans this book!
292 PAGES. 8X10 PAPERBACK. ILLUSTRATED. BIBLIOGRAPHY. $16.95. CODE: **FEH**

ETHER TECHNOLOGY

A Rational Approach to Gravity Control
by Rho Sigma

This classic book on anti-gravity and free energy is back in print and back in stock. Written by a well-known American scientist under the pseudonym of "Rho Sigma," this book delves into international efforts at gravity control and discoid craft propulsion. Before the Quantum Field, there was "Ether." This small, but informative book has chapters on John Searle and "Searle discs;" T. Townsend Brown and his work on anti-gravity and ether-vortex turbines. Includes a forward by former NASA astronaut Edgar Mitchell.
108 PAGES. 6x9 PAPERBACK. ILLUSTRATED. $12.95. CODE: **ETT**

THE TIME TRAVEL HANDBOOK

A Manual of Practical Teleportation & Time Travel
edited by David Hatcher Childress

In the tradition of *The Anti-Gravity Handbook* and *The Free Energy Device Handbook*, science and UFO author David Hatcher Childress takes us into the weird world of time travel and teleportation. Not just a whacked-out look at science fiction, this book is an authoritative chronicling of real-life time travel experiments, teleportation devices and more. *The Time Travel Handbook* takes the reader beyond the government's activities, such as the Philadelphia Experiment—the U.S. Navy's forays into invisibility, time travel, and teleportation—and deep into the uncharted territory of early time travellers including a spate of "UFO" sightings and landings in the 1890s and early 1900s. Childress looks into the claims of time travelling individuals, and investigates the unusual claim that the pyramids on Mars were built in the future and sent back in time. A highly visual, large format book, with patents, photos and schematics. Be the first on your block to build your own time travel device!

316 PAGES. 7x10 PAPERBACK. ILLUSTRATED. $16.95. CODE: **TTH**

UNDERGROUND BASES & TUNNELS
What is the Government Trying to Hide?
by Richard Sauder, Ph.D.

Working from government documents and corporate records, Sauder has compiled an impressive book that digs below the surface of the military's super-secret underground! Go behind the scenes into little-known corners of the public record and discover how corporate America has worked hand-in-glove with the Pentagon for decades, dreaming about, planning, and actually constructing, secret underground bases. This book includes chapters on the locations of the bases, the tunneling technology, various military designs for underground bases, nuclear testing & underground bases, abductions, needles & implants, military involvement in "alien" cattle mutilations, more. 50 page photo & map insert.

201 PAGES. 6X9 PAPERBACK. ILLUSTRATED. $15.95. CODE: UGB

UNDERWATER & UNDERGROUND BASES
Surprising Facts the Government Does Not Want You to Know
by Richard Sauder

Dr. Sauder lays out the amazing evidence and government paper trail for the construction of huge, manned bases offsore, in mid-ocean, and deep beneath the sea floor! Bases big enough to secretly dock submarines! Official United States Navy documents, and other hard evidence, raise many questions about what really lies 20,000 leagues beneath the sea. Many UFOs have been seen coming and going from the world's oceans, seas and lakes, implying the existence of secret underwater bases. Plus, new information on tunneling and cutting-edge, high speed rail magnetic-levitation (MagLev) technology. There are many rumors of secret, underground tunnels with MagLev trains hurtling through them. Is there truth behind the rumors? Underwater and Underground Bases carefully examines the evidence and comes to a thought provoking conclusion!

264 PAGES. 6X9 PAPERBACK. ILLUSTRATED. BIB & INDEX. $16.95. CODE: UUB

MAN-MADE UFOS 1944—1994
Fifty Years of Suppression
by Renato Vesco & David Hatcher Childress

A comprehensive look at the early "flying saucer" technology of Nazi Germany and the genesis of man-made UFOs. This book takes us from the work of captured German scientists to escaped battalions of Germans, secret communities in South America and Antarctica to todays state-of-the-art "Dreamland" flying machines. Heavily illustrated, this astonishing book blows the lid off the "government UFO conspiracy" and explains with technical diagrams the technology involved. Examined in detail are secret underground airfields and factories; German secret weapons; "suction" aircraft; the origin of NASA; gyroscopic stabilizers and engines; the secret Marconi aircraft factory in South America; and more. Introduction by W.A. Harbinson, author of the Dell novels *GENESIS* and *REVELATION*.

318 PAGES. 6X9 PAPERBACK. ILLUSTRATED. INDEX. $18.95. CODE: MMU

THE ENERGY GRID
Harmonic 695, The Pulse of the Universe
by Captain Bruce Cathie.

This is the breakthrough book that explores the incredible potential of the Energy Grid and the Earth's Unified Field all around us. Cathie's first book, *Harmonic 33*, was published in 1968 when he was a commercial pilot in New Zealand. Since then, Captain Bruce Cathie has been the premier investigator into the amazing potential of the infinite energy that surrounds our planet every microsecond. Cathie investigates the Harmonics of Light and how the Energy Grid is created. In this amazing book are chapters on UFO Propulsion, Nikola Tesla, Unified Equations, the Mysterious Aerials, Pythagoras & the Grid, Nuclear Detonation and the Grid, Maps of the Ancients, an Australian Stonehenge examined, more.

255 PAGES. 6X9 TRADEPAPER. ILLUSTRATED. $15.95. CODE: TEG

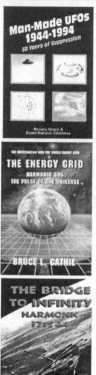

THE BRIDGE TO INFINITY
Harmonic 371244
by Captain Bruce Cathie

Cathie has popularized the concept that the earth is crisscrossed by an electromagnetic grid system that can be used for anti-gravity, free energy, levitation and more. The book includes a new analysis of the harmonic nature of reality, acoustic levitation, pyramid power, harmonic receiver towers and UFO propulsion. It concludes that today's scientists have at their command a fantastic store of knowledge with which to advance the welfare of the human race.

204 PAGES. 6X9 TRADEPAPER. ILLUSTRATED. $14.95. CODE: BTF

THE HARMONIC CONQUEST OF SPACE
by Captain Bruce Cathie

Chapters include: Mathematics of the World Grid; the Harmonics of Hiroshima and Nagasaki; Harmonic Transmission and Receiving; the Link Between Human Brain Waves; the Cavity Resonance between the Earth; the Ionosphere and Gravity; Edgar Cayce—the Harmonics of the Subconscious; Stonehenge; the Harmonics of the Moon; the Pyramids of Mars; Nikola Tesla's Electric Car; the Robert Adams Pulsed Electric Motor Generator; Harmonic Clues to the Unified Field; and more. Also included are tables showing the harmonic relations between the earth's magnetic field, the speed of light, and anti-gravity/gravity acceleration at different points on the earth's surface. New chapters in this edition on the giant stone spheres of Costa Rica, Atomic Tests and Volcanic Activity, and a chapter on Ayers Rock analysed with Stone Mountain, Georgia.

248 PAGES. 6X9. PAPERBACK. ILLUSTRATED. BIBLIOGRAPHY. $16.95. CODE: HCS

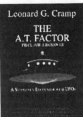

THE A.T. FACTOR
A Scientists Encounter with UFOs: Piece For A Jigsaw Part 3
by Leonard Cramp
British aerospace engineer Cramp began much of the scientific anti-gravity and UFO propulsion analysis back in 1955 with his landmark book *Space, Gravity & the Flying Saucer* (out-of-print and rare). His next books (available from Adventures Unlimited) *UFOs & Anti-Gravity: Piece for a Jig-Saw* and *The Cosmic Matrix: Piece for a Jig-Saw Part 2* began Cramp's in depth look into gravity control, free-energy, and the interlocking web of energy that pervades the universe. In this final book, Cramp brings to a close his detailed and controversial study of UFOs and Anti-Gravity.
324 PAGES. 6X9 PAPERBACK. ILLUSTRATED. BIBLIOGRAPHY. INDEX. $16.95. CODE: ATF

COSMIC MATRIX
Piece for a Jig-Saw, Part Two
by Leonard G. Cramp
Leonard G. Cramp, a British aerospace engineer, wrote his first book *Space Gravity and the Flying Saucer* in 1954. Cosmic Matrix is the long-awaited sequel to his 1966 book *UFOs & Anti-Gravity: Piece for a Jig-Saw.* Cramp has had a long history of examining UFO phenomena and has concluded that UFOs use the highest possible aeronautic science to move in the way they do. Cramp examines anti-gravity effects and theorizes that this super-science used by the craft—described in detail in the book—can lift mankind into a new level of technology, transportation and understanding of the universe. The book takes a close look at gravity control, time travel, and the interlocking web of energy between all planets in our solar system with Leonard's unique technical diagrams. A fantastic voyage into the present and future!
364 PAGES. 6X9 PAPERBACK. ILLUSTRATED. BIBLIOGRAPHY. $16.00. CODE: CMX

UFOS AND ANTI-GRAVITY
Piece For A Jig-Saw
by Leonard G. Cramp
Leonard G. Cramp's 1966 classic book on flying saucer propulsion and suppressed technology is a highly technical look at the UFO phenomena by a trained scientist. Cramp first introduces the idea of 'anti-gravity' and introduces us to the various theories of gravitation. He then examines the technology necessary to build a flying saucer and examines in great detail the technical aspects of such a craft. Cramp's book is a wealth of material and diagrams on flying saucers, anti-gravity, suppressed technology, G-fields and UFOs. Chapters include Crossroads of Aerodymanics, Aerodynamic Saucers, Limitations of Rocketry, Gravitation and the Ether, Gravitational Spaceships, G-Field Lift Effects, The Bi-Field Theory, VTOL and Hovercraft, Analysis of UFO photos, more.
388 PAGES. 6X9 PAPERBACK. ILLUSTRATED. $16.95. CODE: UAG

THE GIZA DEATH STAR
The Paleophysics of the Great Pyramid & the Military Complex at Giza
by Joseph P. Farrell
Physicist Joseph Farrell's amazing book on the secrets of Great Pyramid of Giza. *The Giza Death Star* starts where British engineer Christopher Dunn leaves off in his 1998 book, *The Giza Power Plant.* Was the Giza complex part of a military installation over 10,000 years ago? Chapters include: An Archaeology of Mass Destruction, Thoth and Theories; The Machine Hypothesis; Pythagoras, Plato, Planck, and the Pyramid; The Weapon Hypothesis; Encoded Harmonics of the Planck Units in the Great Pyramid; High Freggency Direct Current "Impulse" Technology; The Grand Gallery and its Crystals: Gravito-acoustic Resonators; The Other Two Large Pyramids; the "Causeways," and the "Temples"; A Phase Conjugate Howitzer; Evidence of the Use of Weapons of Mass Destruction in Ancient Times; more.
290 PAGES. 6X9 PAPERBACK. ILLUSTRATED. $16.95. CODE: GDS

DARK MOON
Apollo and the Whistleblowers
by Mary Bennett and David Percy
•Was Neil Armstrong really the first man on the Moon?
•Did you know a second craft was going to the Moon at the same time as Apollo 11?
•Do you know that potentially lethal radiation is prevalent throughout deep space?
•Do you know there are serious discrepancies in the account of the Apollo 13 'accident'?
•Did you know that 'live' color TV from the Moon was not actually live at all?
•Did you know that the Lunar Surface Camera had no viewfinder?
•Do you know that lighting was used in the Apollo photographs—yet no lighting equipment was taken to the Moon?
All these questions, and more, are discussed in great detail by British researchers Bennett and Percy in *Dark Moon,* the definitive book (nearly 600 pages) on the possible faking of the Apollo Moon missions. Bennett and Percy delve into every possible aspect of this beguiling theory, one that rocks the very foundation of our beliefs concerning NASA and the space program. Tons of NASA photos analyzed for possible deceptions.
568 PAGES. 6X9 PAPERBACK. ILLUSTRATED. BIB & INDEX. $25.00. CODE: DMO

TECHNOLOGY OF THE GODS
The Incredible Sciences of the Ancients
by David Hatcher Childress
Popular *Lost Cities* author David Hatcher Childress takes us into the amazing world of ancient technology, from computers in antiquity to the "flying machines of the gods." Childress looks at the technology that was allegedly used in Atlantis and the theory that the Great Pyramid of Egypt was originally a gigantic power station. He examines tales of ancient flight and the technology that it involved; how the ancients used electricity; megalithic building techniques; the use of crystal lenses and the fire from the gods; evidence of various high tech weapons in the past, including atomic weapons; ancient metallurgy and heavy machinery; the role of modern inventors such as Nikola Tesla in bringing ancient technology back into modern use; impossible artifacts; and more.
356 PAGES. 6x9 PAPERBACK. ILLUSTRATED. BIBLIOGRAPHY. $16.95. CODE: TGOD

VIMANA AIRCRAFT OF ANCIENT INDIA & ATLANTIS
by David Hatcher Childress, introduction by Ivan T. Sanderson
Did the ancients have the technology of flight? In this incredible volume on ancient India, authentic Indian texts such as the *Ramayana* and the *Mahabharata* are used to prove that ancient aircraft were in use more than four thousand years ago. Included in this book is the entire Fourth Century BC manuscript *Vimaanika Shastra* by the ancient author Maharishi Bharadwaaja, translated into English by the Mysore Sanskrit professor G.R. Josyer. Also included are chapters on Atlantean technology, the incredible Rama Empire of India and the devastating wars that destroyed it. Also an entire chapter on mercury vortex propulsion and mercury gyros, the power source described in the ancient Indian texts. Not to be missed by those interested in ancient civilizations or the UFO enigma.
334 PAGES. 6x9 PAPERBACK. ILLUSTRATED. $15.95. CODE: VAA

LOST CONTINENTS & THE HOLLOW EARTH
I Remember Lemuria and the Shaver Mystery
by David Hatcher Childress & Richard Shaver
Lost Continents & the Hollow Earth is Childress' thorough examination of the early hollow earth stories of Richard Shaver and the fascination that fringe fantasy subjects such as lost continents and the hollow earth have had for the American public. Shaver's rare 1948 book *I Remember Lemuria* is reprinted in its entirety, and the book is packed with illustrations from Ray Palmer's *Amazing Stories* magazine of the 1940s. Palmer and Shaver told of tunnels running through the earth—tunnels inhabited by the Deros and Teros, humanoids from an ancient spacefaring race that had inhabited the earth, eventually going underground, hundreds of thousands of years ago. Childress discusses the famous hollow earth books and delves deep into whatever reality may be behind the stories of tunnels in the earth. Operation High Jump to Antarctica in 1947 and Admiral Byrd's bizarre statements, tunnel systems in South America and Tibet, the underground world of Agartha, the belief of UFOs coming from the South Pole, more.
344 PAGES. 6x9 PAPERBACK. ILLUSTRATED. $16.95. CODE: LCHE

ATLANTIS & THE POWER SYSTEM OF THE GODS
Mercury Vortex Generators & the Power System of Atlantis
by David Hatcher Childress and Bill Clendenon
Atlantis and the Power System of the Gods starts with a reprinting of the rare 1990 book *Mercury: UFO Messenger of the Gods* by Bill Clendenon. Clendenon takes on an unusual voyage into the world of ancient flying vehicles, strange personal UFO sightings, a meeting with a "Man In Black" and then to a centuries-old library in India where he got his ideas for the diagrams of mercury vortex engines. The second part of the book is Childress' fascinating analysis of Nikola Tesla's broadcast system in light of Edgar Cayce's "Terrible Crystal" and the obelisks of ancient Egypt and Ethiopia. Includes: Atlantis and its crystal power towers that broadcast energy; how these incredible power stations may still exist today; inventor Nikola Tesla's nearly identical system of power transmission; Mercury Proton Gyros and mercury vortex propulsion; more. Richly illustrated, and packed with evidence that Atlantis not only existed—it had a world-wide energy system more sophisticated than ours today.
246 PAGES. 6x9 PAPERBACK. ILLUSTRATED. $15.95. CODE: APSG

A HITCHHIKER'S GUIDE TO ARMAGEDDON
by David Hatcher Childress
With wit and humor, popular Lost Cities author David Hatcher Childress takes us around the world and back in his trippy finalé to the Lost Cities series. He's off on an adventure in search of the apocalypse and end times. Childress hits the road from the fortress of Megiddo, the legendary citadel in northern Israel where Armageddon is prophesied to start. Hitchhiking around the world, Childress takes us from one adventure to another, to ancient cities in the deserts and the legends of worlds before our own. Childress muses on the rise and fall of civilizations, and the forces that have shaped mankind over the millennia, including wars, invasions and cataclysms. He discusses the ancient Armageddons of the past, and chronicles recent Middle East developments and their ominous undertones. In the meantime, he becomes a cargo cult god on a remote island off New Guinea, gets dragged into the Kennedy Assassination by one of the "conspirators," investigates a strange power operating out of the Altai Mountains of Mongolia, and discovers how the Knights Templar and their off-shoots have driven the world toward an epic battle centered around Jerusalem and the Middle East.
320 PAGES. 6x9 PAPERBACK. ILLUSTRATED. BIBLIOGRAPHY. INDEX. $16.95. CODE: HGA

One Adventure Place
P.O. Box 74
Kempton, Illinois 60946
United States of America
Tel.: 815-253-6390 • Fax: 815-253-6300
Email: auphq@frontiernet.net
http://www.adventuresunlimitedpress.com
or www.adventuresunlimited.nl

ORDERING INSTRUCTIONS

✓ Remit by USD$ Check, Money Order or Credit Card
✓ Visa, Master Card, Discover & AmEx Accepted
✓ Prices May Change Without Notice
✓ 10% Discount for 3 or more Items

SHIPPING CHARGES

United States

✓ Postal Book Rate { $3.00 First Item
50¢ Each Additional Item
✓ Priority Mail { $4.00 First Item
$2.00 Each Additional Item
✓ UPS { $5.00 First Item
$1.50 Each Additional Item
NOTE: UPS Delivery Available to Mainland USA Only

Canada

✓ Postal Book Rate { $6.00 First Item
$2.00 Each Additional Item
✓ Postal Air Mail { $8.00 First Item
$2.50 Each Additional Item
✓ Personal Checks or Bank Drafts MUST BE
USD$ and Drawn on a US Bank
✓ Canadian Postal Money Orders OK
✓ Payment MUST BE USD$

All Other Countries

✓ Surface Delivery { $10.00 First Item
$4.00 Each Additional Item
✓ Postal Air Mail { $14.00 First Item
$5.00 Each Additional Item
✓ Payment MUST BE USD$
✓ Checks and Money Orders MUST BE USD$
and Drawn on a US Bank or branch.
✓ Add $5.00 for Air Mail Subscription to
Future *Adventures Unlimited* Catalogs

SPECIAL NOTES

✓ RETAILERS: Standard Discounts Available
✓ BACKORDERS: We Backorder all Out-of-
Stock Items Unless Otherwise Requested
✓ PRO FORMA INVOICES: Available on Request
✓ VIDEOS: NTSC Mode Only. Replacement only.
✓ For PAL mode videos contact our other offices.

Please check: ☑
☐ This is my first order   ☐ I have ordered before

Name		
Address		
City		
State/Province		Postal Code
Country		
Phone day	Evening	
Fax		

Item Code	Item Description	Qty	Total

Please check: ☑

☐ Postal-Surface
☐ Postal-Air Mail (Priority in USA)
☐ UPS (Mainland USA only)
☐ Visa/MasterCard/Discover/Amex

Subtotal ➡
Less Discount-10% for 3 or more items ➡
Balance ➡
Illinois Residents 6.25% Sales Tax ➡
Previous Credit ➡
Shipping ➡
Total (check/MO in USD$ only)➡

Card Number

Expiration Date

10% Discount When You Order 3 or More Items!

Adventures Unlimited, Pannewal 22,
Enkhuizen, 1602 KS, The Netherlands
http: www.adventuresunlimited.nl